Emerging Themes in Polymer Science

Emerging Themes in Polymer Science

Edited by

Anthony J. Ryan,
Department of Chemistry,
University of Sheffield,
Sheffield, UK.

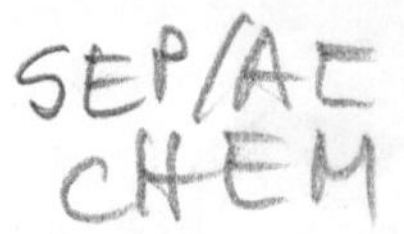

The proceedings of the Macro Group UK meeting entitled Emerging Themes in Polymer Science held at The University of Sheffield 18–20 April 2000.

Special Publication No. 263

ISBN 0-85404-890-1

A catalogue record for this book is available from the British Library

Published by The Royal Society of Chemistry
Thomas Graham House, Science Park, Milton Road, Cambridge CB4 0WF, UK

Registered Charity Number 207890

For further information see our web site at www.rsc.org

Typeset by Computape (Pickering) Ltd, Pickering, North Yorkshire, UK
Printed and bound by MPG Books Ltd, Bodmin, Cornwall, UK

Preface

This book comprises papers presented at the 'Emerging Themes in Polymer Science' discussion meeting held in Sheffield from 18th–20th April 2000. The format of this MACRO Group UK meeting was based on the RSC Faraday Discussions and the authors were asked to provide papers to provoke discussion. In many cases the themes were chosen by the editor (and conference chairman) and the author was not given a choice of subject; the papers you see here have been revised in light of the discussion.

The meeting covered a wide range of subjects from the compelling industrial dimension through polymer synthesis, characterisation, processing, properties and applications. The emerging themes, *i.e.* those that dominated the discussion, where the interactions between biology and polymer science and the use of polymers in electronics and nanotechnology.

A jolly and discursive group of about 70 assembled at Stephenson Hall in the University of Sheffield. The discussion was led by a number of attendees not presenting papers, in particular the humour of Dr John Stanford will be fondly remembered by all who contributed to the sessions on Rheology and Processing. We are not quite sure who started the 'why do block copolymers segregate to interfaces' series of questions, but the editor, Mark Geoghegan and Richard Jones kept it going, and the whole subject of amphiphilic behaviour was discussed without being initiated by a written paper, or properly represented here.

We are very grateful to the MACRO Group UK who sponsored the meeting and to the staff at the RSC's Thomas Graham House for publishing this book. My final vote of thanks goes to Mrs Elaine Shaw who kept everything running smoothly from the inception of the meeting until the final submission of the papers.

A.J. Ryan
Sheffield November 2000

Contents

Molecular Machines

1

Perspective and Summary

Randal W. Richards

INTERDISCIPLINARY RESEARCH CENTRE IN POLYMER SCIENCE AND TECHNOLOGY, UNIVERSITY OF DURHAM, DURHAM DH1 3LE, UK

In 1999 the Royal Society of Chemistry published a compendium book[1] setting out the contributions to society that have been made by chemistry. The title, 'The Age of the Molecule' neatly encapsulated the philosophy that a molecular understanding is the basis for optimum exploitation of new substances or materials provided by chemistry. Much of the material in the book was drawn from 20th Century developments but with evident foundations on the advances made in the latter half of the 19th Century. If the preceding 150 years is the 'Age of the Molecule' then the 20th Century has been 'The Century of the Macromolecule'.

It was not until the last century was some thirty years old that the concept of 'giant' molecules, polymers, was widely accepted. Evidently the science of polymers or macromolecules is relatively young compared to other aspects, *e.g.* quantum theory's beginnings pre-date polymer science by some twenty years. Of course mankind had been using polymers for many centuries before their macromolecular nature was recognised. These were mainly the naturally occurring polymers; cotton, wool, rubber, gums and leather. Driven by the blockade of raw materials during World War I, a process for the production of synthetic rubber was developed in Germany but the high cost of butadiene made the process commercially infeasible in peacetime. It was the rapid development and huge expansion of the petrochemicals industry catalysed by the strategic demands of World War II that led to the widespread use of synthetic polymeric materials by modern society.[2] This use was just preceded by an understanding of the main facets of the molecular nature of polymers,[3] quite unlike other materials, *e.g.* steel and alloys.

From the outset polymer science has involved physicists, chemists, engineers, materials scientists and design engineers. The multidisciplinary nature of polymer science from its earliest days is a feature that is not often exhibited by other fields of natural science until a certain 'maturity' has been reached. The synthetic polymer industry in the UK expanded greatly over the years from *circa* 1950 to *circa* 1995, as did the scope where polymeric materials are used. We rely on polymers to keep us warm and dry (fibres); to preserve and protect our food, housing and transport (packaging, protective coatings); to provide entertainment (recording media); contribute to defence systems and high speed travel (composites in military and civil aircraft); materials that are vital to the exploration of space (light weight antennae and ablative layers for re-entry) and developments over the last eight years presage their use as the next display medium and probably supplanting (eventually) the cathode ray tube. However, we still have much to learn in understanding and manipulating (for beneficial purposes) the macromolecules of life, proteins, polypeptides *etc.* The success of the Human Genome Project will undoubtedly accelerate that understanding.

During the last five years of the twentieth century it was evident that polymer science in the UK was changing both industrially and in academe. The industrial changes were caused by reduction in profit margins for the high volume polymers, polyolefins, polyesters, polyacrylics *etc.* as production plants with significantly lower operating costs elsewhere in the world came on stream. Increased need and use for speciality polymers, *e.g.* polyaromatic ketones, main chain liquid crystal polymers, needed a closer engagement with and understanding of the end-user market. These factors, and others, have led to a fragmentation of the industry with small units being more active in areas where the philosophy is production of smaller volumes of polymer but with more 'high tech' specifications. A rapidly developing area that exemplifies these aspects is electro-active and electroluminescent polymers. In the academic world, the 'bulge' of university expansion in the early to mid 1960s was passing through the system and recruitment of young academics was buoyant. This has resulted in the largest cohort for many years of academics with research interests in polymer science being recruited into British universities, such that at the end of 1999, some 35 academics all with less than 5 years appointment could be identified.

These evident changes and the awareness that others, perhaps more momentous, were on the horizon prompted the Pure and Applied Macromolecular Chemistry Group (MacroGroup UK) of the Royal Society of Chemistry and the Society of Chemical Industry to attempt a 'survey of the landscape'. The subliminal questions to be addressed included 'What's going on now?' 'What aspects are likely to develop in the future?' 'What should we be aware of?' The meeting 'Emerging Themes in Polymer Science' was an attempt to do this survey; the brief was not to attempt to pick winners but to inform, foster debate and discussion, encourage boldness in future efforts. To foster these aspects the format of the meeting was highly structured in one aspect only; all papers were circulated well before the meeting, each invited contributor had ten minutes to 'hit the highlights' and express 'the bottom line' of their paper. Thereafter each paper had at least 20 minutes 'official' discussion time, this format was adopted from the model of the highly successful Faraday Discussions. The responsibility for organising the meeting and choosing the speakers was accepted by one of the newer centres of polymer science in the UK in the form of Professors Tony Ryan (Chemistry) and Richard Jones (Physics) at the University of Sheffield. It is the contributed papers, with some post meeting changes in a few cases, that forms the body of this volume. Their choice of speakers resulted in a meeting where there was true debate and discussion, participants were prepared to put forward radical opinions but at no point were the boundaries of good manners crossed. The themes aired may appear to be eclectic but they emphasise, to some extent, the multidisciplinarity and diversity that is modern polymer science.

The summary overview given here does not cover all aspects exhaustively; rather it is an attempt to capture the scope and contrasts that were expressed in the unscripted discussions. The subjects covered ranged from the philosophy of survival in rapidly changing companies driven by the need to generate revenue to stay viable, to speculations on the probability of producing 'machines' by molecular assembly processes at the nanometer scale, the moving parts being driven by fundamental physico-chemical processes. The common view expressed from industry was that change leads to clear decision-making and 'captures the essence of the evolutionary spirit'. What was not clear was whether this 'change' was in response to demand or a desire to lead the technology market. It is certainly accepted that during a person's working lifetime in industrial R&D, they may need to re-invent or change their expertise base as companies move into new market areas or shed more mature aspects for new, higher added value products generally produced in smaller quantities. Paradoxically, in one of the fastest developing areas already referred to, electroluminescent polymers, it was suggested that continual change might not be a good thing. Rather than wait for the 'next best polymer' to come along, more profitable results may come from optimisation of what is already available. For example, by developing the ability to control precisely the molecular packing in thin films and optimising interfacial area effects. The willingness and ability of synthetic chemists to meet the challenge of producing well defined complex molecular architecture in polymers needs to be balanced by the increased costs incurred and an awareness that sufficient architectural complexity is enough, any more is unnecessary and not cost-effective.

Polymer rheology and processing and colloidal dispersions of polymers (emulsions, latexes *etc.*) have been major components of the application and use of polymers over the past 50 years. The complexity of the systems or the apparent 'black arts' needed to make progress have daunted wide participation at the level of fundamental research. Challenging problems are indeed posed by both these areas and call for lateral thinking and ability to 'mine' apparently unrelated areas for ideas. Evidently the understanding and manipulation of polymer colloids requires a knowledge of surfactant behaviour, an appreciation of the concepts of particle stabilisation and some awareness of the rheology of the suspensions in addition to polymerisation kinetics and thermodynamics of multicomponent systems. This is an area that seems to require consolidation and proper evaluation rather than innovation before further advances are made in an optimised way. Elegant and sophisticated machinery is now commonplace in the polymer processing industry (extruders, moulders *etc.*) but these have been defined and designed by production engineers and the input of polymer scientist appears to be minimal. The question remains to be answered whether more optimal processing or products could be obtained by increased input from the latter group. Rheology is evidently a key aspect to understand in the processing of polymers and the recent availability of numerical flow solvers has greatly aided the 'visualisation' of melt flow during processing (and the use and further development of these solvers will undoubtedly grow). The link with molecular level theories and molecular level architecture is still rather tenuous despite rapid advances in the last few years. The molecular approach to discussing polymer rheology is well discussed herein as are other aspects of modelling polymers.

Although the papers herein covering the theory and modelling of polymer systems are quite specific, the discussions were more philosophical and queried the basic approach. The tone of the discussion is perhaps judged by noting that one approach was described as 'epistemological anarchy'. Evidently, theoreticians are not particularly happy at the apparent evaporation of interest by experimentalists as soon as the theory fits the data! Perhaps theoreticians and experimentalists can agree that theory should entice cutting edge experiments and experiments should challenge theory, not in a combative manner but in the spirit of intellectual curiosity. Certainly, this seems to have been the dynamics of the process that produced the major development in polymer science in the late 1970s and early 1980s, the scaling law description of polymer configuration and dynamics.[4] The major challenge to theoreticians in the field is little different to those in other areas of condensed matter theory; how to scale up detailed ideas at the quantum mechanical level to the length scales of bulk materials. Some advances have been made at meso scales but there is much, much more to do.

Evidently, from the birth of the macromolecular hypothesis in the 1920s, there has been significant development in the understanding, synthesis and application of synthetic polymers. For biopolymers the situation is much more 'patchy' and certainly the level of sophisticated understanding is not as great as for synthetic polymers. There are several aspects covered by the term biopolymers and not all were dealt with here. The use of polymers to generate new tissue and engineer its production at defined sites is evidently a very exciting area for synthetic chemists since new molecules with specific properties are needed. Some caution may be needed since enthusiasm to take on challenging syntheses may run ahead of the defined needs and indeed if there *is* a real need for the molecules. Furthermore, major influences in this area will be ethical issues, the need for regulation and public perception. Misunderstanding or manipulation of incorrect information to produce dramatic headlines (*e.g.* 'Frankenstein' molecules) may be a greater hindrance to developments than the actual scientific problems to be solved. The Royal Society of Chemistry has begun to define a strategy for the role of Chemistry in the life sciences. Perhaps more immediate benefits of polymers in the 'bio' area may result from more (apparently) mundane aspects. The increase in life span and more involvement in 'vigorous' leisure activities (*e.g.* skiing, climbing) has resulted in joint replacements at younger ages and the likelihood that the prosthetic joints will be replaced more than once during a lifetime. This is mainly due to wear of the polymeric components, consequently any improvement in the 'cushioning' of impact or wear of these components would improve the subsequent quality of life of the users. A major area for biopolymers is foodstuffs and polysaccharides. These are abundant; amenable to manipulation, have sufficient natural variants that a range of properties, morphologies and behaviours are accessible. They are generally accompanied by greater or smaller amounts of water and exhibit specific functionalities that may be a boon or a bane depending on what you wish to do with the polymer. Over the last few years, the application of well-developed techniques from other areas (mainly physics) and well-designed experiments has produced much new insight.

Whatever type of polymer is being dealt with, generally the first question to be dealt with concerns the provenance of the material. Is it what you believe it is? Does it have the desired properties, *e.g.* molecular weight, composition? Does it have the organisation that is required? All these questions come under characterisation. Mass spectrometry is being more and more applied to high molecular weight polymers and evidently is extremely useful and informative in the hands of the expert. Questions were asked about applicability to molecular weight ranges more typical of polymers and the observation that under different conditions, different data could be obtained. These questions notwithstanding, it is clear that recent developments in mass spectrometry will make it a key analytical tool for the more 'exotic' polymers that will be needed in, for example, tissue engineering applications. The desire to investigate and understand more complex systems or polymers under actual use conditions will lead to demands for more sophisticated, expensive instrumentation only available at nationally supported facilities. We are on the threshold of new neutron and synchrotron sources that will enable a greater range of experimental investigations, *e.g.* following reactions (of polymers) in real time, simultaneous investigation of molecular level dynamics and organisation when polymers are subjected to stimuli. I suspect we are only limited by our imaginations regarding the investigations that will be possible in the future. One area that has blossomed has been the qualitative and quantitative investigations of polymer surfaces and interfaces. The various probe microscopies are continually developing and are now approaching mesoscopic length scales, and amazing images will not be enough for discerning people when more fundamental information is extractable. Such probe microscopy techniques will be germane to the precise construction of 'molecular machines' either *via* simple reactions or self-assembly routes. However, these 'soft' processes must produce a strong assembly to be useful. The goal is to produce molecular shuttling, *via* local pH changes for example, that can be converted to mechanical work. The realisation of such devices is probably 10–15 years ahead but will rely on the ability to characterise them by some means.

Contained herein is only a small part of the activity in polymer science in the UK but the scope addressed, I believe, epitomises the vibrancy and health of the subject in the UK. Each contribution can be read individually. I suspect that the information 'density' is so high that a 'linear' read from beginning to end would be mentally exhausting, and the reading of individual chapters over a longer period may be more fruitful. I remarked earlier that there are some 30 plus young academics now in the UK university system; there is a greater number of similar age in the UK polymer industry (large and small). Between them they will be responsible for defining future emergent themes and it will be interesting to observe this metamorphosis of the subject. A meeting of this type should not be an annual event, there needs to be time for maturing of new fields, a gestation of new ideas and the necessary discarding of branches that bore no fruit. There is a need for future meetings on a 5–6 year cycle, since it enables an overview to be obtained and a refreshment of perceptions. I look forward to the next one.

References

1 '*The Age of the Molecule*' The Royal Society of Chemistry, London, 1999.
2 A good account of the historical aspects of polymer science is contained in '*Enough for One Lifetime Wallace Carothers, Inventor of Nylon*', American Chemical Society 1996 and '*Polymers, the Origins and Growth of a Science*', H Morawetz, Wiley, NY, 1985.
3 '*Principles of Polymer Chemistry*' P J Flory, Cornell University Press, Ithaca, NY 1953.
4 '*Scaling Concepts in Polymer Physics*' P G de Gennes, Cornell University Press, Ithaca, NY 1979.

The Future of Industry

2

Emerging Trends in Polymer Science

David Bott

SPECIALTY SYNTHETIC POLYMERS DIVISION,
NATIONAL STARCH & CHEMICAL COMPANY

1 Predicting the Future[1]

Actually, it is fairly easy to predict how polymeric materials science can contribute to a better future and improved life. It is 'well known' that mankind would greatly benefit from materials that transport energy without loss and store it harmlessly at enormous levels. And, obviously, human life would be of a dramatic higher quality if the degenerative effects of age and strain on the human body were to be moderated by polymers that help the body replace those parts that have gone bad. Also, clearly, urgent needs exist in the area of water harvesting, purification and distribution. And if we seek continued comfortable life on earth for future generations, sustainability and renewal are prerequisites.

However, reducing these grand strategies to the more mundane level of what we do in the laboratory each day is more difficult. But do we need to answer the question: 'What should be invented?' Virtually by definition, the inventions (from Latin '*inventus*', 'find') that have revolutionised the world, perhaps more often than not have been serendipitous; and, therefore, are unpredictable. Well-known examples of such 'lucky events' have taken place in the past century, such as the discovery of poly(tetrafluoroethylene) [Teflon]. Of course, a mind set and circumstances to recognise the unusual, and the ability to turn findings into use, unlike inventions themselves, can and must be created.

It might be argued that not being limited by profound knowledge of, or belief in the known laws of nature (it is the feebleness of our understanding of those laws that make the 'purely scientific' approach to innovation impractical . . .), it is really the unusual thinkers such as science-fiction writers – people like Jules Verne come to mind – who take the lead in expanding our minds with respect to new materials functions and applications, as well as entirely new concepts for food, shelter, safety, transportation and communication.

Do we think in these terms within the polymer community? Or are we focussed on our own local, short-term problems? When was the last time we watched a science fiction movie (or read a futuristic book) and pondered whether we could do that?

2 Increasing Control of Polymer Architecture, Both Shape and Distribution of Chemical Functionality

When polymers were first made, the scientists who made them were happy just to have made them! As time went by, and with the advances in analytical chemistry that had occurred, it became obvious that there was the possibility for wide variation in chemical structure. At first there was an almost macho drive for high molecular weight, or a narrow molecular weight distribution – something that could be simple and easily measured. Then people started to correlate the performance of the polymers in real world use with their understanding of polymer structure. As relationships in this area were discovered, there was a natural desire to make polymers that more precisely matched the needs of the application. Synthetic polymer scientists set out to make materials that gave precisely the right properties for the most valuable applications – in theory. In fact, they discovered two factors that have bedevilled this custom approach. The first is the fact that many of the desirable properties of polymers are derived from their complex nature. A single molecular weight polymer would, for example, have a much sharper melting point and would be difficult to process – it would lose its 'plastic' nature. It would also tend to have inferior mechanical properties, especially in the areas of toughness and chemical resistance. These are not insurmountable problems and once properly defined, it should be possible to make the polymer material with the correct balance of properties. The problem comes in the definition of the required design. Polymers are strongly influenced by their history – the manner of their synthesis, the thermal and mechanical forces they have been through all play a major role in controlling the distribution of the polymer chains relative to one another in space, and it is this state which largely determines the macroscopic properties of the polymer.

The future of custom synthesis of designer polymers will therefore have to have a much higher input from the processing side of the technology. Although this is strong within the 'materials' end of the polymer spectrum, it is likely also to be true at the more specialist end. Consider as an example spider silk. It is spun out of water to form a water insoluble fibre – a transformation in properties brought about by moving from one kinetically stable state to another thermodynamically (??) stable one. As we design polymers to form micelles, or synthetic enzymes, we polymer scientists will have to understand and embrace the polymer technologists view of the world – perhaps a new discipline will be born – Polymer Engineers?

3 Sustainable Feedstocks

This use of examples from nature and the need (stated in the first paragraph) for a smaller impact on the environment have led to a greater interest in naturally derived building blocks for polymer structures. All the permutations on polymer structure and distribution of chemical functionality can be found in the polymeric systems used in nature – polysaccharides, polypeptides, lipids, *etc.* have complexities by design whereas the majority of the synthetically produced polymers have complexity as a by-product – and often one that is undesirable.

4 Combinatorial Approaches

It happened first in the pharmaceutical industry, where they seem to have enough money to try almost anything and where the size of the returns justifies taking a mixed portfolio of risks. They used a matrix approach to synthesis and designed simple yes–no screens to quickly whittle down the number of potential solutions to a more manageable number. Then it moved across into materials science, first in kinetically stable inorganic materials produced by vacuum deposition, then into catalysts. Here the approach was different. Instead of 10,000 samples, they use 100 samples, and instead of throwing away the samples that failed the screen, they saved them in libraries. This caused some crossover with polymer science, where it was used to screen polymerisation catalysts (I guess the lawyers fees from the metallocenes adventure had sensitised everyone to cost!). Now it is here in polymer science. Admittedly, not everywhere, but in free radical polymerisation and for basic characterisation and some application screens. As with its introduction in other areas, there is the question of how the community will deal with it? Rejection or involvement? It is not uncommon for 'proper' scientists to see the combinatorial approach as a diminution of their 'scientific' input – the insight into systems built up through experience – and investors in research to mistake the different approach as a way of cutting cost.

5 Ever More Specific Applications

The definition of the relationship between polymer structure and application, which has been at the centre of a significant proportion of grant applications or industrial programmes over my scientific lifetime, will need to move to a new level. We claim to understand what the relationship is in some of the more mature bits of the polymer industry – such as injection moulded thermoplastics – but we still seem to encounter problems in end use which run against received wisdom.

Despite what is drawn in scientific papers we do not make nice clean polymers, with only a distribution of molecular weights to challenge the professionalism of the synthetic chemist – and that is before a process engineer has got his/her hands on the recipe.

We need to approach the derivation of structure–property relationships with a wider view. Can we capture the current knowledge and reduce it to simple models that can be used in a practical situation? Can we incorporate the effect of processing to enable us to make exactly the right structure for the end use?

Or can we harness the power of what have now become fairly standard computers to provide more complex models where the approximations do not cause significant error between aim and result?

6 Higher Financial Hurdles

There are other, more practical matters that help shape our industry and its ability to fund and exploit these developments in science and technology.

The money we use to provide the capital and revenue for our industry comes primarily from shareholders. As such we are in competition with other places where they can invest their money. Also, the money does not come directly to us – either in industry or academia. That means that there is a 'supply chain' of money where everyone seeks to 'add value' by helping invest more effectively. Despite the warning that comes with every investment advertisement these days, everyone uses past performance as a guide to future potential.

7 Fashionablity of Re-organisation

One knock-on effect of this desire for demonstrable progress is that the whole industry seems to be in a constant state of re-organisation. Companies acquire other companies, merge with other companies and, increasingly, simply disappear. The driver for this is partly to do with increased efficiencies in those parts of the organisation it is possible to share – with the consequent job losses – but much more to do with assembling an organisation with the full set of skills to deal with the requirements of the chosen market sector or technology. Companies will buy other companies because they believe they have the piece of intellectual property they need to be more attractive to investors. This can be scientific knowledge, manufacturing know-how, customer lists or routes to market. This process is also being carried out at a global level. Companies are no longer tied by where their headquarters have historically been. Shareholders are now spread around the world. Markets and customers are similarly dispersed, and the drive for local manufacturing to lower supply chain costs is placing factories closer to the markets.

The process has not yet finished and there will be more rationalisation of the polymer industry. There are several learned analyses out there that purport to define the likely structures that will result. My own favourite is down to the Arthur D Little organisation.[2] They see roles for four types of companies:

The first, which they define as a Chemicals Utilities company, will have as its value proposition the lowest total cost to produce. Its characteristics will include; an R&D organisation focussed on process improvement/development, being good at manufacturing commodity or speciality chemicals, a low-cost position, large-scale production of a narrow range of products, a highly efficient interface with suppliers and customers, little technical service – trouble-shooting only, and performance driven, command and control culture;

The second, which they call a Solution Partner, will have as its value proposition the provision of the best solution to its customers. Its characteristics will include; being good at proactively understanding markets' needs, working closely with customers, tailoring chemical products to market requirements, frequently outsourcing its production, manufacturing a large number of grades (many of which are customer specific), extensive technical service/application development, and a 'customer is king' culture;

The third, which they call a Chemicals Service Provider will have as its value proposition the best customer-specific result. Its characteristics will include; being good at understanding customer needs and economics, providing chemical-related services to diverse end users, operation of 'outsourced' chemical-related activities within customer site, manufacturing frequently being linked to customers' production, and a service-oriented culture;

The final category, and the one where most research people would see their role is called the Product Innovator. Its value proposition is control of the best product/process. Its characteristics will include; the development and marketing of new patentable chemicals or new patentable processes, being good at exploiting and managing technology, it can be an innovative development company or established company, Licensing out developments, production frequently outsourced, and bright, innovative technology-driven culture.

The source paper for this work is now about a year old and fails to identify the new e-commerce wing of the chemical companies such as ChemConnect, ChemPoint and CheMatch.

The other question to ponder is how these different types of company will address the relationship with academia. The availability of finance and the cost of labour will determine what type of company will be based in the UK, and that in turn will control the interaction between industry and the universities. If the headquarters is in the UK for historical reasons, but the operating divisions and units are in other countries, why would they invest in UK academic polymer science?

8 An Image Problem?

The final challenge that faces the polymer community comes from the products themselves. Polymers are mostly seen as the cheapest option in the materials portfolio. People hear polymers and think cheap, commoditised thermoplastics. So paints are not thought of as polymers – despite the fact that they are a mixture of polymers, surfactants, pigments and so on – although they no longer have an organic solvent! Most products that we use to clean and 'upgrade' ourselves with contain a polymeric component to alter the rheology or final mechanical properties. Hair spray consists of a propellant gas, a polymer and again surfactants. Soap and detergents moved long ago from the cottage industry status to large-scale production, with added perfumes and bactericides, as well as surfactants (increasingly of a molecular weight that would define them as polymers!) to make the foam feel that 'special' way on our skins! Aside from fresh meat and vegetables, there is very little in the supermarket which does not contain a polymer or oligomer to make it attractive or useful. Pastes, creams, syrups, *etc.* use polymers to give them their texture and (often) taste. The whole 'low fat' industry depends on polymeric materials to deliver the other attributes which went when the fat was removed. The revolution in the electronics industry relies on photolithographic masks that are used to define smaller and smaller structures, adhesives that are used to assemble the components and the base materials from which the disks and tapes are made. The Automotive industry relies on polymers to make the interior of the car more habitable, and the components within the engine and suspension lighter and (more and more) to stick the whole thing together! Pharmaceuticals increasingly rely on polymers to deliver the physiological effect that have value for in a controlled manner! Very little of our modern life is untouched by polymers, yet what most people associate with the word 'polymers' is cheapness and pollution – and not a better life. I wonder what we should do about it?

References

1 Stolen from Professor Paul Smith, ETH – An Article on the Future of Materials Science.
2 *The Future of the Chemical Industry*, Gordon Nechvatel, Prism, 1999.

3

The Impact of Environmental Trends in the Coatings Industry on Polymer Design

Steve G. Yeates

AVECIA LIMITED P.O. BOX 42, HEXAGON HOUSE, BLACKLEY, MANCHESTER M9 8ZS, UK

1 Introduction

This paper will discuss the impact of environmental and legislative pressures upon the surface coatings industry and serve to highlight areas where advances in polymer technology are required to meet these developments. Environmental legislation continues to be a major force for change within the surface coatings industry. This is driven both by concern for human health (classifiction/labelling, occupational exposure limits, notification/testing and market restrictions/bans) as well as the environment (reduced emissions, VOC controls, waste minimisation). We are therefore witnessing a move towards cleaner technologies which take a holistic view of the whole coatings process. Technologies are becoming ever more complex in order to deliver comparable or superior in-use performance coupled with an enhanced environmental profile, demanding ongoing advances and developments in polymer chemistry.

1.1 Short-term/High-probability Issues

The move from energy and solvent intensive technologies towards more ecofriendly alternatives is the macro environmental business driver in the surface coatings industry today. Put simply, it is generally taken to mean the move towards VOC compliant or clean technologies such as high-solids (>>65% solids), waterborne, powder and radiation curing. Legislation controlling the release of volatile materials to the atmosphere now exists and when enacted will give criteria for reducing volatile emissions. This will further accelerate the switch to compliant technologies.

Penetration of 'clean technologies' is expected to grow in Europe from 35% in 1994 to around 47% by 2004. Environmental dynamics are different for each market segment due to the differing impact on the environment of the manufacturing processes, of the redundant coating in the waste cycle as well as the need for the coating to perform a defined task. Running counter to this is the continued reduction in costs, which can in the extreme result either in a drift back to older technologies where regulations allow, or geographic relocation to countries where environmental constraints are not so restrictive (Far East). Therefore solvent based technologies are expected to persist in the longer term. In solvent borne technology we are witnessing a move both to higher solids as well as towards the use of more benign solvents. In waterborne coatings we can see the need for greater control of macroscopic architecture (*i.e.* core-shell structures in emulsion polymerisation) for improved film formation and property balance as well as the trend towards co-solvent removal. In solvent free technologies (*i.e.* thermoset powder) there is the move to lower temperature crosslinking coupled with a desire to exploit radiation curing technologies.

1.2 Medium-term/High-probability Issues

Response to changing market and legislative demands, often requires a step change in product design philosophy, increasingly integrating eco-design into product development. Each link in the supply chain will have its own environmental impact and will also contribute to the impact of those links further down the chain. Conversely issues important further down the chain will have design and supply implications in the back specification of materials. The demands of the short term will not necessarily be those that are important 10 years out.

Legislative and market pressures are forcing a number of materials to be phased out of resin and paint formulations. Examples include tin and zinc replacement in antifouling paints and floor polishes respectively. Environmental drivers for the removal of oestrogenic (alkyphenol ethoxylates) and halogenated (chlorinated) materials are necessitating product reformulation. This is a dynamic issue. Formaldehyde free systems have been a topic of debate for a number of years. However, the problem of all proposed alternatives has been inferior performance compared with the excellence and robustness of MF systems. Since such materials tend to be factory applied we are seeing a shift to formaldehyde abatement particularly in the US, which shifts the emphasis to low formaldehyde containing hardeners, and systems which make more effective use of the available cross-linkers.

However, replacement strategies are not straightforward with compromises on either performance and cost having to be addressed. An excellent example of this centres around the chlorine issue. Chlorine imparts a unique set of properties to addition polymers, such as the ability to crystallise on film formation. Vinylidene chloride copolymers have excellent gas, odour and moisture barrier properties impossible to achieve from non-chlorinated waterborne film forming materials. In the barrier packaging area it has a unique position, with proposed replacements being unable to fully match its overall performance whether in terms of overall barrier or flexibility of application. So here we have a paradox, with the environmental lobby striving for the limitation and subsequent removal of chlorinated material but with the downside that greater food spoilage will result. In the end the environmental issue becomes one of compromise, taking into account the overall impact.

Measures such as the Ecolabel for decorative interior paints are being taken with a phased reduction in VOC thresholds out to 2004 and beyond. The concept of examining the inherent risks and associated costs pre legislation is gaining ground and will only be reinforced with the advent of accepted Life Cycle Inventory (LCI) techniques. However, solvent elimination should not be seen as the only way of improving environmental acceptability. Other factors will come into play dependent upon the method of application (*i.e.* factory or on-site), function in-use (*i.e.* durable coating *vs.* 'transitory packaging') and method of disposal (incineration *vs* land-fill).

1.3 Other Longer Term Issues

A major issue for manufacturing processes is the need to minimise the generation of waste, whether it be solid, liquid or gaseous. The reasons for this are two-fold: Firstly the need to operate within agreed regulatory consent limits for emissions and waste disposal, and secondly the increasing cost of waste generation and treatment.

The final part of the picture concerns recyclability of coating materials. Improvements in transfer efficiency to substrate and recycle and reuse of overspray are becoming of increasing importance. The use of super critical CO_2 processing and application is being commercialised and is seen to offer potential benefits over conventional solventborne technologies.

Additionally the fate of the finished article is warranting greater attention with fate analysis being performed upon different resin systems. Thinking is furthest developed in the packaging industry where the transient nature of the coated article, often paper and board, means that disposal is a real issue. Therefore coatings need to not only deliver function in use, but also must not negatively impact on the repulpability (or compostability) of the coated substrate. New market opportunities are being opened up as a consequence of a shift in environmental thinking. The issue is not just important for packaging, but for coatings in general. More durable coatings will have a more favourable impact because of the need to replace on a less frequent basis (decorative paints). The environmental impact of stripping and removal technologies will need to be assessed opposite the new durable technologies.

The use of biosustainable feedstocks is widely discussed, but little if any penetration within the coatings industry has been observed to date (alkyds and shellacs apart). The reason for this has been the generally poor performance when compared against synthetic oil based comparisons. However, performance criteria when weighted against environmental standards may shift, such that renewable resource based products may become practical. It is to be anticipated that this will become a major fruitful area of research within the next decade, with focus upon such feedstocks as starch-dextrin, natural rubber, maize (furans) and novel oils.

2 The Role of Controlled Polymer Architectures

The majority of ongoing developments concern the delivery of functionality to one part or another of the coating in order to perform a specific task, such as enhanced dispersability, interfacial or bulk compatibilisation, adhesion promotion or crosslinking. When considering the available polymeric building blocks, we are to a large extent restricted to those monomer sets which are already EINECS listed. Any polymer which contains >2% of a non-EINECS listed material is required to undergo extensive and expensive notification and testing. Legislation appears particularly biased against European developments *versus* our US, but particularly Far East counterparts. The EEC 7th Amendment of the Dangerous Substance Directive (DSD) has helped ease and harmonise the notification procedure for small volume chemicals, but the effect has been to rationalise the market for chemicals by discouraging the introduction of new ones. For polymers and materials for use in food contact (direct or indirect) the situation is further limited by FDA, The Synoptic VII Food Contact with Plastics Directive (EEC) and national listings. Therefore with limited exceptions we will be required to work within the existing monomer base sets. The situation regarding materials and additives for use at the <2% level however is not necessarily so restrictive, and notification costs (Annex VII) can be phased with volume. Therefore the development of catalysts, additives and comonomers for use at <2% do offer routes to technical differentiation.

Clearly therefore technical advantage will arise to those who can assemble the existing monomer sets in the most effective and elegant fashion. The use of polymer blend strategies, whether through simple physical blends, or sequential polymerisations is well documented but will not be considered further here. However, the ability to additionally impart control at the molecular level is only now beginning to be exploited and exciting opportunities are being opened up.

Through the use of 'controlled' polymerisation techniques it is possible to synthesise polymers, based on existing monomers, which have different properties when compared against conventional 'random' processes. Extended aspects of control include:

- More precise molecular weight development: chain length and distribution.
- Chain architecture: linear, grafts, blocks, stars and dendrimers.
- Functionality distribution: Telechelics.

Step-changes in polymerisation catalysis are expected to be made in the next decade which will enable some of the above to become commercial reality.

2.1 Molecular Weight Control

The ability to synthesise low molecular weight functional (meth)acrylate polymers of narrow molecular weight distribution is increasingly important both in the development of low VOC coatings, and other effect materials. There are a number of reasons for this:

- To achieve low application viscosity at higher solids in solvent-borne coatings it is necessary to have low polymer molecular weight and narrow PDi.
- In thermoset powder coatings in order to obtain the desired melt viscosity profile, within the limits of the required T_g, low molecular weight and narrow PDi are required.
- In waterborne emulsion polymers it is often advantageous to produce low molecular weight polymer in the aqueous phase to improve both film formation and application properties.

Processes for the manufacture of low molecular weight (meth)acrylate polymers are well known and practised. However, there are often issues around high polymerisation temperatures, the need to subsequently remove solvent in the case of solid polymers, and in certain cases the odour associated with high levels of mercaptan based chain transfer agents employed. Catalytic chain transfer polymerisation (CCTP) utilising low spin cobalt(II) complexes is a well documented development which allows the production of low molecular weight methacrylates using conventional free radical techniques. The use of various cobaloximes has been described which have both improved oxidative and hydrolytic stability over the earlier catalysts, as well as enhanced catalytic chain transfer activity to both methacrylates (typically C_s(MMA,bulk) = 2.8×10^4 to 6.6×10^4) and styrene. There still exists the scope for the development of improved catalysts which can be used more effectively in emulsion polymerisation as well as being more tolerant to a wider range of monomers (*i.e.* Acrylates and Vac). This remains an ongoing area of both industrial and academic research. It should be noted that a feature of the CCTP process is that the oligomers so generated have an ω-unsaturated structure. This feature has been exploited in the synthesis of graft and comb block copolymers for such applications as adhesives, dispersants and polymer compatibilisers.

2.2 Chain Architecture

To synthesise polymers with unusual properties from existing basic monomers one needs to place the monomer units in ordered arrays rather than at random. Thus polymer architecture control remains an important area of research. Possible structural elements include block, graft and comb copolymers as well as star and dendritic/hyperbranched topographies. Potential for such structures in the surface coatings and adjacent industries include use as additives (*i.e.* dispersants, polymeric surfactants and impact modifiers) or as main binders having unique rheological or network properties.

The most effective way of controlling polymer architecture in chain polymerisation of vinyl monomers is through living polymerisation. Anionic polymerisation, group transfer polymerisation and screened anionic polymerisation techniques have been developed, but have struggled to find commercial success. Although the advantages are well documented, there are major practical drawbacks such as limited monomer selection along with the need for demanding synthetic conditions, *e.g.* high purity reagents, moisture free and low temperature.

Over the last decade there has been extensively reported work in the area of controlled radical polymerisation which have promised degrees of structure control approaching that of living polymerisation but with manufacturing robustness and economics of conventional radical technology. Various derivations of the concept exist; TEMPO, ATRP and RAFT. In all cases there is a degree of monomer selectivity, less so in the case of RAFT which limits broad application, coupled with a general intolerance to functional groups (acid functionality). For both TEMPO and ATRP the major stumbling block continues to be the level of residual 'catalyst' in the final product; in the case of ATRP up to percentile levels of copper. Consequently strategies which tackle this, either through the need for true catalytic amounts or involving facile cheap removal are required if there is to be broad ranging commercial acceptance. In the case of RAFT there is no catalyst but chain transfer agent is part of the polymer. It should be noted that high added value opportunities may exist where cost sensitivity is less of a issue.

Dendritic polymers continue to attract significant interest representing as they do a high level of structural control. The possibility for higher order terminal functional groups equidistant from the core coupled with a non-Newtonian relationship between viscosity and molecular weight leads to interesting application properties such as favourable rheology and impact modifications. However synthetic strategies are generally complex involving repetitive protection – deprotection strategies which leads to a very rapid rise in material cost with each successive generation. Recent work reported by Courtauld's on a 'one pot' approach shows potential and may have economics consistent with higher end coatings applications. More recently attention has focused upon the synthetically more facile hyperbranched materials. Although they do not have the structural purity of dendrimers they do demonstrate in many instances many of there attractive properties. Reported industrial work has demonstrated a range of effects dependent both upon the available surface functionality, including reduced viscosity, increased toughening and increased reactivity (Perstorp and DSM).

3 Conclusions

Technologies which demonstrate environmental conformance coupled with improved material efficiencies with low energy demand combined with elements of recyclability will win out in the long run. This will lead to environmentally compliant technologies which will have utility outside just the classical surface coatings markets. The conclusion to be drawn from the above is that technical advantage will arise to those who can assemble the existing monomers in the most effective and elegant fashion. We should therefore give serious thought as to how this may come about. Current R&T focus tends to be on polymer blend strategies, whether we mean by this sequential acrylic–acrylic polymerisation, physical blends or composite systems (polyurethane-acrylics or polyester-acrylics). There is no doubt that this will continue to be a fruitful area of activity and will continue to meet many of the environmental targets. However, the ability to manipulate copolymer sequence distribution to manufacture block, star and hyperbranched structures will be important for many applications.

4 The Future

Although the surface coatings industry remains important to UK PLC, that importance as a centre for UK industrial R&T has diminished significantly over the last five years as a consequence of merger, acquisition and rationalisation. Therefore some thought needs to be given to identifying emerging areas which will become commercially important over the next decade, and where UK academe and industry can become engaged.

In the future it is to be anticipated that true speciality polymer opportunities will arise out of the interaction of polymer science with physics and biology. A good example of this are the recently publicised advances in polymeric organic semiconductors and their utilisation as emissive materials in display applications. This is a technology with a bright future, with commercial devices incorporating this technology expected within the next couple of years. Other applications can be envisaged for such materials in areas as diverse as organic lasers to electronic circuitry. These advances have been driven by the close interaction of the solid state physicist with the synthetic polymer chemist. It is an area where UK academia can be said to lead the world and should be identified as an ongoing area of support. The challenge to the UK chemical industry is how to become engaged with these and other exciting opportunities. In doing so one has to rethink the classical business models by which polymers are generally gauged.

There is also a challenge to PhD students within polymer science. More than ever it is important that they develop a deep understanding of polymer chemistry but supported by an ability to work freely across other disciplinary boundaries. An ongoing interest, appreciation and understanding of the recent topical developments and challenges in polymer science, and science in general, is I believe important to help them face the challenges in what is a rapidly changing business environment.

4

Emerging Trends in Polymer Semiconductors and Devices

Donal D.C. Bradley*

DEPARTMENT OF PHYSICS AND ASTRONOMY,
THE UNIVERSITY OF SHEFFIELD, HICKS BUILDING,
HOUNSFIELD ROAD, SHEFFIELD S3 7RH, UK.

1 Introduction

1.1 Applications

Synthetic polymers already find widespread application as replacement structural materials for natural products such as wood and cotton and for traditional man made materials including glass and metal alloys. Their advantages are generally perceived to include their low mass, ease and energy efficiency of processing and their low cost. Recent efforts, nearing commercial fruition, have focused on the application of polymers as active functional materials for electronics, optoelectronics and photonics. Specific areas under current development include light emitting diodes [LEDs],[1] photodetectors and solar cells,[2] TFT arrays and memory elements,[3] amplifiers and lasers.[4] LEDs are the current driving force for commercial development but other areas are now starting to gather momentum.

1.2 Conjugation

In order to be suitable for use in the above applications the polymer in question must contain a significant fraction of conjugated units. Conjugation is the term that describes the existence of a contiguous sequence of double and triple bonds separated by no more than one single bond. The simplest example of a conjugated polymer is *trans*-polyacetylene and we use this historically important example[5] to discuss the bonding found in such materials (Figure 1):

* New Address: Blackett Laboratory, ICSTM, Prince Consort Rd, London SW7 2BW (E-mail: D.Bradley@ic.ac.uk).

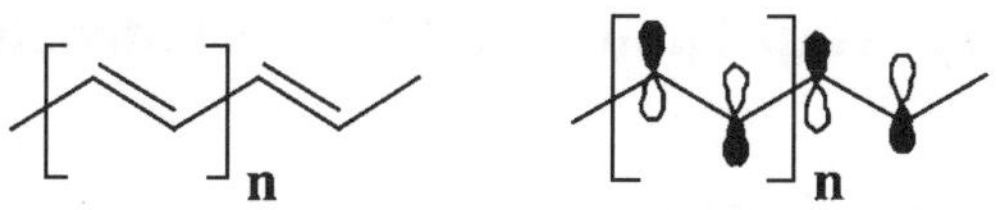

Figure 1 *Chemical structure of* trans-*polyacetylene and a simplified orbital picture showing the p_z orbitals sticking out of the plane formed by three σ-bonds. Only two of these, namely those connecting neighbouring carbon atoms, are shown. There is an additional σ-bond per carbon to a hydrogen atom (not shown)*

Each carbon atom in the polymer backbone has three sp^2 hybrid orbitals that form two σ-bonds with other carbon atoms and one σ-bond with a hydrogen atom. This planar trigonal arrangement of bonds forms the scaffolding that gives the polymer chain its rigidity and mechanical strength. In addition there is one p_z orbital per carbon atom that sticks out of the plane and forms π-bonds by lateral overlap with neighbouring p_z orbitals. These π-bonds can delocalise by mutual interaction along the conjugated polymer backbone. It is the π-bond network that thus forms which is responsible for the desirable electrical propeties since it allows motion of charge carriers and hence conduction. In order to achieve a high conductivity one must also produce a high charge carrier density. The conductivity $\sigma = n\ e\ \mu$, where n is the carrier density, e the charge on the electron and μ the carrier mobility. Typical conjugated polymers have optical gaps of 2 eV or more and hence there is negligible thermal excitation of carriers at room temperature ($kT_{300K} = 0.025$ eV). Extrinsic carriers are introduced by "doping" with redox agents, *e.g.* I_2, AsF_5, Na, Rb *etc.* The dopant has to diffuse into the polymer and undergo a charge transfer reaction. This makes accurate control of doping level (especially in the $\sigma = 10^{-7}$ to 10^{-4} S cm^{-1} range of interest for semiconductor devices) and stability of the doped state very tricky. The delocalised electronic structure also allows optical excitation without the need for bond scission and hence conjugated polymers demonstrate interesting photophysics rather than simply photochemistry. This includes photoluminescence and photoconductivity, both essential processes for a number of devices.

Many conjugated polymers have been synthesised over the years and there is an essentially infinite palette from which to work (see Figure 2). Perversely, this may not be a benefit since it encourages an approach in which the "expected to be better next material" is synthesised rather than making the effort to sort out the purification, processing and stability problems of the current best material. This approach is, however, becoming much less prevalent and a small subset of materials is attracting the bulk of the interest. In particular poly(*p*-phenylene),[6] poly(*p*-phenylenevinylene),[7] poly(thiophene)[8] and poly(fluorene)[9] and their derivatives now very much dominate the activities in the field. Fine-tuning of the electronic properties is achieved by chemical substitution with electron donating or withdrawing groups, by the use of steric interactions to disturb the planar conformation of the backbone

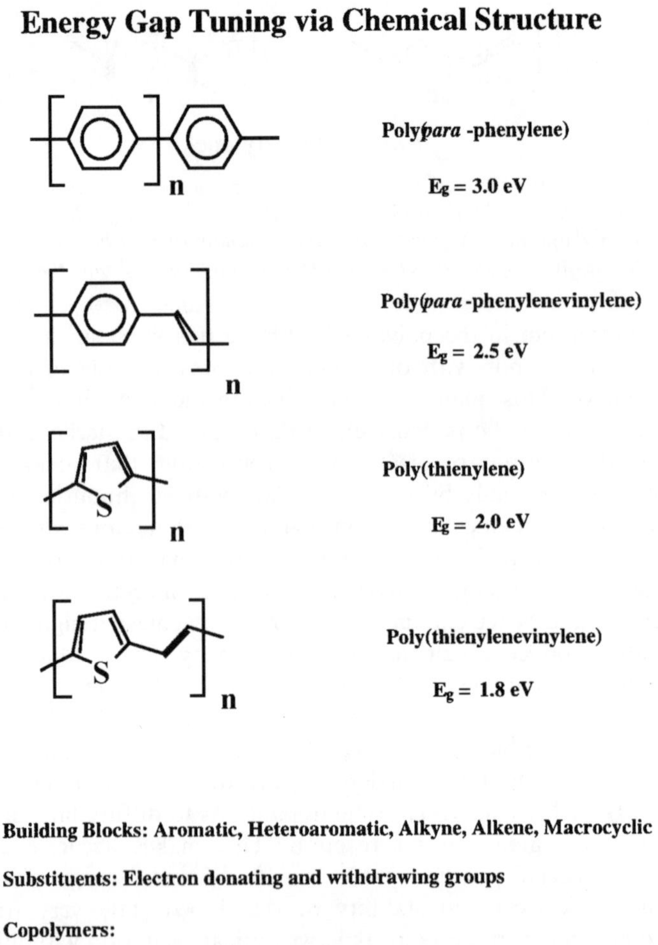

Energy Gap Tuning via Chemical Structure

Poly(*para* -phenylene)

E_g = 3.0 eV

Poly(*para* -phenylenevinylene)

E_g = 2.5 eV

Poly(thienylene)

E_g = 2.0 eV

Poly(thienylenevinylene)

E_g = 1.8 eV

Building Blocks: Aromatic, Heteroaromatic, Alkyne, Alkene, Macrocyclic

Substituents: Electron donating and withdrawing groups

Copolymers:

Figure 2 *Illustration of approaches to energy gap tuning* via *control over chemical structure. The energy gaps of the four polymers are given in the figure.*

(needed for maximal π–π overlap) and by insertion of other moieties within the chain. The latter can be both conjugated units giving rise to conjugated copolymers or saturated units that break conjugation. Through chemical synthesis one can control the fundamental electronic properties of energy gap, electron affinity and ionisation potential. The interactions between molecular sub-units lead to further perturbations of the electronic structure that can have significant influence on light absorption, emission and charge transport and that are very strongly influenced by both chemical and physical structure.

2 Characteristic Features of Molecular Semiconductors

Conjugated polymers are one of several classes of molecular semiconductor. Other examples include organic crystals like anthracene and glassy materials like Alq_3. All molecular semiconductors share a number of common features that distinguish them from more conventional inorganic semiconductors.

2.1 Localised Wavefunctions

Unlike crystalline inorganic semiconductors the electronic wavefunctions of molecular semiconductors are not Bloch wave states delocalised over the whole sample. Instead, the wavefunctions are, more or less, localised to a molecular subunit consisting of a single molecule or a segment of polymer chain. In molecular semiconductors, vibrational deformations of the lattice (typically Raman active modes) couple to the electronic states. The resulting vibronic levels are manifest in the spectral response seen for the absorption and emission of light (see Figure 3) and lead to broad spectral bandwidths that are both potentially attractive (for lasing and amplification) or problematic (for saturated red, green and blue colours in displays). Such lattice coupling is also manifest in the material's response to the addition of charges. Unlike inorganic semiconductors where added charges occupy pre-existing valence and conduction band states, in a molecular semiconductor there is a reorganisation of the local bonding around the charge leading to the formation of polaron-like states. This again affects the spectral response leading to possibilities for novel optical modulators[8] and also to potential losses in electrically pumped amplifiers and lasers.

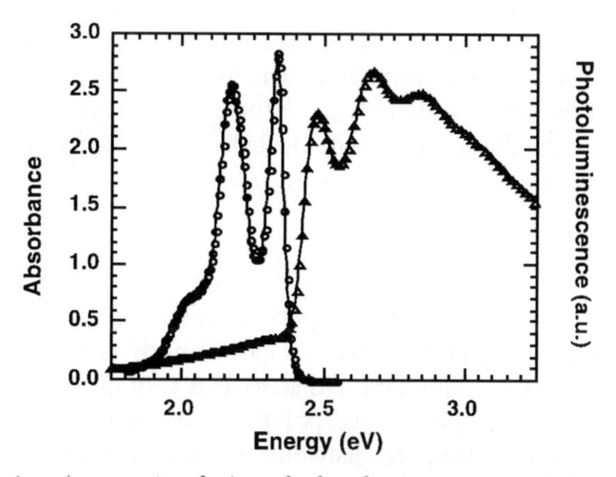

Figure 3 *Absorption (open triangles) and photoluminescence emission (open circles) spectra for a poly(p-phenylenevinylene) film at low temperature (75 K).*

A further consequence of localised wavefunctions is that the interactions between electrons are strong. As a consequence, and re-inforced by the small degree of screening characteristic of the low dielectric constant ($\varepsilon_r \leq 3$–4) typical of molecular semiconductors, the Coulomb and exchange energies are large. A large Coulomb energy leads to excitons that remain tightly bound at room temperature and above. Experimental and theoretical estimates of the binding energy vary substantially but a typical figure would be 0.5 eV. This compares with 0.004 eV for GaAs. A large exchange energy leads to a distinct manifold of triplet states (spin $S=1$) that lie at lower energy than the corresponding singlets ($S=0$). The triplet states are spin-forbidden for dipole transitions to or from the ground state and consequently have long lifetimes. If they do emit light, the emission is described as phosphorescence. In terms of LED performance, the existence of triplets is normally considered to lead to an upper limit for the internal quantum efficiency of 25% of the fluorescence efficiency. This limitation can be tackled by incorporation of materials that efficiently emit phosphorescence or which lead to a mixing of singlet and triplet states. Generally this is achieved through the presence of a heavy atom for which there is a large spin-orbit coupling. An example of materials with strong phosphorescence is Pt-octoethylporphyrin.

2.2 Anisotropy

Inter-chain interactions in conjugated polymers are much weaker than the strong intra-chain π-bonding. To first order the inter-chain interactions are of the van der Waals or weak π–π overlap type. As a consequence, conjugated polymers are intrinsically anisotropic. This is manifest in both optical and electrical responses. Optical anisotropies of up to 100:1 have been measured in absorption for epitaxially oriented poly(diacetylene)s. This anisotropy can be put to good use in polarised backlights for liquid crystal displays and in controlling amplification/lasing thresholds through polarisation control over the allowed modes. A further consequence of the low dimensional electronic structure is that the optical transition elements are very large. Typical values are 10^5 cm^{-1} compared with 10^3 cm^{-1} for a bulk inorganic semiconductor. This is an attractive feature for many optoelectronic devices and also leads to strong optical nonlinearities.

2.3 Disorder

Whilst disorder is not an intrinsic property of molecular semiconductors it is a common feature. Note that this disorder can be both static (introduced during sample preparation) or dynamic (instigated as a response to the environment, *e.g.* ring torsions in phenylene based polymers are thermally activated at room temperature). Disorder leads to distributions in conjugation length (effective delocalisation length for the π-electron system) and hence to distributions in many electronic properties. Spectroscopically it is manifest through inhomogeneous broadening which can smear out the vibronic peaks and lead to broad featureless absorption and emission bands. "Spectral diffusion" is also seen as shown in Figure 3 where the vibronic peak linewidths for the photoluminescence emission spectrum are narrower than those for the absorption spectrum. This is a consequence of the fact that the initially excited population relaxes through the inhomogeneously broadened density of states to occupy states lying lower in energy. The emission then preferentially samples the lower energy states that are better-ordered (lower energy associated with longer conjugation length). Absorption, conversely samples the whole distribution of states and thus shows greater inhomogeneous linewidths.

3 Conjugated Polymer Synthesis and Processing

3.1 Synthesis

As discussed above the desirable (semi)conducting properties of specific polymers derive from their possession of a conjugated backbone structure. However, the presence of conjugation also leads to a rigid chain that, in turn, commonly leads to insoluble and infusible materials for which purification and processing is difficult. Furthermore, the solubility usually decreases rapidly with increasing degree of polymerisation such that the products can often have low molecular weight. Three approaches have been developed to overcome these difficulties:

3.1.1 Precursor-Route Methodologies. Here, a non-conjugated intermediate, or "precursor", polymer is first synthesised. This material is designed to be soluble in common solvents so that it can be readily processed into a desired structure (*e.g.* thin film). The polymer is then subjected to a conversion reaction in which the conjugation is generated by elimination or condensation. The resulting conjugated polymer is insoluble and infusible. Examples of this procedure include the Durham route to polyacetylene[5] and the sulfonium polyelectrolyte route to poly(*p*-phenylenevinylene).[7]

3.1.2 Incorporation of Bound Solvent Moieties. The desired conjugated backbone is laterally substituted by bulky side groups that can act as bound solvents (increasing the entropy of the polymer in solution). This approach is now very widely used. Typical examples include poly(3-alkylthiophene)s,[8] poly(2,5-dialkoxy-*p*-phenylenevinylene)s[1] and poly(9,9-dialkylfluorenes).[9] The substituent groups can have significant effects on the π-electron delocalisation both through steric interactions and through their electron-donating or electron-withdrawing nature.

3.1.3 Vapour Deposition Polymerisation. This is a little studied approach but one that offers significant potential for the fabrication of very thin films and for elaborate multilayer structures. A commercial process has been developed by the Ulvac Corporation in Japan to coat magnetic relay switches with an insulating polyimide layer. A polyamic acid is sythesised by co-deposition of two reactive monomers and is then thermally imidised. The same approach can be used for the condensation polymerisation of poly(azomethine)s,[10] poly(oxadiazoles) and poly(quinoxalines) all of which have been used in LED structures. This approach to polymer synthesis is ripe for further development.

3.1.4 Processing. In order to achieve the promise of cheap device fabrication it will be very beneficial to be able to use solution coating techniques and more particularly techniques adaptable to large area and continuous coating. Batch processing and especially processes that require ultrahigh vacuum are unlikely to allow very low cost device fabrication. Two approaches attracting current interest are reel-to-reel coating and ink-jet printing. The latter offers the attractive promise of a non-lithographic device patterning process. There are, however, significant issues to be addressed before these techniques can be used as general processing methods. Formulation of polymer "inks" to provide the required rheological properties, wetting behaviour on electrode substrates, *etc.* is a challenging problem. Multilayer coating in order to optimise device performance is also difficult since one normally wishes to ensure discrete layers with little intermingling. Film uniformity is a critical parameter in order to ensure uniformity of electrical and optical properties across a device element. For an LED a typical film thickness is only 100 nm and because the device current has a strong field dependence, thickness fluctuations need to be minimised in order to provide a uniform emission.

4 Polymer LEDs

I have chosen to take the polymer LED as a representative example of a polymer optoelectronic device that then allows a more concrete discussion of the role of polymer design, synthesis and processing in optimising device performance. Polymer LEDs are attracting interest for a range of lighting, indicator and display applications. Their attractive features include the fact that they are emissive (viewing quality similar to a standard cathode ray tube), that they can be driven with low voltage DC (for the best green devices 100 cd m^{-2} brightness can be achieved at less than 3 V), that they have high power efficiency (20 lm W^{-1} or higher for green devices), that a full colour range is accessible *via* energy gap tuning and that large area devices should also be available.[1] Additional properties derived from the use of a polymer semiconductor include the possibility of flexible (or conformable) devices, polarised emission devices (see further below) and opportunities for simple processing (as already discussed). The emission of light from a polymer LED requires two key steps, namely exciton formation and exciton decay (see Figure 4).[1] Exciton formation involves carrier injection, carrier transport and carrier combination *via* Coulomb capture. Both singlet and triplet excitons are generated with a 1:3 ratio expected from a simple consideration of spin statistics. Exciton decay relates directly to the photophysics of the material and involves the competition between radiative and non-radiative decay processes (taking into account the influence of the different environment found within an operating device). Phosphorescence from the triplet state is normally spin forbidden and even when weakly allowed through heavy atom effects leads to a long emission decay time that could be a problem for video-rate displays. Optimisation of exciton formation requires efficient and balanced injection, balanced carrier transport and efficient carrier combination. These factors can be combined into a single charge balance factor, γ. The internal quantum efficiency, η_{int}, is then:

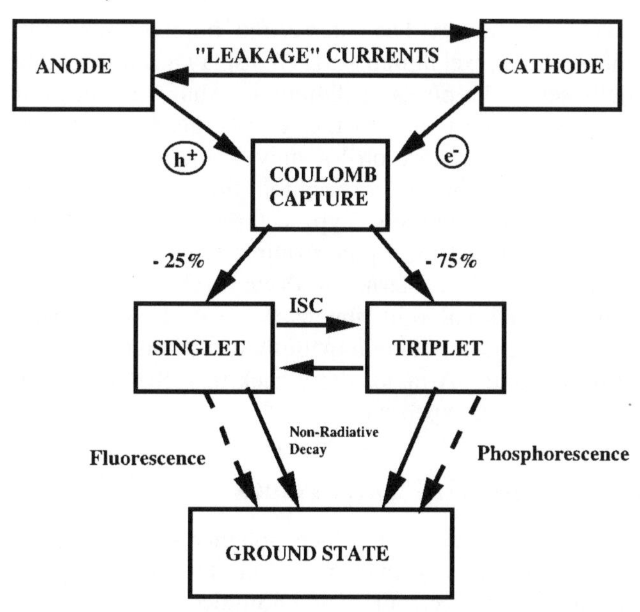

Figure 4 *Flow diagram for bipolar injection electroluminescence.[1] Leakage currents are constituted from those injected charges that pass through the device without undergoing Coulomb capture. ISC stands for intersystem crossing between the singlet and triplet manifolds.*

$$\eta_{int} = \gamma\ \{(\beta\ \phi_{Fl}) + (1 - \beta)\ \phi_{Ph}\}$$

Here β is the proportion of excitons formed in the singlet state and ϕ_{Fl} and ϕ_{Ph} are the solid state fluorescence and phosphorescence quantum efficiencies. In most cases ϕ_{Ph} is negligible but recent interest has focussed on enchancing this value since otherwise up to 75% of the excitons created may be lost without emission of a photon. The external quantum efficiency, η_{ext} , is lower than η_{int} by an amount determined by the out-coupling efficiency of light into the forward (useful) direction. Light that is internally reflected and guided within the plane of the film is lost to the viewer. Devices need also to be power efficient and this requires that a useful brightness (100 cd m^{-2} or more for a typical display device) is achieved with high quantum efficiency but also at a low operating voltage. In terms of polymer design, the focus of interest is on producing materials that have (i) ionisation potentials and electron affinities that match the work functions of preferred electrodes (anodes and cathodes respectively); (ii) high mobilities for electrons and holes so that high brightnesses can be reached and hence that passive matrix addressing may be possible; (iii) high fluorescence and phosphorescence efficiencies and (iv) good colour definition so that red–green–blue colour mixing can be performed to produce full colour images. Some of these issues will be discussed in more detail below. Note that passive matrix addressing is the scheme whereby each pixel of a display is turned on sequentially by applying a voltage to a particular row and column electrode. The intersection point of the row and column selected defines which pixel illuminates. To display a changing image one needs to sequentially select the pixels to illuminate. That this does not appear as a "freeze frame" sequence is due to the refresh rate being faster than the eye response. Individual pixels are on for only a fraction of the time and hence must be driven to higher brightness such that the eye's integration still provides an acceptable average brightness. Typical target peak brightnesses are then a few hundred thousand cd m^{-2} that requires large currents and correspondingly large mobilities. The alternative driving scheme, namely active matrix addressing uses individual thin film transistors to switch each pixel. As a consequence all pixels can be simultaneously turned on or off and no averaging of the brightness occurs. A major draw back to active matrix displays is that they are inherently more expensive.

5 Some Current Materials Issues

There are a number of areas in which polymer semiconductors need to be further developed in order to map more fully the range of potential applications in electronics, optoelectronics and photonics.

5.1 n-Type Polymers with High Mobility

Most conjugated polymers are p-type materials, *i.e.* they have an excess of positive charges that are mobile. This is nothing new for molecular semiconductors since all commercial organic xerographic systems are based on hole transport. Why negative charge carriers are less mobile remains unclear. There is no *a priori* reason in terms of the electronic structure why this should be the case. The most likely explanation is that there are preferential traps for electrons. What these are and why they are ubiquitous to nearly all organic semiconductors is not known with certainty. Molecular oxygen is suggested to be important in this respect. The approach to designing electron transport materials has been to simply increase electron affinity. This means that for a specific trap energy, there will come a point when the electron will prefer to stay on the polymer rather than transfer to the trap. This approach also generally leads to a hole blocking effect since the HOMO level is pushed down and hence the ionisation potential increases. As a consequence holes are hard to inject into the material. High electron affinities are also beneficial for injection from less reactive cathode metals (with higher work functions). Increasing the electron affinity can be achieved by several means. Most common is to add electron withdrawing moieties as substituents.[1] Examples include CN-substituted poly(*p*-phenylenevinylene) derivatives. Another approach is to incorporate electron deficient nitrogen containing aromatic groups,[11] *e.g.* five-membered rings such as triazole,[12] six membered rings such as pyridine[13] and ten membered cycles such as quinoxaline.[14] All of these, and other examples, have been incorporated into polymers that do indeed then have high electron affinities. Note, however, that a high electron affinity is not a sufficient condition for achieving a high electron mobility. To date polymer electron mobilities have generally been very low. Further work is needed to optimise the properties of such polymers. An ability to order the polymer chains would be beneficial in enhancing the mobility. Higher hole mobilities are also desirable and some effort has been expended on this. Fluorene containing copolymers have proven to be especially attractive. Fluorene–arylamine copolymers combine a relatively high hole mobility with a low injection barrier to indium tin oxide.[15] This allows an ohmic contact to be formed.[16]

Another target of interest would be to develop an n-type equivalent of the hole conducting polymer poly(ethylenedioxythiophene) [PEDOT]. PEDOT, when doped with polystyrenesulfonic acid [PSS] forms a transparent conductor that can be used as an injection layer on top of the traditional indium tin oxide anode. This is now used in all state of the art LEDs and promotes efficient injection and also improved device stability. PEDOT/PSS acts as an ohmic contact for many hole transport polymers allowing the observation of trap free space charge limited currents. A corresponding electron injection layer would be very beneficial but n-type doping generally leads to conducting polymers that are unstable in the presence of moisture. The development of a stable n-type conducting polymer is a severe challenge but one that would bring substantial reward.

5.2 High Emission Efficiency Polymers

The same excited states are generated in LEDs as following photoexcitation and hence useful insights can be obtained from photophysics studies. Non-radiative processes that compete with efficient emission include aggregate/excimer/exciplex formation and migration to quenching sites associated with impurities, chemical defects and interfaces. These are strongly influenced by intermolecular interactions and hence by the configuration of the polymer chain. Design criteria adopted for high emission efficiency include ensuring that (i) the lowest singlet is connected *via* a dipole allowed transition to the ground state (avoiding the same parity for ground and excited states (*c.f.* polyenes)), (ii) that excited states have limited mobility and hence largely avoid non-radiative decay sites, (iii) that charge transfer is minimised in the excited states, (iv) that excimer formation is avoided and that (v) aggregation is also avoided (*c.f.* concentration quenching of laser dyes). These requirements can be met in large part by selecting phenylene based polymers that possess segmented backbone structures and a non-planar conformation (often arising from the presence of extended substituent groups). One needs also to consider the effects of the synthesis route through its determination of the end groups, impurities and chemical defects. Furthermore the influence of the processing conditions, thermal and photo-stability and film morphology need to be properly taken into account. When segmented polymers are used it is necessary to separately optimise the composition in respect of LED operation since fluorescence quantum efficiency is not the only factor of importance. Charge carrier mobilities are also critical. One example of a polymer designed as a

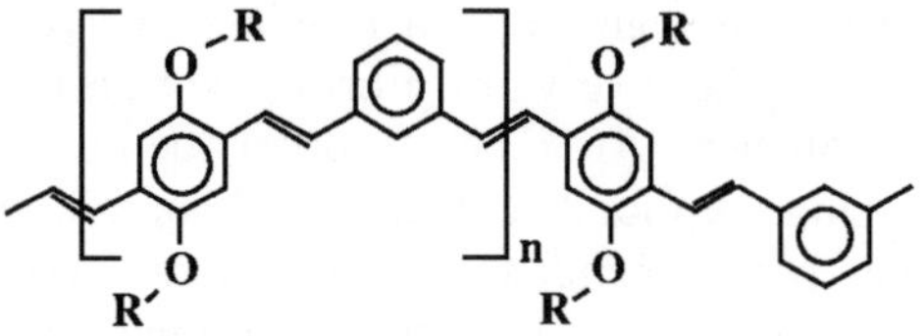

Figure 5 *Chemical structure of the high emission efficiency poly(2,5-dioctyloxy-*p*-phenylenevinylene-co-*m*-phenylenevinylene) ($R = C_8H_{17}$) polymer.*

high emission efficiency material is poly(2,5-dioctyloxy-*p*-phenylenevinylene-co-*m*-phenylenevinylene) (Figure 5).[17] This material contains *meta* linked phenylene rings that effectively limit conjugation without introducing saturated segments. The PL efficiency was found to depend on the fraction of *cis*-vinylene double bonds that are introduced during the Wittig synthesis. Solid state quantum efficiencies lie typically in the range 30 to 70% and lasing can be achieved in polymer solutions.[4] Multilayer LEDs that use this polymer as a green emission material have external quantum efficiencies of up to 1% and a power efficiency of 6 $l\ m\ W^{-1}$.

Additional requirements include high photo, electrochemical and thermal stability materials. These would allow easier access to the operational lifetimes of tens of thousands of hours needed for consumer products. Note that sunlight stability is needed for solar cell applications. In many cases lifetimes are limited by impurities introduced during synthesis or processing, *e.g.* catalyst residues, contaminants from reaction vessels, solvents *etc.* or by chemical defects that are formed by side reactions. Purification is not necessarily straightforward and should be as much a focus for synthesis effort as percentage yields, molecular weights and polydispersities. End groups need also to be carefully considered since in materials of modest molecular weight they can constitute a few weight percent of the total sample. They are then present at the same level as deliberately introduced dyes or other emission materials to which complete Förster energy transfer is often achieved. Encapsulation can also be used to promote stability for both the organic layers and for reactive metal electrode materials. Good barrier films for encapsulation need to exclude both water and oxygen. Finding polymers with suitable properties is not easy and indeed glass remains one of the best barrier materials available. However, for flexible device applications that need to use a polymer substrate, the identification of suitable barrier materials will be a critical component of their development.

6 New and Future Opportunities

The final section of this paper considers several emerging opportunities for the further development of polymer semiconductors. In particular, the focus here is on processing/physical structure aspects that impinge on electronic properties.

6.1 Control over Polymer/Polymer and Polymer/Molecule Phase Separation

Many devices now incorporate multilayer structures in order to optimise the different functions required for efficient operation.[1] In addition polymer/polymer and polymer/molecule blends are widely used to combine different optoelectronic functions within a single layer or to utilise the mutual interaction of the constituents to enhance a specific response. Examples include polymer/polymer multilayers and blends for LEDs[18] and photodiodes/solar cells.[19] Mastering phase separation phenomena in such structures is necessary to promote their operational stability and it also offers an approach to efficient "self-assembly" of specific devices. Both layered (*i.e.* vertically phase separated) and in-plane compositionally modulated structures are of interest. The former is well suited to LEDs where hole and electron injection and transport functions can be separated from an exciton formation and recombination region.[1] Spin coating sequential layers is one approach but there are often difficulties to identify mutually compatible solvent systems that do not dissolve earlier levels of the structure. Phase separation/surface segregation phenomena may alleviate this difficulty. Similarly for photodiodes, exciton dissociation can be optimised independently of electron and hole extraction. Interpenetrating networks can provide a large surface area for exciton dissociation and some of the most efficient photodiodes use blends of this type. Lateral phase separation on a larger length scale can be used to pattern emission regions and thus form arrays of pixels. Again, if sufficient control can be established over the location and uniformity of such structures there is the promise that an efficient, non-lithographic patterning process could be developed. The application of well known principles from polymer physics to the processing of conjugated polymers is rather overdue. To some degree this has been a function of the quality and quantity of materials available. However, the conditions are now ripe to exploit this opportunity.

6.2 Mesogenic Ordering of Conjugated Polymers

Liquid crystalline phases involve the ordering of an array of molecules. If these phases can be oriented into a monodomain and quenched into a room temperature glass they then offer potential for the fabrication of ordered films for use in devices. Order of this type is of interest for several reasons. The first is that there is an intrinsic molecular anisotropy of the electronic structure of conjugated polymers and this anisotropy becomes expressed on micron or higher length scales *via* the mesogenic ordering. It is then correspondingly simpler to investigate the fundamental properties of the material in question since the anisotropy can also be readily probed. Furthermore, one can expect to obtain polarised emission from such samples since the emission dipole moment has a defined alignment with respect to the projection of the polymer backbone. Polarised light sources are of interest as backlights for conventional liquid crystal displays. The ability to avoid the need for an absorptive polariser

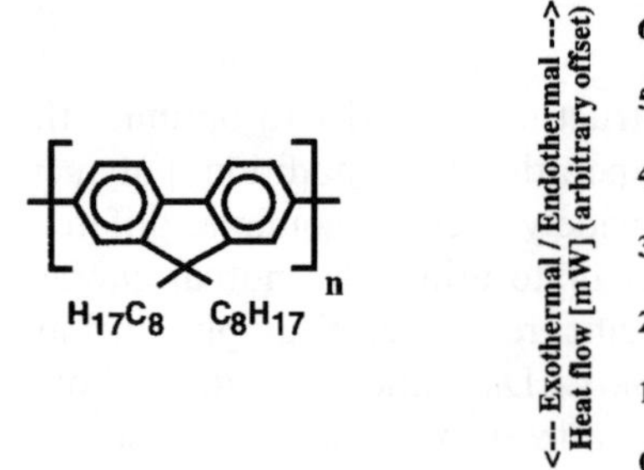

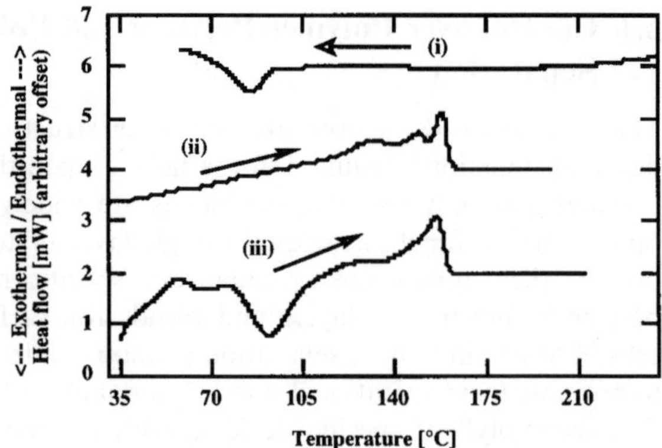

Figure 6 *Chemical structure of poly(9,9-dioctylfluorene) and Differential Scanning Calorimetry measurements of its thermal properties.*[20] *Trace (i) shows the heat flow on slow cooling from the melt, trace (ii) on subsequent heating and trace (iii) on heating a film quenched rapidly from the melt.*

is attractive since it offers significant energy savings due to the removal of a substantial loss element and also offers simplied manufacture. Polarisation ratios need to be at least 12:1 and in some applications the ratio needs to be as high as 200:1. Mesogenic ordering also offers an approach to mobility enhancement. We have previously shown that several polyfluorenes are well-behaved thermotropic liquid crystals and can be quenched (rapid cooling) into a glass or crystallised (slow cooling) from the liquid crystal melt.[20] Figure 6 shows the chemical structure of one such liquid crystalline polymer, namely poly(9,9-dioctylfluorene). Also presented are DSC themograms that show the thermal properties of films prepared in different ways.

Figures 7 and 8 show the effect of mesogenic ordering on both charge carrier transport[21] and on electroluminescence emission.[22] In Figure 7 we see that the

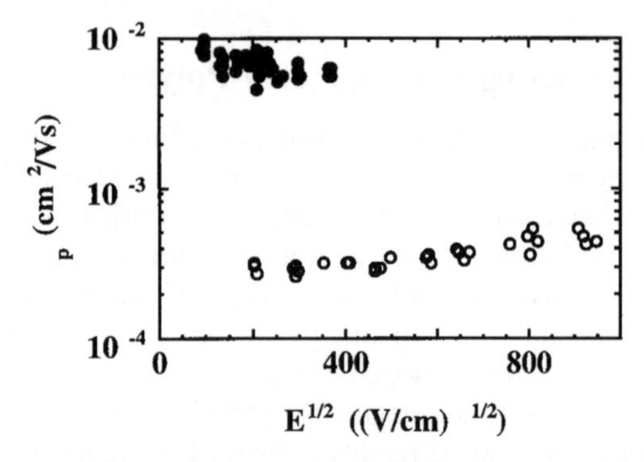

Figure 7 *(left) The electric field dependence of the time-of-flight hole mobility for poly(9,9-dioctylfluorene) films.*[21] *The open circles are data for a film prepared by spin coating. The filled circles are for a sample spin coated on a rubbed polyimide orientation layer, annealed in the nematic phase and then quenched to form a glassy film.*

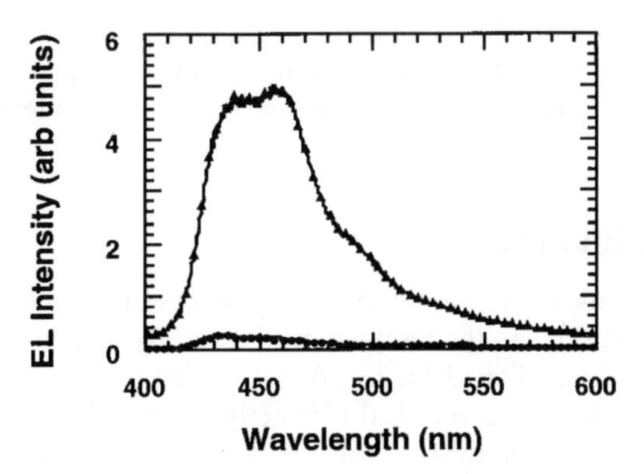

Figure 8 *(right) Polarised electroluminescence spectra from an ITO/rubbed PPV/PFO/Ca LED. The upper curve is for light emitted with polarisation parallel to the orientation direction and the lower curve for perpendicular polarisation. The anisotropy,* $I_{parallel} / I_{perpendicular}$, *has a peak value of 25:1.*

film ordered in the nematic phase has an order of magnitude larger hole mobility than that of a standard spin-coated film. At 10^4 V cm^{-1} the oriented film has $\mu_h = 9 \times 10^{-3}$ cm^2 V^{-1} s^{-1}. In Figure 8 the polarised emission from a polymer LED with an oriented film as the emission layer is shown. The electroluminescence anisotropy reaches a maximum value of 25:1 that would be suitable for application. These LEDs were also reasonably bright and efficient.

6.3 Soft Lithography and Embossing

Another area that offers interesting possibilities, especially for very low cost fabrication, is so-called soft lithography. The approach is to generate a master relief in a silicon or silica substrate by standard microfabrication processes. This is then used to make a stamp using a silicone rubber or similar material. The stamp can then in turn be used to imprint a pattern of material onto the desired substrate. Surprisingly good results have been demonstrated in terms of the possible spatial resolution, the fidelity of the imprint relative to the master and its reproducibility after many stampings. Embossing is another approach that is widely used in the fabrication of micro-optical components, *e.g.* lens arrays.

7 Summary and Conclusion

This paper has presented an overview of some of the emerging trends in polymer semiconductors and devices. It is an extremely rapidly developing field of activity in which many industrial and academic groups are active. Commercial manufacture of LEDs for displays now looks feasible and products are widely anticipated in the next year or so. Other devices are rapidly progressing to a similar level of development. There is, however, still a great deal of research and development to be done, especially in more fully utilising the polymer properties of the materials to allow greater control over physical structure. This looks set to be a thriving field of endeavour for many years to come.

Acknowledgement

I would like to thank the present and past members of my research group for their many contributions to the research outlined above. In particular I wish to acknowledge Paul Lane, David Lidzey, Alasdair Campbell and Martin Grell. I would also like to thank Avecia Ltd (formerly Zeneca Specialities Ltd), The British Council, The Commission of the European Community, Corning Cables (formerly BICC Cables Ltd), The Dow Chemical Company, The UK Engineering and Physical Sciences Research Council, The Higher Education Funding Council of England, The New Energy and Industrial Technology Development Organisation of Japan, The Leverhulme Trust, The Royal Society, Sharp Laboratories of Europe Ltd and Toshiba Corporation for their financial support of the research activities of my group in Sheffield.

References

1 D.D.C. Bradley, *Current Opinion in Solid State and Materials Science* 1996, **1**, 789.
2 R.N. Marks, J.J.M. Halls, D.D.C. Bradley, R.H. Friend, A.B. Holmes, *J. Phys: Condensed Matter* 1994, **6**, 1379.
3 J.H. Burroughes, C.A. Jones, R.H. Friend, *Nature* 1988, **335**, 137.
4 N. Tessler, G.J. Denton, R.H. Friend, *Nature* 1996, **382**, 695; W. Holzer, A. Penzkofer, X. Long, D.D.C. Bradley, *Advanced Materials* 1996, **8**, 974
5 R.H. Friend, P.D. Townsend, D.D.C. Bradley, W.J. Feast, D. Parker, N.C. Billingham, P.D. Calvert, P.J.S. Foot, D.C. Bott, J.N. Winter, *Synthetic Metals* 1987, **19**, 989.
6 G. Grem, G. Leditzky, B. Ullrich, G. Leising, *Advanced Materials* 1992, **4**, 36.
7 D.D.C. Bradley, *J. Phys D: Appl. Phys.* 1987, **20**, 1389.
8 K.E. Ziemelis, A.T. Hussain, D.D.C. Bradley, R.H. Friend, J. Rühe, G. Wegner, *Phys. Rev. Lett.* 1991, **66**, 2231.
9 M. Grell, D.D.C. Bradley, G. Ungar, J. Hill, K. Whitehead, *Macromolecules* 1999, **32**, 5810.
10 M.S. Weaver, D.D.C. Bradley, *Synthetic Metals* 1996, **83**, 61.
11 M.S. Weaver, D.G. Lidzey, T.A. Fisher, M.A. Pate, D. O'Brien, A. Bleyer, A. Tajbakhsh, D.D.C. Bradley, M.S. Skolnick, G. Hill, *Thin Solid Films* 1996, **273**, 39.
12 A.W. Grice, A. Tajbakhsh, P.L. Burn, D.D.C. Bradley, *Advanced Materials* 1997, **9**, 1174.
13 M.S. Weaver, D. O'Brien, A. Bleyer, D.G. Lidzey, D.D.C. Bradley, *Organic and Inorganic Electroluminescence,* Ed R.H. Mauch and H.-E. Gumlich, (Wissenschaft & Technik Verlag, Berlin) 1996 207.
14 D. O'Brien, M.S. Weaver, D.G. Lidzey, D.D.C. Bradley, *Appl. Phys. Lett.* 1996, **69**, 881.

15 M. Redecker, D.D.C. Bradley, M. Inbasekaran, W.W. Wu, E.P. Woo, *Advanced Materials* 1999, **11**, 241.
16 A.J. Campbell, D.D.C. Bradley, H. Antoniadis, M. Inbasekaran, W.W. Wu, E.P. Woo, *Appl. Phys. Lett.* 2000, at press.
17 D. O'Brien, A. Bleyer, D.G. Lidzey, D.D.C. Bradley, T. Tsutsui, *J. Appl. Phys.* 1997, **82**, 2662.
18 L. Palilis, D.G. Lidzey, M. Redecker, D.D.C. Bradley, M. Inbasekaran, E.P. Woo, W.W. Wu, Proceedings ICEL'2, Ed P.A. Lane and D.D.C. Bradley, *Synth. Met.* (2000), at press.
19 J.J.M. Halls, *et al.*, *Nature* 1995, **376**, 498.
20 M. Grell, X. Long, D.D.C. Bradley, M. Inbasekaran, E.P. Woo, *Advanced Materials* 1997, **9**, 798; M. Grell, M. Redecker, K. Whitehead, D.D.C. Bradley, M. Inbasekaran, E.P. Woo, *Liquid Crystals* 1999, **26**, 1403.
21 M. Redecker, D.D.C. Bradley, M. Inbasekaran, E.P. Woo, *Appl. Phys. Lett.* 1999 **74**, 1400.

Polymer Characterisation and Colloids

5

Exploiting Acrylic Polymer Architecture in Surface Coatings Applications

Andrew T. Slark

INEOS ACRYLICS, PO BOX 90, WILTON,
MIDDLESBROUGH TS90 8JE, UK

1 Introduction

The world of surface coatings is diverse. The total value of the global polymer resins market is approximately £20 billion, some of these being acrylic polymers based on the free-radical (co)polymerisation of methacrylate or acrylate monomers. The polymers may be applied as thermoplastic or thermoset coatings, provided that for the latter the polymer is functionalised appropriately and a suitable crosslinking agent is present.[1] The materials are made *via* emulsion, solution, suspension or bulk polymerisation, the relative order of utilisation being emulsion>solution>suspension~bulk. The range of applications is enormous. Surface coatings are applied by a variety of coating methods to different substrates, which include a range of papers, plastics, metals or wood. Typical coating thicknesses are 0.1–100 μm, depending on the application. One approach to generic classification is to segment the area into industrial coatings, paints, electronics, adhesives and inks. Figure 1 illustrates an example, providing further detail within each category.

For the purpose of this paper, most reference will be made to the inks segment but the general principles discussed are relevant to all application fields. The phrase 'polymer architecture' is used in a broad context to include all aspects of chemical and physical composition, not only topology.

Within this paper, the drivers for change in surface coatings industries are considered. These include changes in technology, the influence of environmental legislation on coating technology and the impact of industry dynamics on resources. Expectations are that Research and Development (R&D) projects will become increasingly successful, and this paper considers how this may be achieved by addressing barriers to change. These include recognition that satisfying technical requirements is only one part of a broader picture. Additionally, for a surface coating, it will be emphasised that a polymer is

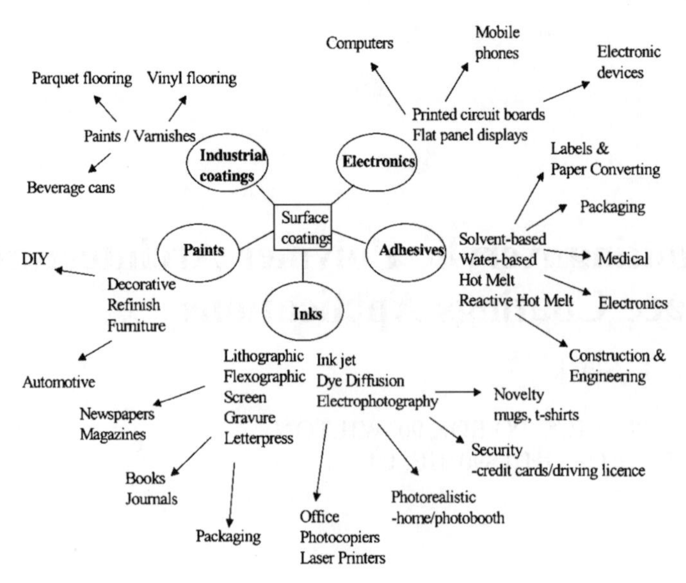

Figure 1 *The diversity of surface coatings applications*

often only one part of the formulation. A particular speciality application will be chosen to emphasise the importance of linking synthesis with architecture, basic properties and applied properties. Only appreciating this whole picture will assist in polymer design, to provide the performance benefits required at an appropriate cost for a particular application. It will be demonstrated that extreme control in polymer architecture may not be required, even for a high speciality application.

Free-radical polymerisation of (meth)acrylate monomers is considered, comparing conventional free-radical polymerisation with controlled 'living' free-radical polymerisation methodologies. These latter techniques are critically compared and their exploitation in surface coatings applications is questioned, although it is recognised that these technologies are at an early stage in their development. A number of disadvantages are highlighted and the lack of knowledge on basic polymer properties is exposed. Using several examples, it is shown that a more pragmatic approach to modifying free-radical polymerisation is possible. It is speculated that this will provide new architectures which have some structural heterogeneity but allow stronger potential for exploitation.

2 Drivers for Change

2.1 The Impact of Changing Technology

More traditional inks can be classified into letterpress, lithographic, flexographic, gravure and screen.[2] These technologies involve different methods of transferring the ink to a substrate under impact. More recently, a number of so-called non-impact methods of printing have emerged, which are related to the explosion in the use of electronic information. The phrase 'digital imaging' can be used to define these technologies which are based on the display of information and images created by an electronic source. These include ink-jet printing, dye diffusion thermal transfer printing and electrophotography.[2] The traditional printing technologies are more mature and polymers are used in relatively high volumes. The cost which end-users are willing to pay for a polymer is moderate, *e.g.* up to £4–5 per kg. However, in the newer, emerging non-impact technologies, polymers are used in smaller volumes where the cost which end-users are willing to pay is higher, *e.g.* greater than £10 per kg. This comparison of traditional and newer technologies reflects a situation apparent in all other surface coatings applications. This is very important because this has a strong impact on defining what polymer chemistry can be done to provide suitable property differentiation at a cost which is tolerable for a particular application. The more specialised the application, the higher the cost which end users are willing to pay for a polymer providing a step-change in properties. However, it should be emphasised that it is much better to satisfy requirements in speciality applications with inexpensive technology, if this is achievable.

2.2 The Impact of Environmental Legislation on Coating Technology

Coatings have conventionally been applied using organic solvents but these are under ever increasing environmental pressure, especially with respect to legislative control of volatile organic content (VOC). Surface coating applicators using organic solvents have the choice of introducing solvent recycling or incineration, moving to higher solids solvent-based systems or using an environmentally friendly coating technology. These include water-based coatings (usually emulsions), powder coatings and UV curable coatings. For water-based coatings, only water needs to be removed, although additional energy may be needed compared to the removal of volatile solvents. Powder coatings and UV coatings represent 100% solids (zero VOC) technologies. For powder coatings, small polymer particles are applied on to a substrate and then heated so that the particles melt and coalesce to form the film. In UV curable coatings, a liquid formulation is applied, typically containing acrylate functionalised oligomers, monomers and a photoinitiator. The entirety of the liquid film applied to a substrate becomes an integral part of the coating after exposure to UV light. Recent developments include combining powder and UV technologies into UV curable powder coatings. The impact of environmental legislation on coatings technology is summarised in Table 1.[3]

This demonstrates that the use of low solids solvent-based coatings will decline rapidly by the end of this decade, due to the impact of environmental legislation. Whilst high solids solvent-based coatings have been used to replace their low solids counterparts in the last five years, their use will also decline although at not such a rapid rate. Water-based, powder and UV curable

Table 1 *European Growth Rates and Volumes for Various Coatings Technologies.*[3] *Volumes for a Given Year Sum to 100%*

Coating Technology	*Growth rate / %*			*Volume / %*		
	1995–2000	2000–2005	2005–2010	2000	2005	2010
Low solids solvent-based	−3.2	−11.2	−4.0	30	15	7
High solids solvent-based	6.8	−3.2	−3.0	12	10	9
Water based	4.0	9.3	2.2	26	34	38
Reactive systems	3.5	4.5	2.0	15	17	18
Powder coatings	5.8	10.5	2.5	12	17	20
UV cured coatings	7.3	10.2	6.5	5	7	8

coatings will grow and prosper, all with very healthy average growth rates approaching 10% per annum during the next 10 years. Similar trends will apply worldwide. Although not considered above, supercritical CO_2 is used in some coatings applications, *e.g.* in the preparation of pigmented powder coatings or spin coating for electronics applications.[4] This activity is also likely to expand.

2.3 The Impact of Industry Dynamics on Resources

There is no doubt that mergers, acquisitions and restructuring in the chemical industry worldwide during the last decade has had a profound effect on the structure of the polymer and surface coatings industries. Over the next ten years, it seems that nothing is more certain than further change. It has been said that 'in 2010 as many as half of today's leading chemical companies may no longer be around, at least not in the form they take today'.[5] There are growing concerns that the restructuring is focusing resources on short-term results and stifling longer term R&D. It is certainly true that there are enhanced expectations of projects delivering a successful outcome in order to meet the ambitious growth targets which many companies demand. Companies can only thrive by growth, part of which involves developing successful new products to deliver sustainable competitive advantage. However, developing a pipeline of successful new products is a challenge; the risks are high but so too are the rewards. It is sobering to recognise historical evidence that for every 100 new ideas or concepts, 10 are researched, only three enter development and, of those, only one succeeds.[6]

3 The Reality of Creating New Products Containing New Polymers

3.1 Maximising the Chances of Success

Some companies attempt to accelerate the process from idea to product launch by employing a new product scheme or a 'stage-gate' process.[6] The main principle behind this is to manage risk. As a project proceeds, costs increase exponentially and the purpose of the stage-gate process is to simultaneously ensure that the risk reduces exponentially, thereby maximising the probability of a successful outcome. Typically, the process includes a number of consecutive stages, such as idea generation, preliminary investigation, detailed investigation, product development including testing and validation, and finally full production and market launch. Between each stage, there will be a review against certain criteria and a decision then made whether (or not) to proceed to the next stage.

3.2 Criteria: More than Technical Performance

In order for a product to be developed successfully, proving a concept purely from a technical viewpoint is quite often the easiest part of the process! However, actually developing a product which satisfies customer needs at an appropriate cost, whilst satisfying the constraints of intellectual property, safety health and environment (SHE) and manufacturing, is the most intensive part. With intellectual property, one has to be concerned with competitive patents. Do I have to obtain a license? Can I patent? Should I patent to gain exclusivity? With SHE, there are two aspects. One is concerned with product liability, ensuring the safety of people by assessing the risk of injury or exposure 'from cradle to grave', *e.g.* in R&D, manufacture, transport, use and disposal. The other element concerns regulatory compliance. In Europe and USA, this is covered by the European Inventory of Existing Chemical Substances (EINECS) and the Toxic Substances Control Act (TSCA), respectively. EINECS requires all raw materials constituting a polymer to be registered, whereas TSCA also has criteria which are polymer specific. For an industrial company, the method of polymer synthesis has to be consistent with its assets. Toll manufacture is possible but requires sufficient high volume/high speciality to justify. All of these factors constitute constraints and expense which must all be addressed in additional to raw technical performance.

Over the next ten years, it is very likely that the majority of companies will use a new product scheme. Many companies will have a strong market orientation, striving to provide unique, superior products, which provide differentiation and value to the customer in attractive markets. Successful companies will place more effort in the 'homework' predevelopment phase, by improving proof of concept from all angles and not solely from a technical viewpoint. Projects will be very sharply defined before entering the cost-intensive development stage. This will be assisted by better definition of the link between applied properties and basic properties, thereby enabling improved material design (see later for further detail). It is also vital that product innovation is based on the company's existing in-house strengths, capabilities and resources. Any technology which is considered step-out is likely to be obtained by licensing, acquisition or joint venture. Organisations will be focused and, despite some negative views, this is constructive provided that it is balanced rationally on the short, medium and long term. Short-term pressures should not be an excuse for speed at the expense of quality R&D work. With respect to the longer term, improved forward vision is required to identify or anticipate future problems which are worthwhile solving. Provided that industrial R&D continues to improve its hit rate of successful new products over the next few years, confidence will build, thereby justifying further resources to be placed on the medium and long term. A further key to success is to have a system in place which allows successful project management, but which is not so bureaucratic as to stifle innovation at the front end. Certainly, compared to the use of these systems today, judgement needs to be suspended further up to the point of detailed investigation (*i.e.* overall proof of concept) so that ideas with higher risk can be nurtured.

4 Seeing the Whole Picture

4.1 More than Polymers

It is rare for a surface coatings application (in particular a speciality application) to use a polymer as an isolated material. Many coatings consist of a formulation, where the polymer is a primary component used for its film forming properties. Taking an ink as an example, the formulation is likely to include a polymer (or blend), dyes or pigments and plasticisers.[1] Fillers may also be used to influence other properties. While the polymer itself has a significant influence on performance, the final balance of properties is governed by the structure of the solid coating (after the solvent or water is removed, if present). Therefore, the role of the polymer in influencing the interactions with the other components becomes paramount. This principle has been explored well in thermal transfer printing, one of the new, speciality non-impact printing technologies. The printing process involves the sequential transfer of yellow, magenta and cyan dyes from a donor ribbon to a receiver sheet.[7] The dye diffuses from a polymer coating in the donor ribbon (dye-donor) to an acceptor polymer coating in the receiver sheet (dye-acceptor).

Systems involving strong specific acid–base interactions between dye and dye-acceptor have been investigated. Infrared spectroscopy was used to probe the interactions of multifunctional hydroxyl dyes with polymers containing a wide variety of electron donating functional groups. Good understanding was obtained on the nature and strength of interactions as a function of dye functionality, concentration and polymer environment.[8,9] For a fixed dye solute, a model was then developed to correlate dye diffusion with dye and dye-acceptor polymer characteristics.[10] As demonstrated in Figure 2, excellent correlations have been established for dye transport as a function of polymer T_g, dye-polymer solubility parameter difference and dye-polymer specific interaction as characterised by the infrared shift of dye solute -OH vibration.[11] It has been demonstrated that these factors have a controlling influence on colour reproduction, a key applied property which is critical to the adoption of the technology.

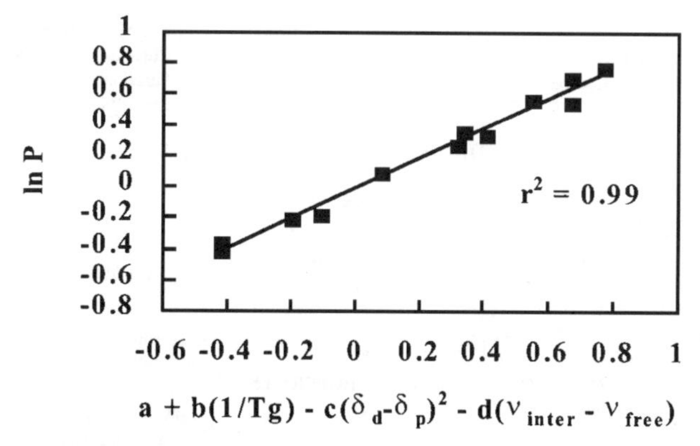

Figure 2 *The transport of a multifunctional hydroxyl dye to various dye-acceptor polymers during thermal transfer printing.* [11] *P is the effective permeability coefficient; Tg is the glass transition temperature of the polymer;* δ_d *and* δ_p *are solubility parameters of dye and polymer, respectively;* ν_{inter} *and* ν_{free} *are infrared absorptions estimating the dye-polymer specific interaction*

4.2 More than Synthesis

These principles were established using a variety of both addition and condensation polymers. All polymers were linear with broad molecular weight distributions and a statistical placement of different functional groups. Therefore, polymers which are fit-for-purpose have been designed for this high speciality application with specific knowledge of the physical chemistry of the coatings based on balancing the polymer T_g, its solubility parameter and the functional group capable of specific interaction with the dye. *Despite the application being high value and able to withstand the cost of making controlled polymer architectures, this is not actually necessary* This emphasises the point that if polymer structure is going to be modified or the architecture controlled, what is the purpose of doing it? Novel chemistry must have a purpose, if it is to be exploited. Polymer synthesis is put into context in Figure 3, which demonstrates the link between synthesis, architecture, basic properties, applied properties and technical differentiation.

This can be viewed either starting from the top left (technology push) or the bottom (market pull). Starting with *synthesis*, why alter polymer composition? Control of polymer chemistry and synthesis leads to a defined polymer *architecture*, *i.e.* the chemical and physical composition of a polymer chain. This *architecture* influences the *basic properties* of bundles of polymer chains, which then correlate with *applied properties*. The polymerisation process may affect the polymer architecture produced and the coating process will influence both the basic and applied properties. Basic properties are independent of the application whereas applied properties may be application specific. For surface coatings applications, the overall performance will be influenced by other raw materials in the formulation, depending on the nature of the individual components and the interaction between them. Improvement in specific

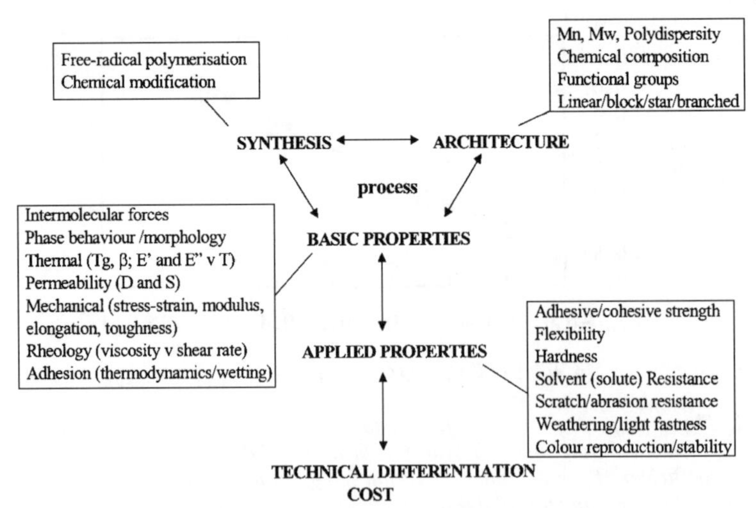

Figure 3 *The fundamental importance of linking synthesis, architecture, basic properties and applied properties. A project will only be successful if technical differentiation is achieved at an appropriate cost for a particular application*

applied properties will lead to enhancement in performance. If this *technical differentiation* is valued by the customer and is cost effective then delivering this improvement will lead to a successful new product or technology. Taking the thermal transfer printing example, polymerisation leads to a conventional molecular weight distribution and statistical placement of specific functional groups (architecture). The basic properties are influenced by the polymer T_g and the dye-polymer interaction (global and specific interaction). These basic properties then control the colour reproduction and image stability. Whilst this has been represented as technology push, problem solving was actually achieved by an appropriate balance of strong market pull. At the time of project conception, there were problems in the marketplace with an imbalance in colour reproduction, which led to the concept of controlling dye delivery. A broader view of this model for various surface coatings applications is also provided in Figure 3.

It is possible to screen new ideas by hopping straight from synthesis to applied properties. In order to reduce costs, polymer suppliers may develop products with little knowledge of the applications performance, relying on the customer as a testbed. This strategy can be flawed, since the quality of feedback depends on the quality of the relationship with the customer, who may be suspicious because he knows that the polymer supplier has interest in supplying to his competitors. It should be noted that the supply chain between resin producer, coating formulator and end user can be quite convoluted, with each relationship contributing resistance to change. For the surface coating world to prosper further in the next ten years, understanding this whole picture will become more important, so that polymers can be better designed which are fit-for-purpose. This could be achieved by better collaboration between polymer suppliers and coatings formulators, so that specific detail concerning applications and performance requirements can be fed back to polymer design. An alternative strategy will be for polymer manufacturers and coating formulators to become integrated *via* mergers and acquisitions. As exemplified above, even in a high speciality application, sophisticated control in polymer architecture may not be required.

5 Exploiting Polymer Architecture

5.1 Conventional Free-radical Polymerisation

Classically, free-radical polymerisation proceeds *via* a chain growth mechanism, involving initiation, propagation and termination.[12] Dead polymer can be created by two different termination mechanisms (combination and disproportionation) or by chain transfer. There is wide-ranging worldwide industrial exploitation of free-radical polymerisation since it is economical and much easier to perform than other polymerisations. Other polymerisation processes such as cationic, anionic, group transfer and coordination polymerisation can be used to obtain better control of polymer structure but they have limited application industrially. This results from the stringent requirements of high purity reagents/solvents, anhydrous conditions, low temperatures and application to a limited number of monomers.[13] Free-radical polymerisation is robust and tolerant to reaction conditions such as impurities, solvents, moisture and temperature. It can be used to homopolymerise or copolymerise a wide range of monomers (both functional and non-functional) to high conversion by a number of inexpensive processes. Bulk, solution, suspension and emulsion polymerisations are all operated industrially. Solution polymerisation can be used to provide a polymer in a solvent, which can then be directly applied as a surface coating. Suspension polymerisation produces polymer beads (typically 100 μm) which can be readily dissolved in solvents or monomers for subsequent surface coatings applications. Emulsion polymerisation provides polymers dispersed in water on a colloidal scale (typically sub-micron), which can be applied as surface coatings in an environmentally friendly fashion. Molecules containing the thiol functional group have high chain transfer constants[14] and are often used for the control of molecular weight. Alternatively, catalytic chain transfer (CCT) is efficient at producing low molecular weight polymers, the majority of chains containing a terminal unsaturation.[15] Cobalt porphyrin complexes can be used at low levels since the chain transfer constants are high. Using conventional free-radical polymerisation, a wide range of (co)polymer compositions, structures, molecular weights and properties may be produced to satisfy applications in a facile and versatile, cost effective manner. A variety of linear and crosslinked polymers can be made in addition to graft copolymers from macromonomers.[15] However, precise control over molecular weight distribution, end groups, monomer incorporation sequence and polymer architecture is difficult. Consequently, there has been recent renewed interest in controlling the polymerisation *via* 'living' free-radical polymerisation methodologies.

5.2 Controlled 'Living' Free-radical Polymerisation

Precise control of free-radical polymerisation has been considered difficult for a long time because of the high reactivity and low selectivity of the growing radical species. Excellent progress has been made in the last 15 years, since several new technologies have been developed which are aimed at precisely controlling free-radical polymerisation. In chronological order, these are nitroxides (mid 1980's),[16] Atom Transfer Radical Polymerisation (ATRP, mid-1990s)[17–21] and Reversible Addition Fragmentation Transfer (RAFT, 1998).[22,23] Nitroxides and ATRP are based on the reversible formation of reactive propagating radicals from dormant covalent species. The equilibrium between dormant and active species for nitroxides is thermally stimulated. The use of nitroxides provides excellent control for styrene but controlled polymerisation of methacrylates and acrylates has been more difficult, although new derivatives are being developed.[24] For ATRP (see Figure 4(a)), the reversible deactivation of propagating radicals is stimulated chemically, using an organic halide initiator in combination with transition metal complexes. For methyl methacrylate (MMA), a suitable initiating system is 2-bromoisobutyrate with a ligated CuBr complex. Various ligands are recommended such as bipyridyl and 2-pyridinecarbaldehyde imines. Other metal complexes can also be used, including those of Ruthenium,[25] Iron[26] and Nickel,[27] *etc.* The claims in the open literature are that polymers of a wide molecular weight range with a narrow molecular weight distribution can be made. Using appropriately

(a)

$$P_n\text{—}X \;+\; M^n/L \;\underset{k_{deact}}{\overset{k_{act}}{\rightleftharpoons}}\; P_n\bullet \;+\; X\text{—}M^{n+1}/L$$

($P_n\bullet$ + monomer, k_p)

(b)

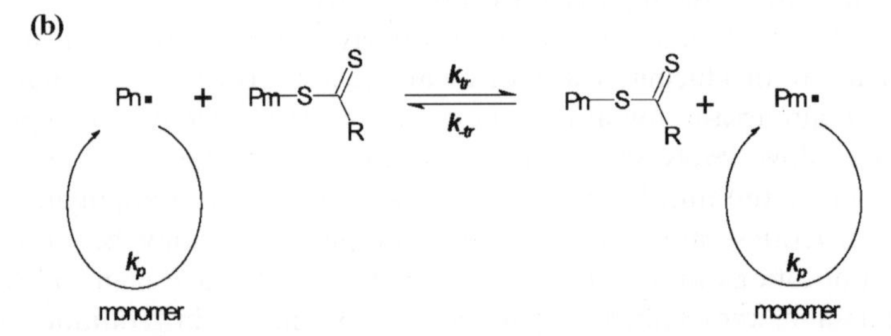

Figure 4 (a) *Schematic representation of ATRP, where X is* –*Cl or Br; M is a metal halide, e.g. Cu, Ru, Ni, Fe; L represents ligands.* (b) *Schematic representation of RAFT using thioesters*

functionalised macroinitiators, it is also possible to produce controlled block, (hyper)branched, graft and star (co)polymers.[28]

RAFT is the newest methodology and is performed by adding a thioester compound to a conventional free-radical polymerisation, as shown in Figure 4(b). The mechanism of RAFT polymerisation is envisaged to involve a series of addition-fragmentation sequences. Polymers with narrow molecular weight distribution can be made, and block or star polymers are also possible.

5.3 Reality: Can these Methods be Exploited?

For any new controlled free-radical polymerisation methodology to be exploited, the following criteria need to be considered:

1. How good is the control of molecular weight and polymer architecture?
2. Is the methodology repeatable? Can control still be achieved on a non-laboratory scale?
3. For each step, can the controlled architectures be achieved with high conversion of monomer to polymer (>95%, preferably>98%)?
4. Are raw materials such as initiators, catalysts and chain transfer agents commercially available and in a form which are easy to use?
5. Can it polymerise a range of monomers, including, *e.g.* direct copolymeriation of functionalised acid monomers?
6. Can current commercial processes (emulsion, solution, suspension & bulk) be used? Is the methodology robust, tolerant to water and achievable under nitrogen (with oxygen not rigorously excluded)?
7. Will the new polymer be regulatory compliant? Will the raw materials (*e.g.* initiators, chain transfer agents, monomers) or the polymer composition need to be registered?
8. How many steps are required to make the material required?
9. Is the product contaminated and do impurities need to be removed?
10. What are the performance advantages compared to materials currently used?
11. Considering all of the above and intellectual property issues, is the cost-benefit balance appropriate to the application? (*i.e.* will end-users pay the price for the performance?).

Table 2 compares conventional free-radical polymerisation, ATRP and RAFT against these criteria. It must be emphasised that this is a current snapshot, where the new technologies are at a relatively early stage in their development. Conventional radical polymerisation scores well on most criteria (2–9). The two issues are that control of molecular weight, chemical composition and topology is poor and that new performance improvements are required. ATRP, in its current form, does not fare so well. There are a number of industrial laboratories where difficulty has been encountered reproducing some of the published chemistry. Typically, molecular weights above 50,000 do not have narrow polydispersities and when using an ATRP derived polymer as

Table 2 *Comparison of Conventional Free-radical Polymerisation, ATRP and RAFT using Various Exploitation Criteria.*

Criteria	*Conventional radical*	*ATRP*	*RAFT*
1. Control of molecular weight and topology	**Low**	High, but see point 2	High
2. Repeatable in different laboratories; control with scale up	Excellent	**Difficult to make Mn >50,000 with narrow PD**	Unproven
3. Conversion of monomer to polymer	High	**Macroinitiators require <80% conversion**	High
4. Initiator or chain transfer agent	Many available, including low odour	Unusual, complicated system	**Novel thioesters strong colour and odour**
5. Monomer range	Excellent	**Not acid comonomers (directly)**	Excellent
6. Commercial processes	All processes	**Bulk and solution**	All processes
7. Regulatory compliance	Good	Poor?	**Poor**
8. Number of process steps	1–2	1–3	1–2
9. Contamination	None	**High concentration of metal**	**Colour, requires further step for removal**
10. Performance advantages	**Improvements required**	**Unproven**	Unproven
11. Cost-benefit balance	Acceptable for some applications	Unknown, likely to be Niche applications only	Unknown, likely to be Niche applications only

a macroinitiator for block copolymers, conversion needs to be limited to 80% for successful regrowth (*i.e.* 20% of monomer is wasted or needs to be recycled). For making architectures such as star branched copolymers, coupling or crosslinking occurs if the polymer concentration is too high, so realistically these structures can only be made at low conversion of monomer to polymer,[29,30] which makes them difficult to exploit.

There is the issue of using an unfamiliar, highly coloured transition metal complex, which can be toxic, and needs to be removed. Interesting attempts have been made to address this issue *via* polymer supported catalysts.[31,32] This removes the colour issue substantially (not completely), but polymerisation control is then diminished resulting from diffusion problems. The chemistry appears to be limited to bulk and solution processes, with suspension and emulsion polymerisation not being as successful to date (and these methods contribute substantially to the number of surface coatings products). Acidic monomers cannot be copolymerised directly since the catalysts are rendered inactive. This is very important, since adhesion to many substrates is provided by acid functional groups in the polymer. RAFT appears to look more promising. There are no complex transition metals to deal with and control of molecular weight or architecture can be achieved in all polymerisation processes taken to high conversion. However, it should be emphasised that this technology is more recent and requires further assessment within industrial laboratories before a preliminary judgement is made. Issues which need to be considered are that novel thioesters are used (regulatory compliance) which have a very high odour, much greater than mercaptans used as a chain transfer agent. They are also highly coloured and may require removal from the polymer before use. Although this may be achieved easily in a laboratory, an extra processing step may be needed and then residues from the chemical modification may require removal from the polymers in a further step.

Although both ATRP and RAFT are relatively new, one criticism of these fields to date is the lack of detailed characterisation on the architecture of the polymers made and the lack of knowledge on the basic physical properties which these materials confer. Importantly, comparisons need to be made with more conventional polymers, so that technical benefits can be quantified. It may be academically useful to control molecular weight, molecular weight distribution and topology but if these methods are to be exploited, what is the benefit? Despite the body of knowledge being generated on ATRP, the author is only aware of a few publications where material properties have been investigated in detail. In one case, interfacial toughening was probed using a variety of statistical, block and graft copolymers of styrene and 4-hydroxystyrene.[33] Despite the careful preparation of a number of precisely controlled architectures *via* controlled radical polymerisation (using ATRP and nitroxides), the random copolymers produced the best performance! Recently, the physical properties of PMMA-block-PBA-block-PMMA copolymers generated by ATRP have been investigated, compared to PS-block-polybutadiene-block-PS prepared by anionic polymerisation.[34,35] Despite apparently good phase separation, the mechanical properties of the ATRP polymer were comparatively poor, part of which results from some broadening of the outer PMMA blocks during extension of the PBA macroinitiator. Therefore, at this moment in time, it appears that property benefits of ATRP derived or RAFT derived polymers have not been demonstrated. The challenge is there to do so: what is the advantage of this extreme control?

Systematic relationships are required between synthesis, architecture and basic properties. The array of basic characterisation tools now at our disposal should also be used to pin down the architectures in more detail. For example, fractionating a block copolymer across the molecular weight distribution followed by detailed analysis would determine how controlled these polymerisation methodologies really are. What will happen if the criteria in Table 2 are not met? It is possible that controlled free-radical polymerisation could remain largely an academic curiosity with relatively few examples of exploitation. Hopefully, this will not be the case as these technologies develop and mature over the next 5–10 years. Certainly, the points raised previously demonstrate that controlled polymerisation methodologies are only likely to be adopted by those applications which can tolerate the cost of a number of polymerisation steps and purification. Therefore, any exploitation, if it happens, seems likely to be confined to niche, speciality applications. Paradoxically, these applications may not require such degrees of control, as demonstrated earlier. Even if they do, polymers with a very high quality standard are required.

5.4 The Middle Ground: Statistical or Pragmatic Modification of Free-radical Polymerisation

A more pragmatic approach to modifying free-radical polymerisation can still be achieved by influencing the processes of initiation, propagation, termination and chain transfer. It is possible for monofunctional polymethacrylate chains to be synthesised by using a combination of a functionalised initiator and a functionalised chain transfer agent. Any termination which does occur relies on disproportionation predominating. Polymethacrylate chains can be made with a polydispersity of 1.5. The monofunctional nature of the chains has been established by chemical end-group analysis, NMR spectroscopy and more recently by MALDI, with no evidence of bifunctionality.[36,37]

Polyfunctional mercaptans can be used to impart some control to free-radical polymerisation. For methacrylate monomers, it has been demonstrated that polymers with polydispersities as low as 1.25 can be made.[38] Unlike conventional radical polymerisations, this proceeds with an increase in molecular weight with time. The reason for this is that the thiol groups react sequentially as polymerisation proceeds. The polymerisation is statistically more controlled.

With respect to branched polymer architectures, recent work has been performed using polymeric chain transfer agents.[39,40] Methacrylate copolymers have been prepared containing statistically placed thiol groups. This polymeric chain transfer agent has then been used in a second free-radical polymerisation using further monomer. Branched polymers are produced *via* reaction of the pendant thiol groups. The polymerisation is modified in a pragmatic fashion to yield a non-crosslinked, branched polymer. Inevitably, some linear polymer is produced but this is a minor fraction compared to the target branched polymer produced. Both free-radical polymerisation processes are taken to high conversion and average molecular weights increase with time. Higher molecular weight polymers result if the polymeric chain transfer agent has a higher molecular weight or an increasing thiol concentration. The level of branching increases simultaneously, as illustrated in Table 3.

Table 3 *The Effect of Polymeric Chain Transfer Agent Composition on Branched Polymer Structure.*[39,40] *All solution polymerisations were performed using MMA at 80 °C, where PCTA is the polymeric chain transfer agent. [MMA] = 40% (w/w), [initiator] = 0.4% (w/w), [toluene] = 47.6% (w/w) for samples 3 and 4, and 55.6% (w/w) for samples 5 and 6. Characterisation was performed using triple detector GPC; α is the Mark-Houwink constant*

Sample	*PCTA Mw / g mol⁻¹*	*PCTA -SH /no. per chain*	*[PCTA] % (w/w)*	*Product Mw / g mol⁻¹*	*Product α*
1	–	–	0	46,200	0.71
2	–	–	0	149,900	0.75
3	9,790	2	12	44,500	0.49
4	9,980	4	12	54,500	0.33
5	9,066	2	4	65,100	0.59
6	37,300	8	4	312,100	0.39

5.5 Is Statistical or Pragmatic Modification the Answer?

These three examples can be considered to enhance traditional free-radical polymerisation. There is an improvement over the structures produced compared to conventional radical polymerisation, but not to the extreme of controlled 'living' radical polymerisation. However, the pragmatic methods are facile and versatile, allowing strong potential of exploitation. If conventional free-radical polymerisation and controlled 'living' free-radical polymerisation are considered as two extremes, statistically or pragmatically modified free-radical polymerisation occupies the middle ground, as shown in Figure 5. Here, it is possible for moderate control to be obtained in a manner which can be exploited industrially.

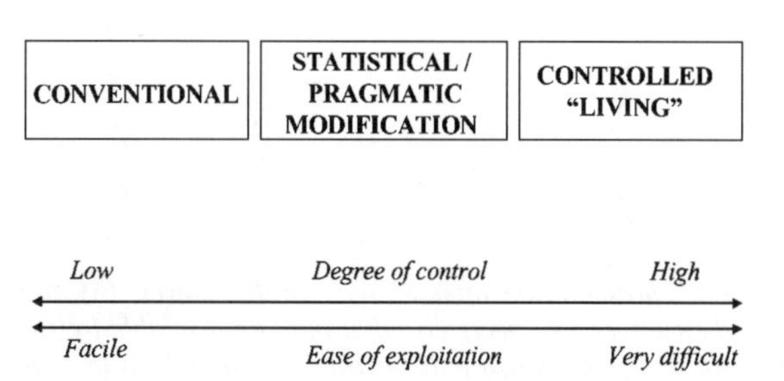

Figure 5 *Comparison of conventional free-radical polymerisation, controlled 'living' free-radical polymerisation and the middle ground of statistical / pragmatic modification*

It is now interesting to speculate what structures could be classified into this middle ground. Figure 6 illustrates telechelics with polydispersities of 1.5 which are monofunctional or bifunctional with respect to end-groups. Methods are available to achieve this, *e.g.* using a functional dimer macro-

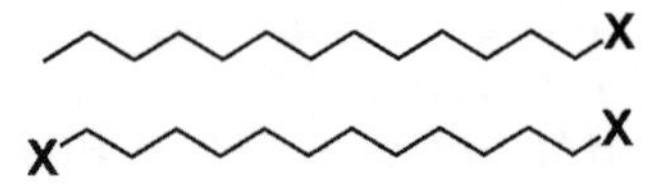

Figure 6 *Schematic illustration of monofunctional or bifunctional telechelics, where X is,* e.g., *-OH, -COOH,* etc

monomer derived from CCT in a radical addition-fragmentation reaction.[41] Newer, more facile methods may be found to achieve these structures.

A number of non-linear architectures may be available, *e.g.* star polymers shown in Figure 7 (a). Unlike the very regular star polymers illustrated in Figure 7(b), these pragmatic stars are likely to have a polydispersity of ~1.5 with different arm lengths. However, the basic structure will be present and this will modify the Tg and the relationship between molecular size and molecular weight. Related to this, randomly branched polymers may also be feasible, as illustrated in Figure 7(c). In comparison to their more controlled counterparts (Figure 7(d)), the branch lengths, the number of branches and the branch distribution will vary. In addition, dendritic features (branches off branches) could also be feasible. Optimising the rheological and other physical properties could be achieved by blending polymers with different architectures.

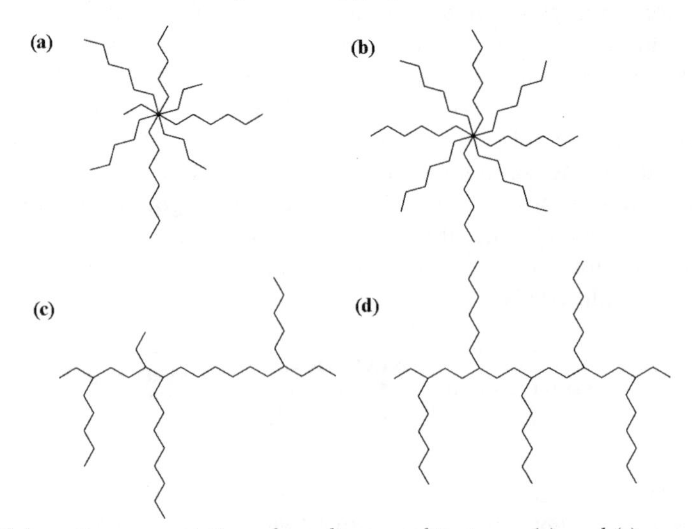

Figure 7 *Schematic representation of non-linear architectures.* (a) *and* (c) *are star and branched structures, respectively, that may be available by statistical/pragmatic modification of free-radical polymerisation.* (b) *and* (d) *are controlled analogues which could be made by 'living' free-radical polymerisation*

Block copolymers may also be possible by facile, cost effective routes and they will contain some compositional heterogeneity. Taking a triblock copolymer as an example, this compositional heterogeneity may result from some block polydispersity (1.2–1.5) or minor fractions of non-targeted molecules such as homopolymer, diblocks or tetrablocks. Branched block copolymers could also be possible, where non-linear architectures are produced with segmentation of A and B blocks. As shown in Figure 8(a), each block will contain some polydispersity, and branch lengths will be statistical. However, completely new structures could be possible, some of them combining addition and condensation polymers. These pragmatic structures will differ from their more controlled analogues (which are shown in Figure 8(b)), the latter having uniform block lengths and uniform branch lengths, with the branches being regularly spaced.

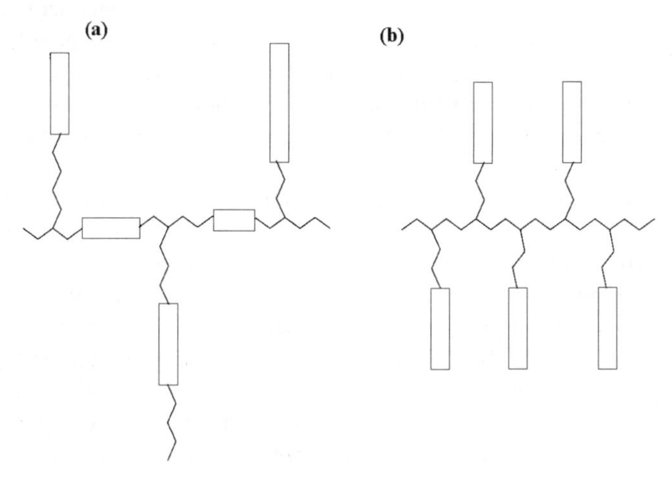

Figure 8 *Schematic representation of* (a) *branched-block copolymers which could be available by statistical/pragmatic modification of free-radical polymerisation.* (b) *represents a controlled structure available from 'living' free-radical polymerisation*

Some of the pragmatic architectures are likely to offer substantial property benefits compared to what is currently achievable with conventional free-radical polymerisation. They will be capable of being produced using one or two steps by current solution, suspension and emulsion processes taken to high conversion. The second step may involve chemical modification. It is possible that these architectures might be produced with readily available raw materials, be able to polymerise all functional monomers, and be unlikely to suffer the problems of high odour, extreme colour and toxic residues. They will clearly not be highly controlled structures with tightly defined molecular weight distributions, branch lengths or block lengths, and in some cases there will be a minor population of structures present which are not targeted. Understanding these architectures in detail, and the relationship between basic physical properties and applied properties will be the test of success. If step change improvements in properties can be realised at a cost suitable for a given application, then the use of polymer architectures to deliver product differentiation will grow and prosper.

6 Conclusions

The diverse world of surface coatings is likely to thrive, with new speciality applications continuing to emerge. The impact of environmental legislation on coating application is likely to accelerate the demise of both low solids and high solids solvent-based coatings. Water-based coatings, powder coatings and UV curable coatings will all grow at their expense. Company mergers and acquisitions will continue over the next ten years. R&D departments are likely to continue to be focused, but the successful ones will ensure a healthy balance of short, medium and long-term work. The latter will prosper further as companies improve their hit-rate on successful new products, provided that systems used to minimise risk are not so bureaucratic as to stifle front-end innovation.

There is likely to be a growing need for acrylic polymers to provide further property differentiation in a wide variety of applications, with a greater realisation that the role of the polymer in the properties of formulations requires better understanding. This will lead to improved knowledge of the relationship between applied physical properties and basic materials properties, thereby enhancing the design of target polymer architectures to provide the leading edge performance differentiation demanded by the marketplace. This needs to be supported by both better understanding of specific short/medium term market requirements, and improved vision of the longer term in order to identify future problems worth solving.

There may be some exploitation of controlled 'living' free-radical polymerisation methods, such as ATRP and RAFT, most likely in the form of block copolymers. The applications will probably need to be speciality to tolerate the cost, so that these controlled methodologies are likely to be confined to niche areas. This may occur, provided that polymers of high standard with a very low level of colour and impurities are produced. Exploitation of RAFT could be a better option, provided that it proves to be robust in industrial environments, and the issues of regulatory compliance, odour and colour are addressed. Improvements to these methods or new controlled radical polymerisation methodologies may also emerge. For many applications (including speciality applications), the extreme control offered by these methodologies could be proven unnecessary. In order to be exploited, an enhanced understanding of the physical properties which these controlled materials confer must be achieved, while satisfying all of the good points of conventional free-radical polymerisation.

It is probable that many applications (including speciality) will be satisfied by pragmatic modification of polymer architecture. Improvements over conventional free-radical polymerisation will be possible, providing new structures which have some heterogeneity. These will be significant improvements over those architectures currently available and they could provide a step-change in property enhancements. All of this is likely to be achieved in a cost-effective manner using current solution, suspension or emulsion technology, with all the advantages of current conventional radical polymerisation.

Acknowledgements

The author would like to thank Mike Chisholm (Ineos Acrylics), Alan Butters (ICI Imagedata), Bob Tooze (ICI Synetix) and Chris Borman (Du Pont) for a number of useful discussions.

References

1 P. Oldring and P. Lam. 'Waterborne & Solvent Based Acrylics and their End User Applications', John Wiley & Sons / SITA Technology Ltd, London, 1996, Vol. 3.
2 R.H. Leach and R.J. Pierce. 'The Printing Ink Manual', Chapman and Hall, London, 1993, 5th Edition.
3 J.P. Howard. 'RadTech Europe '99 Conference Proceedings', 1999, p.13.
4 S. Howdle and A.I. Cooper, private communication.
5 'Royal Society of Chemistry Industrial Affairs Division News Bulletin', Royal Society of Chemistry, 2000, **02,** 1.
6 R.G. Cooper. 'Winning at New Products', Addison-Wesley publishing, New York, 1993, 2nd edition.
7 R.A. Hann and N.C. Beck, *J. Imaging Technol.*, 1990, **16,** 238.
8 A.T. Slark and P.M. Hadgett, *Polymer*, 1998, **39(10),** 2055.
9 A.T. Slark and P.M. Hadgett, *Polymer*, 1998, **39(17),** 3977.
10 A.T. Slark and P.M. Hadgett, *Polymer*, 1999, **40,** 1325.
11 A.T. Slark and P.M. Hadgett, *Polymer*, 1999, **40,** 4001.
12 G. Moad and D.H. Solomon. 'The Chemistry of Free Radical Polymerisation', Elsevier Science Ltd, Oxford, 1995.
13 G. Odian. 'Principles of Polymerisation', John Wiley & Sons, New York, 1991, 3rd edition.
14 K.C. Berger and G. Brandrup in 'Polymer Handbook', John Wiley & Sons, New York, 1989, 3rd Edition.
15 T.P. Davis, D. Kukulj, D.M. Haddleton and D.R. Maloney, *Trends in Polymer Science*, 1995, **3(11),** 365.
16 D.H. Solomon, E. Rizzardo and P. Cacioli, US 4581429, *Chem. Abstr.*, 1985, **102,** 221335q.
17 M. Sawamoto and M. Kamigaito, *Trends in Polymer Science*, 1996, **4(11),** 371.
18 J.S. Wang and K. Matyjaszewski, *J. Am. Chem. Soc.*, 1995, **117,** 5615.
19 K. Matyjaszewski and J.S.Wang, *Macromolecules*, 1995, **28,** 7901.
20 D.M. Haddleton, C.B. Jasieczek, M.J. Hannon and A. Shooter, *Macromolecules*, 1997, **30,** 2190.
21 D.M. Haddleton, D. Kukulj, D. Duncalf, A.M. Heming and A.J. Shooter, *Macromolecules*, 1998, **31,** 5201.
22 J. Chiefari, B.Y.K. Chong, F. Ercole, J. Krstina, J. Jeffery, T.P.T. Le, R.T.A. Mayadunne, G.F. Meijs, C.L. Moad, G. Moad, E. Rizzardo and S.H. Thang, *Macromolecules*, 1998, **31,** 5559.
23 E. Rizzardo, J. Chiefari, B.Y.K. Chong, F. Ercole, J. Krstina, J. Jeffery, T.P.T. Le, R.T.A. Mayadunne, G.F. Meijs, C.L. Moad, G. Moad, S.H. Thang, *Macromolecular Symposia*,1999, **143,** 291.
24 P. Tordo, presentation at 'EPF Workshop on Controlled Radical Polymerisation', 1999.

25 M. Kato, M. Kamaigaito, M. Sawamoto and T. Higashimura, *Macromolecules*, 1995, **28,** 1721.
26 G. Moineau, Ph. Dubois, R. Jerome, T. Senninger and Ph. Teyssie, *Macromolecules*, 1998, **31,** 545.
27 G. Moineau, M. Minet, P. Dubois, Ph. Teyssie, T. Senninger and R. Jerome, *Macromolecules*, 1999, **32,** 27.
28 S.G. Gaynor and K. Matyjaszewski, *Polymer Preprints*, 1997, **1,** 758.
29 S. Angot, K. Shanmugananda Murthy, D. Taton and Y. Gnanou, *Macromolecules*, 1998, **31,** 7218.
30 J.C. Norman and S.C. Moratti, unpublished results.
31 G. Kickelbick, H. Paik and K. Matyjaszewski, *Macromolecules*, 1999, **32,** 2941.
32 D.M. Haddleton, D.J. Duncalf, D. Kukulj and A.P. Radigue, *Macromolecules*, 1999, **32,** 4769.
33 B.D. Edgecombe, J.A. Stein, J.M.J. Frechet, Z. Xu and E.J. Kramer, *Macromolecules*, 1998, **31,** 1292.
34 C. Moineau, M. Minet, P. Teyssie and R. Jerome, *Macromolecules*, 1999, **32,** 8277.
35 J.D. Tong, G. Moineau, Ph. Leclere, J.L. Bredas, J. Lazzaroni, and R. Jerome, *Macromolecules*, 2000, **33,** 470.
36 A.T. Slark and J.V. Dawkins, *Polymer*, to be submitted.
37 A.T. Slark and G. Webster, *Polymer*, to be submitted.
38 M.C. Yuan and G. Di Silvestro, *Macromol. Chem. Phys.*, 1995, **196,** 2905.
39 J.H. Houseman, J.V. Dawkins and A.T. Slark, *Polymer Preprints*, 1999, **40(2),** 167.
40 J.H. Houseman, J.V. Dawkins and A.T. Slark, *Polymer*, to be submitted.
41 D.M. Haddleton, C. Topping, J.J. Hastings and K.G. Suddaby, *Macromol. Chem. Phys.*, 1996, **197,** 3027.

6

Mass Spectrometry of Polymers

David M. Haddleton

DEPARTMENT OF CHEMISTRY, UNIVERSITY OF WARWICK,
COVENTRY CV4 7AL, UK

1 Introduction

One of the most important parameters the synthetic chemist wants to find out about a reaction is the mass of the products. This would normally be achieved by mass spectrometry. Although mass spectrometry has been available for the majority of the last century, the development of mass spectrometry for polymer analysis is relatively immature.[1] This is due to a number of contributing factors which in essence are the high mass of polymers and, by definition, the complex mixture of molecular species which constitute polymeric materials. Until recently most mass spectrometers are mass, or mass/charge, limited to, at most, a few thousand mass units and more typically less than approximately 2000 amu. Although for monodisperse high mass molecules this can sometimes be overcome by having more than one charge unit per molecule, polydisperse systems lead to very complex mixtures of peaks which usually preclude analysis. Thus the simple fact that most, if not all, polymers have higher mass than 2000 g mol^{-1} results in mass spectrometry not being an appropriate analytical technique. One of the first things undergraduate chemists are taught in mass spectrometry classes is that fragmentation occurs inside mass spectrometers and indeed it is this fragmentation which can give structural information as well as the absolute mass values. In the majority of cases the mixture of macromolecular entities is already so complex that fragmentation is often, but not always, undesirable in polymer mass spectrometry.

2 The Mass Spectrometry Experiment

Although it is not the place here to go into detail about mass spectrometry, it is important to note that the mass spectrometry experiment consists of two equally important processes. Firstly, ionization is the process that takes the analyte, polymer molecule, into the gas phase. More traditional techniques such as chemical and electron impact ionization are not appropriate due to the high level of fragmentation induced. Fast atom bombardment (FAB) has been used for polymers but even this relatively mild form of ionization does lead to fragmentation. The impressive advances in more mild ionization methods, *e.g.* field desorption, laser desorption (with and without matrix assistance), plasma desorption and electrospray ionization, so called soft-ionization methods have opened up mass spectrometry to the polymer chemist. Indeed it is probably fair to say that matrix assisted laser desorption ionization (MALDI) is the ionization method that has caught the attention of polymer chemists more than anything. Even these more mild techniques have associated problems. The second part of the experiment is the mass analyser, or detector, which is just as important as the ionization, with a range to choose from, all with associated advantages and disadvantages. Many detectors scan over a given mass range, common examples including magnetic sector and quadrupole. This scanned mass range, or mass limit, is usually relatively low with 2000 or 4000 amu being typical. A time of flight analyser requires ions to be produced in bundles, is particularly useful for pulsed desorption instruments and is finding widespread use in mass spectrometry of polymers in conjunction with MALDI. A detector which is increasing in popularity is ion cyclotron resonance (ICR) which, when used with Fourier Transform techniques, has impressive resolution at high mass. The important factors to bear in mind in relation to the mass analyser are the upper mass limit, the resolution of the detector and that the measurements will be mass sensitive, *i.e.* lower resolution at increasing mass.

When considering the use of mass spectrometry or when analysing data from mass spectrometry it is important that both the ionization and the mass analyser are considered, as both will have limitations in terms of mass discrimination and structural discrimination. This needs to be taken into account when trying to rationalize and analyse data.[2]

Although FAB has been used in polymer analysis,[3] problems with fragmentation and the relatively low mass limit has made this less popular as new techniques have emerged. Plasma desorption has been used successfully but this too has waned in popularity with commercial spectrometers not really readily available. To a large extent polymer mass spectrometry equates to MALDI time-of-flight and the remainder of this article will bear this in mind.[4,5] However, the use of electrospray ionisation (ESI)[6] will be considered in conjunction with either quadrupole detectors or ion cyclotron resonance (ICR) *N.B.* ICR detectors can also be used with MALDI, as this is important and probably not as widely used as it could be.

3 Polymer Types Amenable to Mass Spectrometry

This is considered first as this is quite an important area and is largely ionization type independent. The mass spectrometry experiment requires a molecule to be (i) in the gas phase and (ii) ionized, which is usually achieved by the complexation of a cation which can be a proton, metal ion or organic cation. This results in all polymers that contain heteroatoms such as poly-(ethers), poly(amides),[7] poly(esters),[3] poly((meth)acrylics),[8] poly(siloxanes),[9] poly(carbonates),[10] all being observed by MALDI and ESI. It is noted that the nature of polymer-cation interaction is not a trivial matter and is the scope of extensive study even for the apparently simplest case of poly(ethers). In addition to these, non-polar polymers[11] which contain π-bonds are also ionized under appropriate conditions, *e.g.* poly(dienes), poly(styrenes) and even poly-(isobutylene)s which contain terminal unsaturation. Thus probably the only polymer types which are not routinely observed are poly(olefins) such as poly(ethene) and poly(propene) which is arguably due to either a lack of a site for cationisation or excessive phase separation on sample preparation. However, it is noted that early advertisements for MALDI TOF mass spectrometers did often featured spectra of poly(propene) thus conditions might already be known to overcome this possible limitation.

The nature of the ionization is worth considering at this point. If one considers that poly(isobutylene) (PIB) with an unsaturated terminus can be ionized, and thus observed, but PIB with a saturated end group cannot. Clearly this has implications if the mode of termination in cationic polymerisation is the nature of the research project. This is an important consideration when considering the quantitative nature of the information. Thus how does a change in structural composition alter the relative ionization ability? Cationisation of poly(styrene) in MALDI is usually by either Ag^+ or Cu^+,[12] alkali metal ions proving problematic, overlap of d-orbitals with π-bonds is often used to rationalize this observation. However, ESI of poly(styrene) with transition metals gives no signal whereas K^+ results in excellent spectra.[13] This seems difficult to rationalize at the present time.

Thus most polymers can be both taken into the gas phase and ionized allowing detection by mass spectrometry. There are some polymers that are not routinely observed and although many generalities regarding ionisation, by either MALDI or ESI, can be made it is unadvisable at this point to extrapolate conditions between polymer types. The polymer chemist is now quite likely to obtain a mass spectrum of their sample to complement NMR, GPC, IR, DSC data, *etc.* It is necessary to consider what this information is, how it can be used and does it render other analytical techniques redundant?

4 What Information Do Polymer Chemists Want from Mass Spectrometry?

In general terms there are two types of information available;

- Information regarding mass averages
- Structural information
  - End-group information
  - Sequence distribution
  - Information regarding co and ter-polymerisation

5 Mass Averages from Mass Spectrometry

One of the first ways that MS impacted on polymer chemistry was a claim that the mass spectrometer would replace SEC as the method of choice for routine mass determination of polymers. Gone would be the days of having to use copious amounts of solvent in order to obtain a relative mass. Mass spectrometry heralded absolute mass averages irrespective of mass or polymer structure. It is probably fair to say that this has not yet happened. This is entirely due to mass discrimination in both ionization and detection and structure discrimination in ionization. ESI is so mass discriminate, in both ionisation and detection, that to date it is unlikely that it has been used this way. However, many researchers have published on the use of MALDI TOF for mass measurement[14–16] and many synthetic chemists quote mass averages from MALDI TOF now where SEC proves problematic. On close examination, it is found that most studies that compare SEC with MALDI do so with narrow mass standards where it is unlikely those major deviations will occur. In some relatively early work[12] we looked at the correlation between MALDI and SEC for a series of poly(styrene) and poly(methyl methacrylate) SEC standards with PDi less than 1.10 up to over 10 000 g mol^{-1}; this type of study has been repeated by many groups.[17] It is important in these comparisons to remember to compare the MALDI spectrum with the number distribution, as derived from the chromatogram by dividing once by the calibration curve and once by the molar mass.[12] In these cases a visual inspection reveals that in all cases the curves overlap with impressive consistency. It is also worth noting that the chromatogram, the weight distribution and the number distribution all have similar shapes and do not change in appearance on transformation. However, a closer inspection reveals that MALDI TOF has a decreasing relative sensitivity with increasing mass. Figure 1 shows the error in M_n for poly(styrene) on increasing mass. This trend was identical for PMMA and also holds for M_w and M_z in both cases.

This lack of agreement between the mass distribution from SEC and from MALDI TOF is a recurring problem. It is apparent for a range of polymer types over wide mass ranges. For example, Figure 2 shows the overlay for a series of poly(siloxanes).[18] It is apparent that as higher masses are analysed the errors become larger. The mass spectrometry experiment has a non-linear response with respect to an increase in mass. This discrepancy is more marked for polymers with molar mass distributions greater than about 1.20, *i.e. most polymers*. The effect is hidden to a certain extent by the polymer chemists usual window on a mass distribution, the chromatogram or sometimes the weight distribution. Indeed the number distribution is so unused that much SEC software, including the one routinely used by our group, does not automatically transform data and spreadsheets are usually required. An example to illustrate this is shown in Figure 3 where the number distribution from a RI detector has been calculated and overlaid with the number distribution as measured with a UV detector from SEC.[19] The M_n for this polymer is 20 170 with M_w of 39 700. The most intense peak from the MALDI TOF spectrum

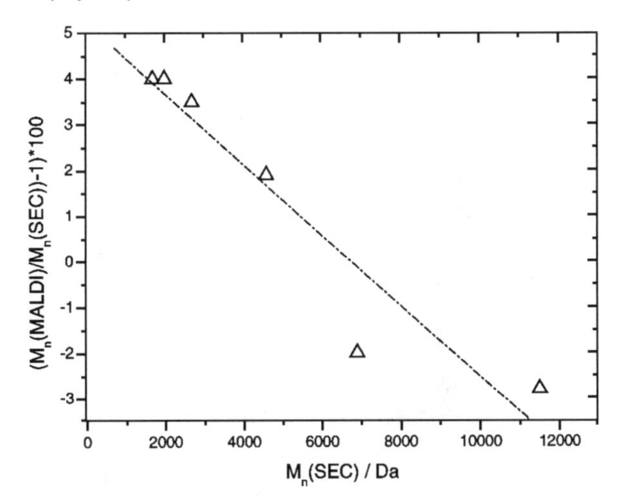

Figure 1 *Percentage difference between number-average molar masses of poly(styrene) narrow PDI samples derived from SEC and from MALDI TOF as a function of the average molar mass (from SEC)*

appears at 1377 amu that coincides surprisingly almost exactly with the M_p from the number distribution from SEC. However, there are no peaks observable in the MALDI TOF above m/z 4000. It is here where mass averages from MALDI TOF become increasingly erroneous. Indeed I have never observed a PDI greater than 1.5 from a MALDI TOF spectrum! The problem is almost simple to state; when collecting a number distribution each molecule counts as one detection unit whereas a molecule 100 times the mass of another will register 100 times more with a mass detector. As polymer properties are dominated by the high mass molecules present just missing a few percent at this end of the envelope will give spurious results.

There are many reasons why mass spectrometry misses the high masses. One of these is that the detectors measure in a linear mass mode which soon loses small *numbers* of molecules in the background noise, contrast this with SEC which collects logarithmically with mass. The detectors are often mass sensitive and this can be corrected to some extent by applying a data manipulation function.[20] Other factors which need to be taken into consideration are loss of low mass regions due to either volatility or ionization problems. This is particularly apparent when looking at condensation polymers or acrylics from catalytic chain transfer polymerization.[21] There are also effects on the mass distribution due to the laser power used thus the minimum laser power is often required but not always applied.

One way around these problems is to use the MALDI TOF as a SEC detector.[22–25] If very narrow PDI fractions are taken and analysed the errors in mass averages will be small, this data can subsequently combined with data from a concentration detector to reconstitute the mass distribution. This is effective but very time consuming and certainly does not lead to the replacement of SEC as routine characterisation. An alternative approach is to use the

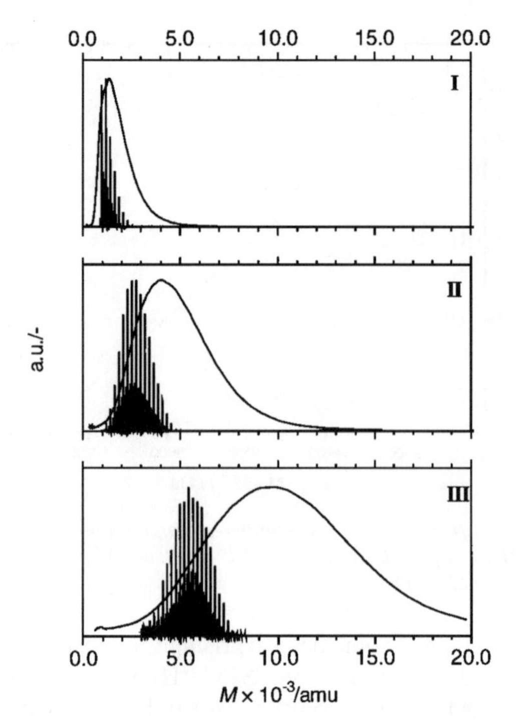

Figure 2 *Overlay of MALDI TOF spectra and the calculated differential number weight distributions obtained* via *SEC analysis of poly(dimethylsiloxane)s I, II and III: M_n (SEC) = 1920, 5000, 10900 M_w (SEC) = 2470, 5920, 12700: M_n (MALDI TOF) = 1290, 2720, 5350, M_w (MALDI TOF) = 1460, 2890, 5530 g mol^{-1} respectively*

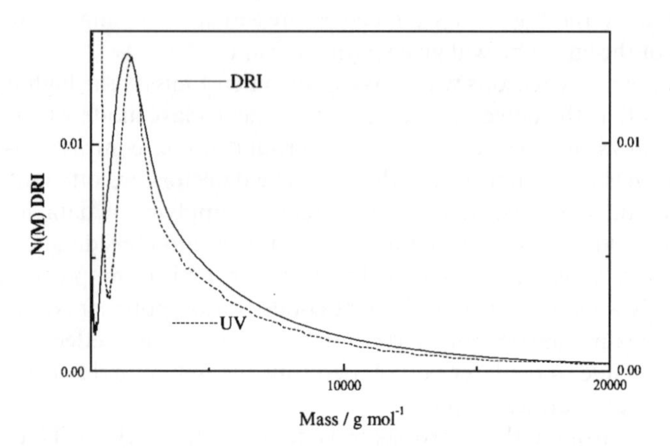

Figure 3 *Dual channel DRI and UV/Visible detection SEC traces for PMMA from catalytic chain transfer polymerisation*

mass spectrometer to construct a calibration curve for SEC by relating hydrodynamic volume to mass; this approach has been used by Simonsick with ESI.[25]

Suffice to say that at present we would never use mass spectrometry to obtain mass-average information at Warwick.

6 Sample Preparation in MALDI TOF

Sample preparation in MALDI is almost an art form with many different methods available. We have found that this is probably one of the major sources for non-representative data from samples. For example, the most often used method is syringe spotting; we have found that if *known* mixtures of PEG's are analysed spotting can lead to completely different spectra from spot to spot and even within the same sample spot. In some cases one component of the known mixture does not appear at all. Many different reasons for this will be found in the mass spectrometry literature but the only rationale we have is that phase separation occurs within the sample on concentration during solvent evaporation. This is possibly a reason why hydrophobic polymers are difficult to analyse due to separation form the, usually, polar matrix. We have found that in order to achieve a degree of quantatisation giving representation of the whole sample a spraying technique is appropriate where the phase regions of the samples are of the same order of magnitude as the laser spot. Electrospray deposition can be used[26] or more simply we have employed an artist's spray gun, available at low cost from hobby shops.[27] Figure 4 shows the results of various functionalised PEG mixtures prepared by this method, it is noted that commercial devices are now available based on an aerosol principal. An alternative simple method is to build up a thick layer of salt and deposit the analyte on to this salt-bed. Again it is apparent that there are a wide range of sample preparation techniques being used in the literature. This makes comparison of results and attempts to reproduce conditions for specific polymer types often difficult to achieve.

7 Mass Spectrometry for Structural Analysis of Polymers

Structural analysis of polymers is the power of mass spectrometry for polymer characterisation and is where the future lies. There are many possible examples that could be used to illustrate this; a few will be chosen to illustrate the point.

7.1 Dendrimer Analysis

Dendrimers are monodisperse high mass molecules. MALDI TOF is ideal for following reactions and can even replace TLC to follow deprotection steps for example. If the synthesis has been carried out correctly a single peak corresponding to the exact mass will be present. This is particularly powerful when used in conjunction with SEC, which can help distinguish synthetic

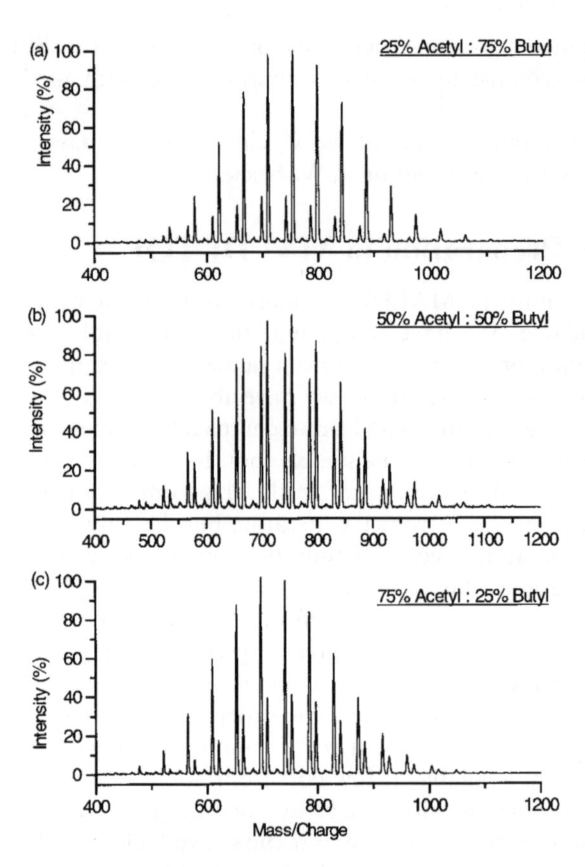

Figure 4 *MALDI-TOF mass spectra of functionalised PEG mixtures as prepared by air-spraying with the compressed air applicator*

imperfections from possible fragmentation. Indeed it is difficult to see any other technique which can conclusively give information regarding the effectiveness of a dendrimer synthesis.[28] It is apparent that many of the dendrimers reported in the early stages of this emerging area where complex mixtures and the absence of mass spectrometry data precluded this being known.

7.2 End Group Analysis

Polymer end group analysis gives information regarding both initiation and termination mechanisms. Even in the simplest cases of the free radical polymerisation of MMA and styrene the exact nature of the ratio of termination by disproportionation and recombination is not known accurately. Textbook estimates for methacrylonitrile range for 35 to 100% combination! These discrepancies are due to the fact that techniques other than mass spectrometry give information on the bulk sample. Looking at polymer end groups above about 3 000 g mol^{-1} is a difficult experiment. MALDI TOF has been applied to this problem and not only gives an easy route into end group analysis but also allows chain length dependence of termination to be examined as each degree of polymerisation is observed separately.[29] Indeed these experiments allow the probing of some fundamental aspects of free radical polymerisation. The use of high-resolution detectors such as ICR allows for a polymer, which has either a terminal unsaturation or saturation (a difference of two mass units), to be observed with ease, even at masses over 10 000 g mol^{-1}. We have been able to identify the disproportionation of primary AIBN radical fragments into methacrylonitrile that is able to copolymerise using high resolution ESI mass spectrometry. Thus new insight that is available has not been possible to investigate without this instrumentation.

End group analysis of living polymerisation methods is very enlightening. For example, intra-molecular cyclisation of PMMA in both anionic and group transfer polymerisations may be followed kinetically.[30,31] Thus the mode of termination cannot only be identified but synthetic limitations found which are necessary to know when designing block copolymer synthesis. Again this mechanistic information is available as a function of chain length, propagation time, which is inconceivable by any other technique. Figure 5 shows the

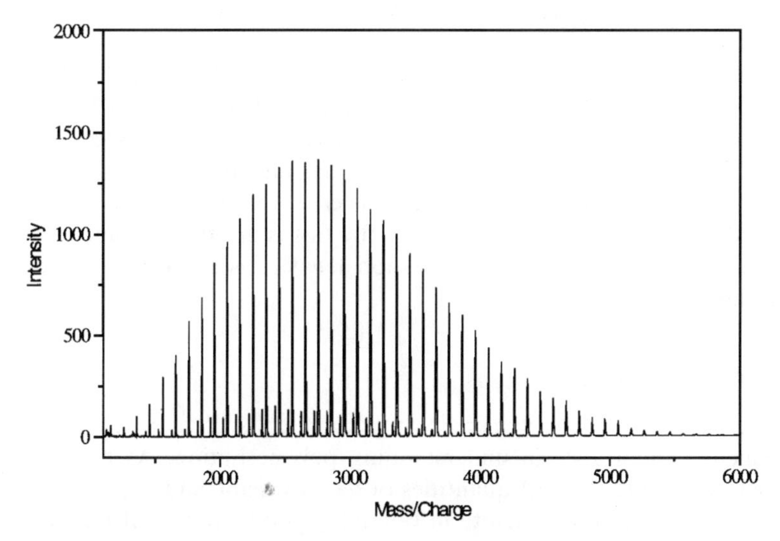

Figure 5 *MALDI-TOF spectrum of GTP prepared hydroxyl functional PMMA hydrogen terminated*

MALDI spectrum of a hydroxyl functional polymer prepared by GTP: the level of the impurity is easily seen and the success of this reaction clearly observed.[32] However, this is not a universally applicable technique. Perhaps the most topical living polymerisation at present is transition metal mediated radical polymerization. This typically gives a polymer with a tertiary halide terminal group. This group, as has nitroxide, has been found to be very labile in the mass spectrometer leading to fragmentation.

Even at threshold laser intensity for tertiary halide terminated polymers there is evidence for cleavage of the carbon halogen bond. The peak at 1045.7

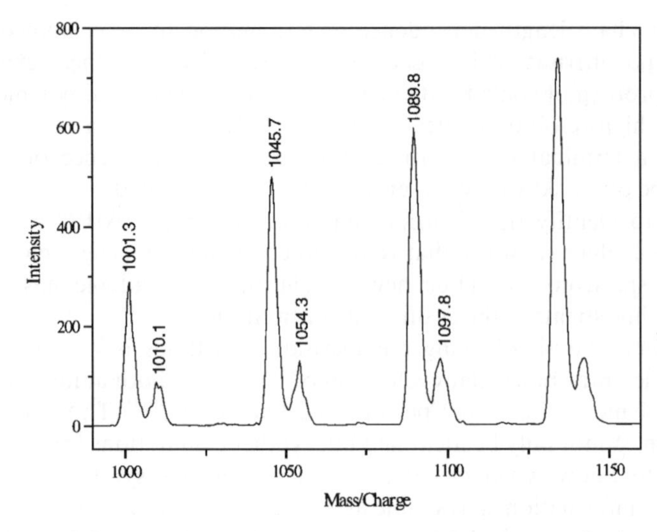

Figure 6 *Expanded region (995-1105amu) MALDI-TOF spectrum for bromoisobutyryl PEG 1000 in reflectron mode at threshold laser intensity*

in Figure 6 represents the sodium adduct of a bromoisobutyryl derivatised PEG 1000 (a model compound for living radical polymers) with degree of polymerisation approximately = 12 (calculated 1043.2). The second smaller series, *i.e.* 1010.1 represents polymer which has undergone bromine end group loss, *i.e.* 1089.8 − 79.9 = 1009.9.[32,33] In this case we are confident that this has occurred in the mass spectrometer. Usually this is not the case and as soon as secondary series are observed the question always arises: is it the synthesis or is it the mass spectrometry?

7.3 Copolymerisation

The measurement of copolymerisation parameters such as reactivity ratios is of immense importance in understanding polymerisation. As most polymer chemists are aware the vast quantities of data available in the literature are at best flawed, due to the difficulty in obtaining good data. MALDI-TOF mass spectrometry was used to obtain bivariate distributions of composition and chain length for a series of copolymers of methyl methacrylate (MMA) and n-butyl methacrylate (BMA) produced by free radical catalytic chain transfer polymerization. The relationship between the bivariate distributions of copolymers and the kinetics of the copolymerization reactions from which they were formed were derived using the terminal copolymerization model. For this copolymerization the reactivity ratios, (catalytic) chain transfer coefficients, and initiator selectivity were determined to be $r_{MMA} = 1.09$, $r_{BMA} = 0.77$, $C_{S,MMA} = 17900$, $C_{S,BMA} = 6150$ and $S_{MMA} = 0.535$ respectively. The reactivity ratios for these polymerizations were also determined from ^{1}H NMR to be $r_{MMA} = 0.75$ and $r_{BMA} = 0.98$ using an error-in-variables-model (EVM) treatment. The parameters determined from the mass spectra were compared with both the NMR values and literature values, where available and general agreement was shown.[21] The work illustrated the power of mass spectrometry as a tool for the characterization of synthetic polymers that are relatively complex such as statistical copolymers.

In a separate study we utilized ESI quadrupole mass spectrometry to analyse MMA/methacrylic acid copolymers.[34] For each degree of polymerisation the distribution of monomer units are seen giving again chain length dependence information. This is extremely powerful application of mass spectrometry applied to polymers. Mass spectrometry has even been used for the measurement of k_p for MMA polymerisation in conjunction with PLP.

7.4 Polymer Identification

An area that finds increasing application is the identification of an unknown polymer to obtain not only repeat unit information but also on the nature of the chemistry used to synthesise the molecule. This has been used with polyethers identifying the initiating alcohol/phenol and with acrylics made by living polymerisation. This is a potential area in which to carry out competitor analysis and in intellectual property protection.[35]

8 The Future

The opinion of the author is that the future of mass spectrometry in polymer chemistry is to obtain structural information. This gives additional and complementary information to the polymer chemist. Information on initiation, propagation, chain transfer and termination are all accessible. We will see the powerful high-resolution spectrometers become more accessible in terms of cost and ease of use. Mass spectrometry will become, if it has not already, an additional tool for the polymer chemist and information will become available that is not from any other technique. Most respectable polymer synthesis groups already have access to mass spectrometry and this will become the norm. The range of ionization techniques will increase in use, in particular ESI which has been largely ignored by polymer chemists. MALDI TOF will benefit from improved, automated sample preparation. However, I cannot see the mass discrimination problems being overcome and I do not see MS becoming the method of choice for mass average determination. It is noted that this is not the view of all practitioners of mass spectrometry of polymers and I hope I am wrong on this point. In reality mass spectrometry offers new data but care has to be taken with data analysis and an appreciation of what is in the "box/spectrometer" or how the data is collected is needed. This is obviously true for all analysis and the mass spectrometer cannot be viewed as a *magic box*. These aspects will improve as manufacturers of spectrometers become more familiar with polymer chemistry needs and polymer chemists become more familiar with the pros and cons of the experiments. There are vast improvements already happening rapidly, *e.g.* delayed extraction, ion-mirrors, new detectors, *etc.* that improve resolution and sensitivity. New matrices for MALDI will continue to be developed and this will move from an art to a more exact science. In particular MALDI TOF is a very easy experiment to perform and this will become ever more common and necessary in our field.

In summary, mass spectrometry is a technique to be ignored at your peril. Data interpretation and sample preparation are not always trivial and the best results will be obtained by polymer chemists and mass spectrometry experts working in collaboration.

Acknowledgments

I would like to thank Professor Peter Derrick for continued collaboration in this area, Professor Tom Davies for learning lessons and benefiting from our joint naiveté with me along the way. I would also like to thank the efforts of all my past co-workers who have worked hard in this area, in particular Drs Carl Waterson and Kevin Suddaby.

References

1 G. Montaudo, *Trends in Polym. Sci.*, 1996, **4**, 81.
2 C. A. Jackson and W. J. Simonsick, *Curr. Opin. Sol. State & Mat. Sci.*, 1997, **2**, 661.
3 S. C. Hamilton, J. A. Semlyen, and D. M. Haddleton, *Polymer*, 1998, **39**, 3241.
4 H. S. Creel, *Trends Polym. Sci.*, 1993, **1**, 336.
5 B. S. Larsen, W. J. Simonsick, and C. N. McEwen, *J. Amer. Soc. Mass Spectrom.*, 1996, **7**, 287.
6 C. H. McEwen, W. J. Simonsick, B. S. Larsen, K. Ute, and K. Hatada, *J. Am. Soc. Mass Spectrom.*, 1995, **6**, 906.
7 G. Montaudo, M. S. Montaudo, C. Puglisi, and F. Samperi, *J. Polym. Sci. Polym. Chem.*, 1996, **34**, 439.
8 P. O. Danis, D. E. Karr, W. J. Simonsick, and D. T. Wu, *Macromolecules*, 1995, **28**, 1229.
9 G. Montaudo, M. S. Montaudo, C. Puglisi, and F. Samperi, *Rap. Commun. Mass Spectrom.*, 1995, **9**, 1158.
10 C. Puglisi, F. Samperi, S. Carroccio, and G. Montaudo, *Rap. Commun. Mass Spectrom.*, 1999, **13**, 2260.
11 P. O. Danis, D. E. Karr, Y. S. Xiong, and K. G. Owens *Rap. Commun. Mass Spectrom.*, 1996, **10**, 862.
12 P. M. Lloyd, K. G. Suddaby, J. E. Varney, E. Scrivener, P. J. Derrick, and D. M. Haddleton, *Euro. Mass Spectrom.*, 1995, **1**, 293.
13 C. B. Jasieczek, A. Buzy, D. M. Haddleton, and K. R. Jennings, *Rapid Commun. Mass Spectrom.*, 1996, **10**, 509.
14 M. S. Montaudo, C. Puglisi, F. Samperi, and G. Montaudo, *Rapid Commun. Mass Spectrom.*, 1998, **12**, 519.
15 G. Montaudo, E. Scamporrino, D. Vitalini, and P. Mineo, *Rapid Commun. Mass Spectrom.*, 1996, **10**, 1551.
16 M. S. Montaudo, C. Puglisi, F. Samperi, and G. Montaudo, *Macromolecules*, 1998, **31**, 3839.
17 C. M. Guttman, P. O. Danis, and W. R. Blair, *Abs. Amer. Chem. Soc.*, 1997, **214**, 90.

18 D. M. Haddleton, S. A. F. Bon, and K. Robinson, *Macromol. Chem. Phys.*, 2000, **201**, 694.
19 D. M. Haddleton, C. Topping, J. J. Hastings, and K. G. Suddaby *Macromol. Chem. Phys.*, 1996, **197**, 3027.
20 J. Axelsson, E. Scrivener, D. M. Haddleton, and P. J. Derrick, *Macromolecules*, 1996, **29**, 8875.
21 K. G. Suddaby, K. H. Hunt, and D. M. Haddleton, *Macromolecules*, 1996, **29**, 8642.
22 D. J. Aaserud, L. Prokai, and W. J. Simonsick, *Anal. Chem.*, 1999, **71**, 4793.
23 G. Montaudo, *Abs. Amer. Chem. Soc.*, 1996, **211**, 261.
24 G. Montaudo, *Abs. Amer. Chem. Soc.*, 1996, **211**, 261.
25 W. J. Simonsick, D. J. Aaserud, and M. C. Grady, *Abs. Amer. Chem. Soc.*, 1997, **213**, 208.
26 J. Axelsson, A.-M. Hoberg, C. Waterson, P. Myatt, G. L. Shield, J. Varney, D. M. Haddleton, and P. J. Derrick, *Rapid Commun. Mass Spec.*, 1997, **11**, 209.
27 D. M. Haddleton, C. Waterson, and P. J. Derrick, *Euro. Mass Spectrom.*, 1998, **4**, 203.
28 H. S. Sahota, P. M. Lloyd, S. G. Yeates, P. J. Derrick, P. C. Taylor, and D. M. Haddleton, *JCS Chem. Commun.*, 1994, 2445.
29 M. D. Zammit, T. P. Davis, D. M. Haddleton, and K. G. Suddaby, *Macromolecules*, 1997, **30**, 1915.
30 D. R. Maloney, K. H. Hunt, P. M. Lloyd, A. V. G. Muir, S. N. Richards, P. J. Derrick, and D. M. Haddleton, *JCS Chem. Commun*, 1995, 561.
31 K. H. Hunt, M. C. Crossman, D. M. Haddleton, P. M. Lloyd, and D. P. J, *Macromol. Rapid Commun.*, 1995, **16**, 725.
32 D. M. Haddleton, C. Waterson, P. J. Derrick, and A.-M. Hoberg, unpublished results.
33 D. M. Haddleton, C. Waterson, P. J. Derrick, C. Jasieczek, and A. J. Shooter, *Chem. Commun.*, 1997, 683.
34 D. M. Haddleton, E. Feeney, A. Buzy, C. B. Jasieczek, and K. R. Jennings, *JCS Chem. Commun.*, 1996, 1157.
35 D. J. Aaserud and W. J. Simonsick, *Progress in Organic Coatings*, 1998, **34**, 206.

7

An Overview of Current and Future Themes for Polymer Colloids Research

Peter A. Lovell

MANCHESTER MATERIALS SCIENCE CENTRE,
UNIVERSITY OF MANCHESTER AND UMIST,
GROSVENOR STREET, MANCHESTER M1 7HS, UK

1 Introduction

Polymer Colloids is a generic term encompassing all stable colloidal dispersions of polymers in aqueous or non-aqueous media for which the polymer particle size may be conveniently expressed in nanometres.[1,2] For almost all synthetic and naturally-occurring polymer colloids the mean particle size falls in the 100–2000 nm range, but most commonly is 100–500 nm.

In Europe the synthetic polymer colloids industry produces around three million wet tonnes per annum, with world-wide production of the order of 10 million wet tonnes per annum.[3] Industrial applications range from those involving direct utilisation (for which surface coatings, inks and adhesives are important examples), to the manufacture of plastics and rubbers, and also include more specialised applications, such as those in the biomedical and photographic industries.[1,2] The need for new water-borne polymer colloid coating systems continues to increase with environmental pressures driving the move away from solvent-borne coatings. These needs bring with them major scientific challenges in both academic and industrial research. There also is growing use of polymer colloids in specialised, high added-value applications, such as diagnostic and other biomedical applications.[1] The extensive range of applications for polymer colloids is summarised in Figure 1.

Given the size and scope of the field of polymer colloids, an exhaustive review of the important themes for polymer colloids research is not feasible in the context of this paper. Instead, an overview of some of the more significant issues for current and future polymer colloids research is given. Important themes for research are presented in relation to industrial needs, specifically in terms of water-borne polymer colloids.

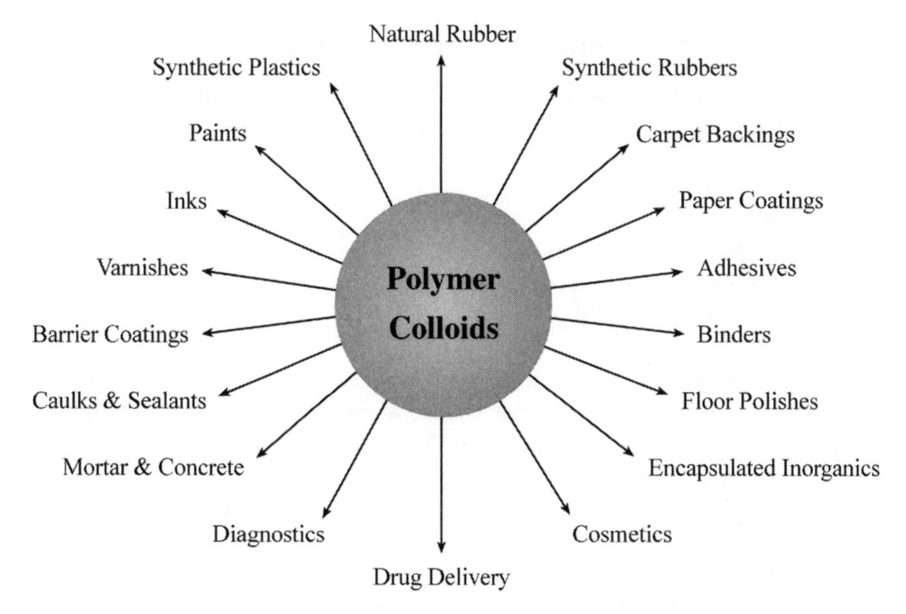

Figure 1 *Diagram showing the wide diversity of applications for polymer colloids*

2 The Principal Disciplines in Polymer Colloids Research

The most important feature of polymer colloids research is that it is both interdisciplinary and multidisciplinary. Indeed, most polymer colloids research programmes necessarily require integration of skills from several different areas of science and engineering. At present, the disciplines of most importance are:

- organic synthesis
- polymer chemistry
- chemical engineering
- colloid and interface science
- rheology
- polymer physics

Each of these disciplines can be divided into sub-categories that are individually of importance for progress in the field of polymer colloids. These will be considered both individually and in an integrated manner where appropriate in the following sections, highlighting essential links between research in each of these areas, with most emphasis placed on research driven by polymer chemistry.

3 Conventional Routes to Polymer Colloids

There are several routes to polymer colloids that may be considered *conventional* in that they have existed for many years, the most important being:

- emulsion polymerisation
- dispersion polymerisation
- miniemulsion polymerisation
- microemulsion polymerisation
- dispersion of preformed polymers

The first four routes are characterised by the use of conventional free-radical polymerisation to form the polymer *in situ*, whereas the latter does not involve polymerisation and was developed in order to produce aqueous dispersions of polymers formed by polyaddition or polycondensation.

Even though the subject of heterogeneous polymerisation can rightly be considered mature, there remain many unresolved issues and the need for fundamental research is substantial, both in more completely understanding existing processes and in the development of new routes to polymer colloids.

3.1 Research on Emulsion Polymerisation

Emulsion polymerisation is a complex heterogeneous polymerisation that at present is the most widely used and important route to synthetic polymer colloids.[1,2,4] Although the scientific understanding of simple emulsion polymerisations is well developed, a massive amount of fundamental research by *polymer chemists* is required if a predictive understanding of the more complex systems which find commercial application is to be achieved. The needs range from measurements of fundamental quantities (such as rate coefficients) for use in modelling and testing of theory, to control and accurate theoretical prediction of particle nucleation, polymer molar mass distribution, copolymer composition distribution, particle morphology and the location of functional groups in functionalised latexes, to the effects of high conversions and VOC (volatile organic compound) reduction on polymer properties.

The importance of gaining a deeper understanding of emulsion polymerisation is particularly evident when considering process modelling. Although there exist several good models of emulsion polymerisation, in relation to commercial processes they generally are incomplete or specific to a particular polymerisation.[2,4] This gap between the needs of the industry and the present predictive capability of general process models needs to be bridged and provides great opportunities for research by *chemical engineers*. Such research will feed from knowledge of accurate values of rate coefficients and monomer partitioning functions generated from the fundamental studies by *polymer chemists*. The aim must be to develop more general accurate predictive process models that take into account all factors of importance in commercial processes, including non-chemical factors such as heat flow and mixing effects.

Present understanding and control of the effects of monomer feeding strategies, free monomer, monomer partitioning, and concentration, heat and shear rate gradients is incomplete, as is the control of particle nucleation and growth in polymerisations that involve several monomers with different water solubilities. Research into process modelling should also be aimed at innovations in reactor and process design, especially with a view to achieving industrial targets for improved product control and faster, cleaner processes.

Attention has rightly been drawn[5] to the fact that highly-carboxylated latexes have received relatively little attention from academics, despite their commercial importance and the potential they offer for innovative research at the boundary between dispersion and solution behaviour.

3.2 Functionalised Polymer Colloids

The need to reduce or replace the use of conventional surfactants, particularly those based on alkyl phenols, in the preparation of polymer colloids has existed for many years and a considerable amount of research has been carried out world-wide with this objective. The development of 'inisurfs' and 'transurfs', which are, respectively, initiators and chain transfer agents that react to yield surfactant-like radicals, was an ill-conceived strategy for providing built-in stabilisation because they are limited to one or two surfactant groups per polymer chain (at the chain ends), much too low a concentration to be effective. The alternative strategy of developing polymerisable surfactants is far more realistic since the concentration of surfactant groups in the polymer chain can be controlled by the amount of polymerisable surfactant used. This topic has been the subject of intense effort over the past decade,[6,7] but despite the vast amount of work there are relatively few successes. More research is needed because the goal is highly desirable and realistic.

The 'polymerisable surfactant' approach will undoubtedly feature more generally in future research. For example, it can be envisaged as a means for anchoring of biocides to particle surfaces and of achieving the control of particle functionality that is required to support growth in the use of polymer colloids for more specialised applications, such as in the diagnostics, biomedical and catalysis fields where post-functionalisation of particles is likely to feature prominently. Research of this kind will be important for the development of smart polymer colloids, such as those that can respond to electrical or magnetic stimuli and those that can produce self-repairing coatings. At present, there are relatively few applications which take advantage of the high specific surface area available in polymer colloids (*e.g.*, $\sim 10^4$ m^2 kg^{-1} for non-porous particles of 200 nm diameter at 40% solids) and there is great scope for developments in this direction. Thus, the general topic of designed monomers and chemistries for introducing specific functionalities to the surfaces of particles is one that will continue to grow in importance and will be most effectively approached through close collaborations between *organic chemists*, *polymer chemists* and *biochemists*.

Although a very large number of different crosslinking systems have been developed for use with water-borne polymers,[2,8,9] none are entirely satisfactory and there is a need for more research into the development of new crosslinking chemistries. In addition to improving the performance characteristics of polymer latex coatings in established applications, such crosslinking systems are crucial to successful replacement of solvent-borne coatings with water-borne systems. One-part crosslinking systems (that require no further additions to the latex after preparation) are especially important and present a major scientific challenge because although crosslinking should not occur during storage of the latex or in the early stages of film formation, the crosslinking reaction ideally should proceed as the coating dries and consolidates at room temperature. Joint research by *organic* and *polymer chemists* will provide the most efficient way forward in achieving the innovations necessary to take forward this commercially-important field of fundamental polymer colloids research.

Thus *organic chemists* have a great deal to contribute in the synthesis of new monomers for use in preparation of functionalised and smart polymer colloids. The challenges are substantial given that, in most cases, the monomers must have low toxicity and sufficient chemical robustness that they do not degrade in the presence of the highly reactive species involved in polymerisations.

4 Unconventional Routes to Polymer Colloids

The dominant position of conventional emulsion polymerisations in the preparation of polymer colloids is unlikely to change. However, studies of new routes for preparation of polymer colloids are essential for growth of the industry and to meet the increasing demand for replacement of solvent-borne coatings. Thus, there is great scope for the development of less conventional routes to polymer colloids and the application of newer methods of polymer synthesis to preparation of polymer colloids will undoubtedly be a fruitful area for research.

4.1 Controlled Polymerisation in Dispersed Media

The field of controlled radical polymerisation (CRP) has seen rapid growth during the past decade,[10–12] growth that in more recent years has begun to embrace work on CRP in dispersed media.[7] The two-phase nature of heterogeneous polymerisations, however, imposes severe constraints on what can be achieved and restricts the scope for extending the established methods of CRP to heterogeneous systems.

Nitroxide-mediated CRP has been investigated for use in emulsion polymerisation systems, but with mixed the results.[13,14] Atom transfer radical polymerisation (ATRP) offers greater scope than nitroxide-mediated CRP in that it is less discriminating in terms of the monomers that can be used. However, there are few reports of work on adapting ATRP to heterogeneous systems.[15–17] Given the established requirements for control of these polymerisations, both nitroxide-mediated CRP and ATRP suffer from problems brought about by partitioning of reactants between the various phases present in a heterogeneous system. New strategies are required if the potential for control of molar mass, molar mass distribution and chain architecture is to be achieved in heterogeneous media. Both routes can benefit from use of miniemulsion polymerisation rather than emulsion polymerisation, since this largely eliminates the issues of reactant transport that exist with typical emulsion polymerisation methods.[18,19] New nitroxides for CRP and ligands for ATRP need to be developed that are less susceptible to partitioning and which are to able to constrain the active species to the loci of polymerisation. For ATRP, the incompatibility with acidic monomers (*e.g.*, acrylic acid) and the need for subsequent removal of the metal (most commonly copper) from the polymer particles will undoubtedly be serious issues that may well prove to be the ultimate factors restricting the use of ATRP in heterogeneous media.

Thus, there are several obstacles to implementation of nitroxide-mediated CRP and ATRP in dispersed media. Nevertheless, the concept of producing polymers of controlled molar mass and narrow molar mass distribution by heterogeneous polymerisation remains attractive. Success in achieving this aim may be easier to achieve by adapting the reversible addition-fragmentation-transfer (RAFT) process[20,21] to the preparation of polymer colloids. Again miniemulsion polymerisation is best suited to such polymerisations. Catalytic chain transfer processes have already been adapted for use in emulsion and miniemulsion polymerisation,[22–25] and use of a RAFT-like process termed MADIX (macromolecular design through interchange of xanthate) has recently been reported as being successfully adapted to emulsion polymerisation of butyl acrylate.

Clearly, CRP in heterogeneous media will be an important theme for future polymer colloids research. So far, most of the studies have been carried out by *synthetic organic chemists* with a strong interest in polymers. As the research progresses there will be a growing need to understand the physical chemistry of the polymerisations,[2] in particular the polymerisation kinetics, a traditional strength of the *polymer chemist*. In addition to the studies of CRP, there have been some studies of other chain polymerisations, including ring-opening metathesis polymerisation, coordination polymerisation and anionic polymerisation, though these have been quite limited and much more can be done, especially given that such polymerisations give access to control of polymer microstructure.

4.2 Alternatives to Polymer Dispersion

The growing commercial use of polymer colloids whose preparation involves step polymerisations, hybrid systems and/or new routes to dispersion of polymers has not yet spawned an equivalent level of activity in academic studies. The success of the Avecia (*ex* Zeneca Resins) technology for preparation of polyurethane dispersions is a good example of the importance of such developments.[26,27] There also is the attraction of using conventional solvent-borne coating vehicles (*e.g.*, alkyd resins) in aqueous media and hybrid systems consisting of combinations of polymers produced by conventional radical polymerisations with alkyds have obvious attractions. As for CRP in dispersed media, reactant transport issues make conventional emulsion polymerisation unsuitable for clean preparations of hybrid latexes, whereas miniemulsion polymerisation offers great potential due to the self-contained nature of the reaction loci. There is also the distinct advantage that the thermodynamic effects of having pre-formed (water-insoluble) oligomer or polymer present in the miniemulsion droplets lead to much better control of particle number and final particle size and its distribution. Thus, the use of step polymerisations, hybrid systems and dispersion of polymers is in need of substantial academic research activity aimed at providing the fundamental understanding of existing systems that is essential for further exploitation of these polymer colloids.

4.3 Other Developments

The continuing developments in the preparation of dendrimers and hyperbranched polymers[28–30] have moved this field into consideration by the polymer colloids community. As dendrimers and hyperbranched polymers get ever larger, their dimensions become sufficient to consider their 'solutions' as molecular polymer colloids. Inevitably, therefore, research into such polymers will feature in new directions for polymer colloids research, both as polymer colloids in their own right and as additives for modification of the properties of existing polymer colloids. Such research will be most effective if carried out through collaborations of *organic* and *polymer chemists* with *colloid scientists.*

Encapsulation of inorganic particles is likely to be of increasing importance and requires more research. Initial studies have focused on encapsulation of pigments and magnetic particles for use as colloids, but further research is necessary and could be extended to non-colloid applications for encapsulated particles. Additionally, the growing development of ink-jet routes for rapid prototyping and production of 3-dimensional ceramics articles can be expected to generate a demand for new research into the preparation of water-borne polymer-encapsulated ceramics particles, where the polymer acts firstly to stabilise the dispersion and then to provide coherency to the 'green' ceramics article before and during its sintering.

Given the recent emphasis on learning from nature and the fact that many polymer colloids are produced naturally (not least natural rubber latex), it is remarkable that there has been virtually no research into mimicking or developing biological routes to polymer colloids. This must surely be rectified by future research initiatives and will bring the need for new collaborations with *biological scientists.*

Thus there is still great scope for research into new strategies for producing polymer colloids. The resulting demands for new reactants that are specifically designed for use in heterogeneous media can only be satisfied through collaborations between *organic chemists* and *polymer chemists.*

4.4 Process Modelling

Process models for most of these newer routes to polymer colloids do not exist, but will be needed to support the growth in use of these alternative routes. For example, miniemulsion polymerisation has great potential as an effective means of performing heterogeneous CRP, but its commercialisation is restricted by the lack of control arising from the high monomer concentrations in the droplets and the relatively low solids contents. Thus, if the full potential of miniemulsion polymerisation is to be realised, *chemical engineers* have a major role to play in developing new approaches to carrying out the reactions that overcome these problems. Thus the development of these newer routes for the preparation of polymer colloids into commercially-realistic processes provides an opportunity for three-way collaborations between *organic chemists, polymer chemists* and *chemical engineers.*

5 Characterisation and Properties of Polymer Colloids

The complexity of polymer colloids is evident from the wide of range of methods needed for their characterisation.[1,2] These encompass both classical and newer methods of colloid and polymer analysis, and are essential for support of polymer colloids activities at all levels from fundamental research to commercial processes.

5.1 Colloidal and Interfacial Characterisation

Polymer colloids present a rich field of research for *colloid* and *interface scientists*, providing at one extreme well-defined colloids for testing of theory and at the other complex colloids for characterisation.

Understanding how to control the colloidal stability of polymer colloids is critical, whatever route is used for their preparation. Predictive knowledge of simple systems is well established, but more complex stabilisation systems that combine several approaches are less well understood. Additionally, research into colloidal stability under extremes of temperature, pH, ionic strength and shear rate is important and will require detailed studies of the particle interactions and dynamics in well-characterised polymer colloids. The results from such studies will also be important for achieving process control.

Thorough colloidal/surface characterisation is fundamental to the success of research on polymer colloids. A wide range of complementary techniques are available for colloidal/surface characterisation of polymer colloids and access to several is necessary since no single technique can provide full characterisation. There is an ongoing need for experimental and theoretical work on improvements to existing methods and on development of new techniques to support the needs of research. Additionally, the necessary improvements in process modelling will naturally lead to a demand for advances in on-line analysis to support feedback loops for process control and manufacturing. Thus, further developments in on-line methods for measurement of particle size and its distribution, surface tension, shear rate, viscosity, and unreacted monomer concentrations are necessary and could be integrated into providing 3-dimensional maps of the reactor contents for use in more advanced process control systems. Studies of this kind will provide good opportunities for combining the expertise of *colloid scientists* and *chemical engineers*.

5.2 Rheology

The rheology of polymer colloids is crucially important for coating applications and better control of rheology is sought by the polymer colloids industry. The development of theoretical treatments that model the particles realistically (*i.e.*, not as hard spheres) is paramount.

The realisation of very high solids contents in commercial latexes through inventive control of particle size distribution has created a need for both theoretical and experimental research into the rheological behaviour of polymer colloids at high particle volume fractions.

Recent developments in rheology control through use of associative thickeners has prompted significant academic research activity.[31–33] Much more research, however, is needed, particularly with a view to understanding and controlling the dynamics of the interactions that give rise to the thickening effects. These studies will need to include the effects of the hydrophilic surface layers that arise from use of water-soluble monomers and/or protective colloids. The use of microgels for modification and control of the rheological properties of polymer colloids is another topic worthy of much greater research,[34] as is the extension of such approaches to use of even smaller crosslinked particles.

5.3 Polymer Physics

Even though the process of film formation from latexes has been thoroughly researched for conventional polymer latexes,[2,35] there is still much work to be done on the dynamics of the process and the effects of the many chemical and physical factors that influence film formation and film properties for commercial latexes. Additionally, research into the factors affecting film formation and coating properties for the newer polymer colloids are required if they are to be fully developed. In hybrid systems, the effects of phase separation processes will be particularly important and are yet to be seriously studied.

Although toughness is often an important requirement for protective coatings, little fundamental research into coating toughness has yet been done. This contrasts quite starkly with the large volume of research into toughening of bulk polymers.[36–41] Hence, there is great scope for both theoretical and experimental research aimed at enhancing coating toughness and understanding the effects of multiphase particles and microgels on the mechanical properties of thin films. Similar studies aimed at gaining a predictive understanding of the factors that control the properties of water-borne soft adhesives (*e.g.*, pressure-sensitive and contact adhesives) are of crucial importance if the polymer colloids industry is to meet the challenges posed by the environmental drive to move away from solvent-borne adhesives. These studies will necessarily require substantial contributions from *polymer chemists*, *polymer physicists* and *colloid and interface scientists*.

6 The Way Forward

The themes for polymer colloids research presented in outline in the previous sections highlight the wide range of research that is necessary to take the field of polymer colloids forward in relation to industrial needs, and emphasise the importance of collaborative research across several disciplines. The principal disciplines that will be essential to advancing future research in the field of polymer colloids are shown in Figure 2. An increasing contribution from the disciplines of biochemistry and inorganic chemistry can be anticipated as newer, more specialised applications develop.

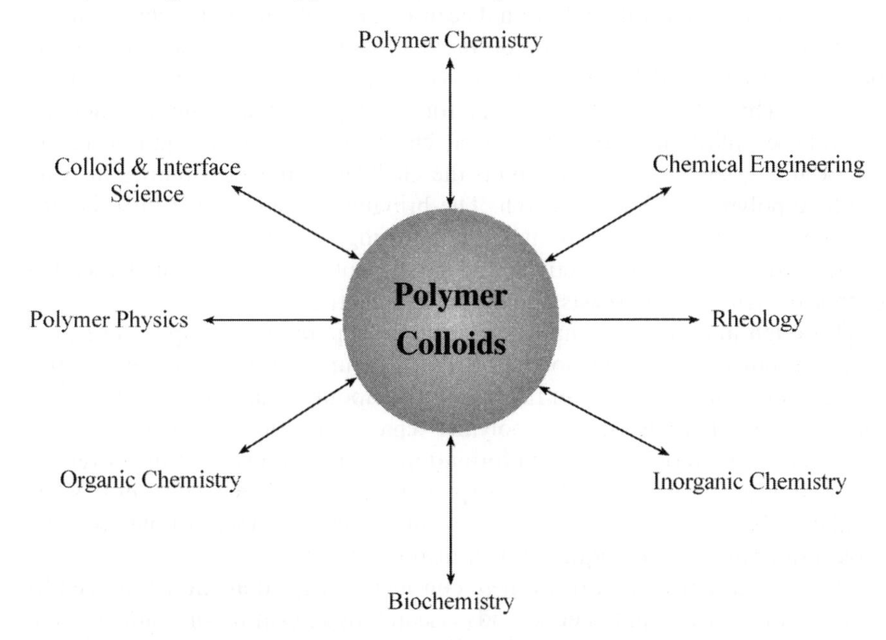

Figure 2 *Diagram highlighting the principal disciplines involved in polymer colloids research. Interactions between scientists from two or more of these disciplines will be essential to success in each of the directions for future developments in polymer colloids research*

Quite clearly, collaborations are the way forward for polymer colloids research, but such research is inhibited in the UK by the dispersed locations of the academics active in the field. The UK Polymer Colloids Forum, which was set up in 1993 to act as a focus for the UK polymer colloids community, embracing both industry and academe, has recently published a booklet giving profiles of the UK academics with research interests and activities in the field of polymer colloids. There are 36 academics represented in the booklet and, if considered collectively, the amount of UK polymer colloids research is substantial. Furthermore, the contributions by UK academics to polymer colloids research are held in high regard internationally. However, the 36 academics are distributed amongst no less than 22 universities and 26 departments. The greatest concentration of polymer colloids activity is at Bristol University, which is represented by four academics, all from one department (though one of the academics formally retired a few years ago).

Thus a serious issue to be addressed in taking polymer colloids research forward in the new millennium is how to effectively bring about stronger research collaborations. Those academics who are active in polymer colloids research and have complementary skills that could beneficially be combined, could be encouraged to collaborate more by implementation of appropriate initiatives, but would still be subject to the constraints imposed by geographical separation. Even if such collaborations were to be effectively instigated, there will be a natural limitation because the number of polymer colloids academics from the disciplines of organic synthesis, chemical engineering, polymer physics and biochemistry is insufficient to meet the needs of future research. This, of course, leaves scope for existing polymer colloids academics to enthuse colleagues from their local chemistry, chemical engineering and physics/materials departments about the challenges and possibilities in collaborative polymer colloids research. The bringing of new people into the field can only be beneficial, especially in allowing long-standing issues to be considered from new perspectives, and should not just be targeted at chemists, chemical engineers, physicists and materials scientists.

Thorough analysis and characterisation is of paramount importance in all research, but research into polymer colloids requires an unusually wide range of techniques, including those for organic compound characterisation, colloid and surface characterisation, polymer characterisation, measurements of rheology, characterisation of film formation and mechanical testing. Several of the techniques require significant capital investment and so are not widely available. In fact, none of the locations of current polymer colloids research have a full range of techniques at their disposal.

The obvious conclusion from these considerations is that there is a need for consolidation of UK polymer colloids research by relocation of academics into more focused groups which would facilitate close collaborative work and make funding of a wide range of equipment more realistic. Precedents for such strategies exist in other countries. For example, substantial groups of academics working together in single locations on polymer colloids research have been established at Lehigh University in the USA and at Eindhoven University of Technology in The Netherlands, the Lehigh activity having been prominent for around 30 years. Another long-standing, focused polymer colloids activity exists in the CNRS at Lyon in France, though this is on a smaller scale than those at Lehigh and Eindhoven. Much more recent are the focused activity set up at the Max Planck Institute for Colloids and Interfaces in Germany and the creation of the Key Centre for Polymer Colloids at Sydney University in

Australia. A common feature of these polymer colloids research groups is that they were created through a desire to enhance research activity in recognition of the importance of the polymer colloids industry and have subsequently been sustained in part by financial support from both the local and world-wide polymer colloids industry. An initiative of this kind is desired by the UK polymer colloids industry, as voiced through the various meetings of the UK Polymer Colloids Forum, and without doubt would lead to major advances in fundamental polymer colloids research that would service the needs of the polymer colloids industry. This remains a laudable goal for the UK, but cannot be achieved without real commitment from the academics, the industry and the research councils.

References

1 R. M. Fitch, *Polymer Colloids: A Comprehensive Introduction*, Academic Press, London (1997).
2 P. A. Lovell and M. S. El-Aasser, eds., *Emulsion Polymerization and Emulsion Polymers*, John Wiley & Sons, Chichester (1997).
3 Data acquired through the UK Polymer Colloids Forum.
4 R. G. Gilbert, *Emulsion Polymerisation: A Mechanistic Approach*, Academic Press, London (1995).
5 J. C. Padget, Papers presented at the 34th High Polymer Research Conference (Moretonhampstead, UK) 1994, and the Polymer Colloids Symposium (Runcorn, UK) 1999.
6 K. Holmberg, *Prog. Org. Coatings*, 1992, **20**, 325.
7 A. Guyot, *Colloids and Surfaces, Physicochemical and Engineering Aspects*, 1999, **153**, 11.
8 E. S. Daniels and A. Klein, *Prog. Org. Coatings*, 1991, **19**, 359.
9 R. J. Esser, J. E. Devona, D. E. Setzke and L. Wagemans, *Prog. Org. Coatings*, 1999, **36**, 45.
10 C. J. Hawker, *Trends Polym. Sci.*, 1996, **4**, 183.
11 T. E. Patten and K. Matyjaszewski, *Adv. Mater.*, 1998, **10**, 901 (16 pages).
12 T. E. Patten and K. Matyjaszewski, *Accounts Chem. Res.*, 1999, **32**, 895.
13 S. A. F. Bon, M. Bosveld, B. Klumperman and A. L. German, *Macromolecules*, 1997, **30**, 324.
14 C. Marestin, C. Noel, A. Guyot and J. Claverie, *Macromolecules*, 1998, **31**, 4041.
15 S. G. Gaynor, J. Qiu and K. Matyjaszewski, *Macromolecules*, 1998, **31**, 5951.
16 J. Qiu, S. G. Gaynor and K. Matyjaszewski, *Macromolecules*, 1999, **32**, 2872.
17 K. Matyjaszewski, *Macromol. Symp.*, 1999, **143**, 257.
18 T. Prodpran, V. L. Dimonie, E. D. Sudol and M. S. El Aasser, *Polym. Mat. Sci. Eng.*, 1999, **80**, 534.
19 P. J. MacLeod, B. Keoshkerian, P. Odell and M. K. Georges, *Polym. Mat. Sci. Eng.*, 1999, **80**, 539.
20 J. Chiefari, Y. K. Chong, F. Ercole, J. Krstina, J. Jeffery, T. P. T. Le, R. T. A. Mayadunne, G. F. Meijs, C. L. Moad, G. Moad, E. Rizzardo and S. H. Thang, *Macromolecules*, 1998, **31**, 5559.
21 R. T. A. Mayadunne, E. Rizzardo, J. Chiefari, Y. K. Chong, G. Moad and S. H. Thang, *Macromolecules*, 1999, **32**, 6977.

22 K. G. Suddaby, D. M. Haddleton, J. J. Hastings, S. N. Richards and J. P. Odonnell, *Macromolecules*, 1996, **29**, 8083.
23 D. Kukulj, T. P. Davis, K. G. Suddaby, D. M. Haddleton and R. G. Gilbert, *J. Polym. Sci., Polym. Chem.*, 1997, **35**, 859.
24 D. Kukulj, T. P. Davis and R. G. Gilbert, *Macromolecules*, 1997, **30**, 7661.
25 D. M. Haddleton, D. R. Morsley, J. P. Odonnell and S. N. Richards, *J. Polym. Sci., Polym. Chem.*, 1999, **37**, 3549.
26. R. Satguru, J. McMahon, J. C. Padget and R. G. Coogan, *J. Coatings Tech.*, 1994, **66**, 47.
27 R. Satguru, J. McMahon, J. C. Padget and R. C. Coogan, JOCCA, *Surf Coatings Int.*, 1994, **77**, 424.
28 J. Roovers and B. Comanita, *Adv. Polym. Sci.*, 1999, **142**, 179.
29 C. J. Hawker, *Adv. Polym. Sci.*, 1999, **147**, 113.
30 A. Hult, M. Johansson and E. Malmstrom, *Adv. Polym. Sci.*, 1999, **143**, 1.
31 W. H. Wetzel, M. Chen and J. E. Glass, *Adv. Chem. Ser.*, 1996, **248**, 163.
32 M. R. Tarng, Z. Y. Ma, K. Alahapperuma and J. E. Glass, *Adv. Chem. Ser.*, 1996, **248**, 449.
33 T. Svanholm, F. Molenaar and A. Toussaint, *Prog. Org. Coatings*, 1997, **30**, 159.
34 C. Raquois, J. F. Tassin, S. Rezaiguia and A. V. Gindre, *Prog. Org. Coatings*, 1995, **26**, 239.
35 J. L. Keddie, *Mat. Sci. Eng. Rep.*, 1997, **21**, 101.
36 C. B. Bucknall, *Toughened Plastics*, Applied Science, London (1977).
37 A. J. Kinloch and R. J. Young, *Fracture Behaviour of Polymers*, Applied Science, London (1983).
38 C. K. Riew, ed., *Rubber-Toughened Plastics*, Adv. Chem. Ser., Vol. 222, American Chemical Society, Washington, D.C. (1989).
39 I. K. Partridge in *Multicomponent Polymer Systems*, I. S. Miles and Rostami S, eds., Longman, Harlow, United Kingdom (1992), Chapter 5, p. 149.
40 C. K. Riew and A. J. Kinloch, eds., *Toughened Plastics I: Science & Engineering*, Adv. Chem. Ser., Vol. 233, American Chemical Society, Washington, DC (1993).
41 C. K. Riew and A. J. Kinloch, eds., *Toughened Plastics II: Novel Approaches in Science & Engineering*, Adv. Chem. Ser., Vol. 252, American Chemical Society, Washington, DC (1996)

Biomaterials and Tissue Engineering

8

Biomaterials: Tissue Engineering and Polymers

Steve Rimmer

DEPARTMENT OF CHEMISTRY, UNIVERSITY OF SHEFFIELD, SHEFFIELD S3 7RH, UK

1 Introduction

A biomaterial is generally regarded as any material that is used to fabricate a device to be used in a medical application. These applications are many and varied and include demanding applications such as artificial veins, *etc.* through to disposal use-once applications such as surgeon's gloves and blood bags. Perhaps the fastest growing aspect of biomaterials is the emerging field of tissue engineering. In tissue engineering one attempts to apply the principles of engineering design to the production of an artificial biological tissue. The design process, of course, is reliant on sound empirical data that can be extrapolated to the desired application. In the case of designing a new tissue this data is generally biochemical.

2 The State of the Art: A Materials Chemist's Perspective

Several variants of tissue engineering can be identified (see Figure 1). In the first method a polymeric scaffold is produced in the shape of the organ to be replaced. Cells pertinent to the target tissue are then seeded onto this article. The cells are then grown *in vitro*. This growth process eventually produces a tissue, *i.e.* an organized functional assembly of cells. As the cells grow the scaffold degrades, hopefully at a controlled rate, releasing benign materials, until eventually only the tissue is left behind. Thus in a perfect system a new tissue will have been produced from a small sample of cells. This tissue might then be implanted as a replacement for a diseased tissue or in theory it could be combined with other tissues to produce an organ. In a second variant a scaffold is first seeded with appropriate cells. This scaffold is then implanted so that further cell growth is determined by the bodies natural control mechan-

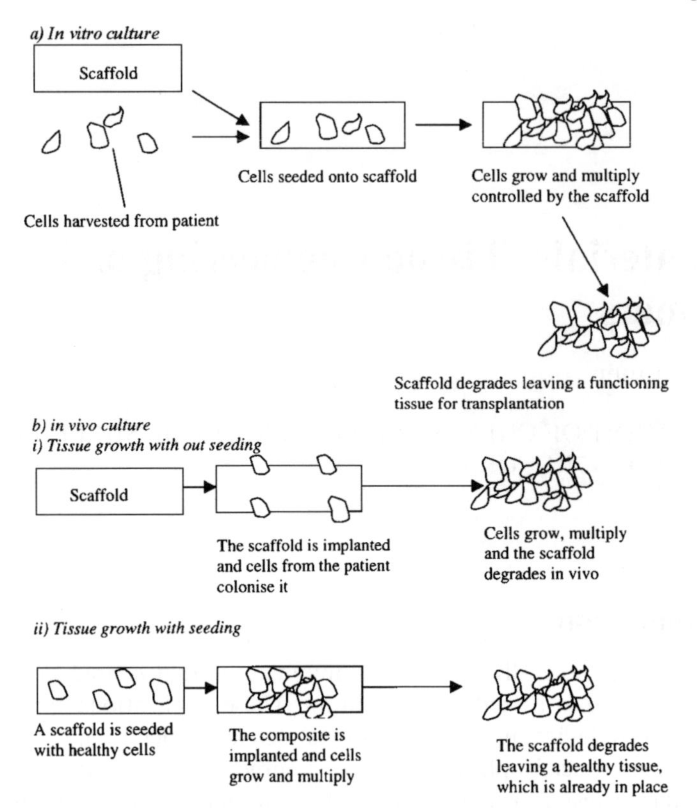

Figure 1 *Tissue engineering procedures*

isms. Thus the cells grow and proliferate under the influence of the body's own growth factors and other cell signaling molecules. Once again as the cells grow and the tissue develops the scaffold degrades. In this varient it is even more vital that the degrading scaffold does not produce toxic materials during degradation since these toxins could now travel to other organs. Another way of carrying out tissue engineering *in vivo* is to implant the scaffold and then allow the patients own cells to colonise it, grow and multiply. Another, but little explored, method involves the use of non-biodegradable scaffolds. This technique will be most applicable to the generation of tissue that needs to withstand mechanical stress, such as cartilage. In this case the polymeric scaffold can be used to enhance the mechanical strength of the construct. The final material is thus a cell/synthetic hybrid material. The cells' function here would be to mediate the biological response to the device.

3 Scaffold Materials

Polymeric biomaterials play the vital roles of guiding/controlling the growth of the new tissue and protecting the immature tissue from environmental stresses, such as abrasion. Several materials are available including natural and synthetic polymers that degrade at varying rates.

3.1 Natural Polymers as Scaffolds

Natural protein containing polymers include those based on collagen, albumen and silk. Generally collagen and albumen based materials need to be modified to give mechanically stable structures. The simplest way to do this is to use glutaraldehyde[1] cross-linking but activated PEGs have also been used.[2] Modification is also necessary in order to control the degradation rate of the scaffold.

Polysaccharide based materials include glycosaminoglycans (GAGs)[3] and modified cellulose,[4] chitin[5] and starch.[6] Once again such materials require cross-linking to provide mechanical stability and control of swelling in water. Typically PEGs may be used *via* reaction with free amino groups on the GAGs.[7] It is also possible to cross-link these materials with non-covalent electrostatic interactions provided by incorporation of polyvalent cations.[8]

3.2 Synthetic Polymers as Scaffolds

The most widely used biodegradable polymers used are the poly(α-hydroxy acids) especially poly(D,L, lactide) and polyglycolic acid. The main reason for their use is that they degrade to natural metabolites. Large accumulating concentrations of lactic acid can however be problematic since this compound can induce changes in the ECM and at sufficiently high levels may become cytotoxic.[9,10] Although other evidence shows that the associated changes in pH are insignificant.[11] The polymers are however hydrophobic: a property that is known to affect cell adhesion, morphology and growth and probably effects the composition of the ECM. Degradable polymers that are related to the lactide/glycolides include the polycarbonates and the poly(dioxanones) and copolymers of these with lactide and glycolide repeat units. The success of these polymers is derived from a compromise between bioproperties and processing. That is the degradation products from these materials are generally acceptable to biological systems and the polymers can be processed into useful forms by for example fibre spinning. Tyrosine-derived polycarbonates are used in orthopaedic applications. They are strong and bioabsorbable and can be prepared as amphiphilic variants. Polyanhydride polymers are also biodegradable and they have been used in both drug delivery and hard tissue engineering applications. However, they can be difficult to process. Another class of carboxylic acid derived degradable polymers of use as degradable scaffolds are the polyorthoesters. These materials tend to have low moduli and load bearing capacity.

In general all of these materials are hydrophobic although some hydrophilic modifications are known. Commercial hydrophilic variants, known as polyactive® and shown in Figure 2, are composed of blocks of poly(butylene teraphthlate) and poly(ethylene oxide).

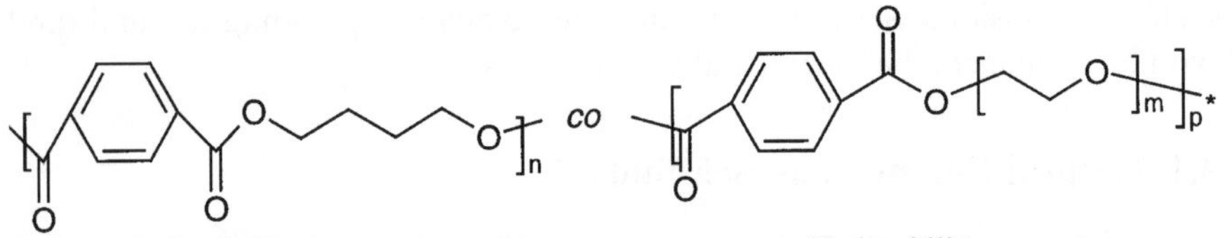

Hydrophobic segement Hydrophilic segment

Figure 2 *The segmented amphiphilic structure of Polyactive®*

These polymers are of increasing utility and are good examples of the cellular requirement for water in tissue engineering scaffolds.[13–15] Polyphosphazenes are degradable polymers that have been functionalized with a variety of biomolecules such as peptides and growth factors.[16]

3.2.1 Hydrogels. Hydrogels based on copolymers of hydrophilic vinyl monomers are well known biomaterials and they can have uses in tissue engineering. Their utility stems from the incorporation of large but controllable amounts of water in the material. Cells of course live in an aqueous environment so it is not too surprising that successful tissue culture often requires a hydrophilic polymer surface. Currently the most important hydrogel polymers are those based on polyethylene glycols.[17–20] Since this polymer is water soluble at physiological temperatures it must be modified either by cross-linking or by addition of hydrophobic blocks. Radiation cross-linking and photopolymerization have both produced useful scaffold materials. PVA has also found some limited use.[21] Cross-linked natural hydrophilic polymers such as alginate[22–25] and gelatin[26,27] have recently been used as scaffolds as have sophisticated blends of various natural polymers, which aim to mimic the ECM.[28,29]

4 Challenges in Tissue Engineering

The challenges that face us can broadly be set out into (1) biochemical/cell biology challenges and (2) materials challenges.

4.1 Biochemical Challenges

The response of cells to the artificial environment of a tissue engineered construct is of course governed by the complex interactions of adhesive proteins, integrins and growth factors as well as the interplay of surface energies and surface topology. The understanding of all of these interactions will be vital if the field is to grow to its full potential. Very few reports are available that detail experiments designed to assess the interplay of all of these factors. A good example of this is the often quoted use of arginine-glycine-aspartate (RGD) sequences of the adhesive proteins, fibronectin, vitronectin and laminin, which clearly do induce cell adhesion on otherwise non-adhesive surfaces.[30,31] However, few experiments that compare adhesion and spreading on RGD functional surfaces with non-RGD containing surfaces of high surface energy have been reported and yet surface energy of the scaffold clearly plays a role in adhesion and spreading. An interesting recent series of experiments by Tija *et al.*[32] has clearly shown that fibronection adsorbed to a hydrophobic surface, poly(lactide-*co*-glycolide) (PLG), is less biologically active than the free protein and that this effect is due to conformational changes following adsorption. Even the observation that larger amounts of fibronectin adsorped to PLG than to glass did not overcome this lowering of activity. So that addition of adhesive fragments may not be sufficient to induce migration and adhesion. These same authors have also shown that secretion of cellular fibronection to PLG coated with collagen I was a more effective strategy. Since fibronection is known to interact *via* the 70 kDa N-teminius and cell surface receptors a future ambitious strategy might be to attempt to build a hydrophilic polymer, which should be resistant to non-specfic interactions, that also contains a species capable of interacting with this N-terminal sequence. In this way one might maintain the confirmation of the RGD sequences and thus make them available for cellular interactions.

Other peptide sequences have been studied in the context of tissue engineering but the results can often be contradictory. For example, the IGSR sequence found in laminin is an integrin receptor ligand that has been shown to promote spreading and proliferation of arterial cells[33] but results from experiments involving fibroblasts are contradictory.[34] There is also now evidence that RGDS attachment while enhancing cell spreading and adhesion decreases the amount of ECM produced during cell growth.[34] This is a serious issue since as the scaffold degrades it must be replaced by the ECM and changes in its composition should be avoided. Currently it is very difficult to predict the effect a scaffold might have on the composition and extent of the ECM. The main source of this difficulty is the lack of data on well-defined systems. For this reason work on model systems prepared irrespective of cost, *in vivo* performance or practicality of large-scale production is vital. Perhaps, none of these models would ever see commercial success but the value of data gathered in well designed biochemical experiments on well defined models is incalculable.

Components of cells that are important for cell ECM adhesion include the proteoglycans (PG) and glycosaminoglycans (GAG). Heterodimeric proteoglycans known as integrins are particular important for attachment of cells to the ECM as are cell surface proteoglycans such as heparin sulfate, keratin sulfate, *etc.* Interactions of all of these ECM and cell wall components can influence the formation of engineered tissue. Clearly total characterization of the biochemical response to a material is difficult but if complete understanding is to be gained all of these factors will in the future need to be investigated. Techniques that can generate data quickly from the whole biological milieu would provide a powerful addition to the knowledge base and even qualitative but general techniques will be useful in obtaining a broad view of the biochemical response. Mass spectrometry fits this requirement. In particular MALDI-TOF and electrospray mass spectrometry alone and coupled to other separation techniques, such as capillary electrophoresis and chromatography are ideal techniques for examining the whole biochemical system in a few analytical steps. Also these approaches have the advantage that one may discover hitherto unknown but important biomolecular species by serendipity. Application of mass spectrometry to biomaterials research has been pioneered by Leize *et al.*[35] and Greisser *et al.*[36–39] and is common in many other biochemical spheres.

Huge advances in molecular biology will in the future also open up the possibility of using designer cells, that is cells specifically designed to interact favourably with the scaffold.[40] In the current climate this aspect will undoubtedly attract publicity. The possibilities, for this kind of gene modification of the seeded cells, appear to be almost boundless. Not only could cells be modified to match the scaffold: it might be possible to prepare improved tissues: for example an improved liver or heart certainly appear to be worthwhile goals.

4.2 Materials challenges

4.2.1 Chemistry and Biochemistry. Several materials are available for scaffold formation but since the biochemical understanding of how cells interact with materials is still in its infancy it is not a simple matter to design a new scaffold. Scaffold design is often empirical although some rational chemical design can be employed. For this reason parallel approaches to the synthesis of series of libraries have the potential to unlock some still controversial issues. Often polymeric materials have several diverse structural features all of which may influence the physical and biological properties of the material. For example segmented block polymers contain diversity in: segmental structure; mole ratio of the segments; water content; cross-link density and charge density and type. This diversity coupled with the complexity of the biochemical response makes establishment of structure/bioproperty relationships using conventional linear techniques extremely time consuming so that such studies are very rarely carried out. A more powerful method of assessing biomaterials properties is to produce rational libraries of materials: that is arrays of materials in which the structural parameters vary in a sequential manner. Kohn *et al.* have pioneered this approach in their examinations of several ranges of biomaterials libraries[41–47] and an example of this kind approach, from our laboratories, is shown in Figure 3.

These data, which are concerned with the adsorption of immunoglobulin G to a series of tricomponent hydrogels, show the danger in examining small samples of materials that are able to display multivariate diversity. Clearly, several naive correlations can be made by accessing sections of these data. For example the correlation shown by the tie lines, which connect hydrogels of constant monomer feed composition and variable cross-link density, could easily lead to several false conclusions. Driven by the pharmaceutical industry, new advances in parallel synthesis are now occurring at a significant rate and the application of these methods to polymer synthesis will be the key to

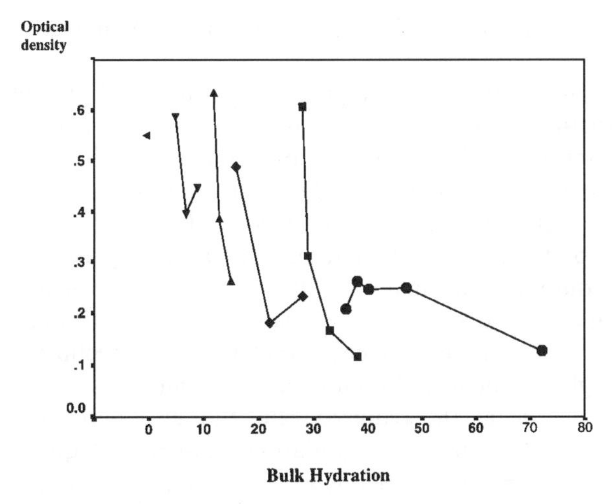

Figure 3 *Adsorption IgG to a series of poly(lauryl methacrylate-co-glycerol methacylate-co-ethylene glycol dimethacrylate)s: Results from immuno assay, optical density is directly proportional to amount of adsorbed protein*

unlocking the secrets of the synthetic polymer/biological properties paradigm. It should also be possible to couple the fast analytical capability of techniques such as mass spectrometry and LC-NMR with experiments carried out on parallel arrays of materials.

As covered, above design of new polymer scaffolds is not a simple process since the lack of good structure-property relationship data hinders theoretical prediction of the type of polymer structures that we should be aiming at. We are therefore left with a rather empirical approach to the area: for example we do not have a theory that would have predicted the utility of polyactives® as a scaffold and yet van Blitterswijk *et al.* have clearly shown that these materials do have potential. Probably, the ultimate guiding principle for scaffold design is to mimic the ECM. The ECM is the material in which cells reside and is the medium by which they communicate and gain nutrients. Perhaps the most important feature of the ECM is that it contains mainly water. The presence of large amounts of water is important since this allows the cells to migrate into the material and allows for free passage of nutrients. Also, of course, most biochemical processes occur in water. The ECM is also essentially a cross-linked gel in which often the GAG hylauronic acid acts as the hydrophilic component and collagens act as cross-linking, structural and recognition units. Therefore, hydrogels would appear to be suitable candidates for new scaffolds. However, suitable hydrogels will need to possess several properties that this class of materials do not generally display. That is high water content hydrogels are not generally tough materials. This is an important aspect since a successful tissue manufacturing facility will require that scaffolds can be handled by non-specialist, maybe even non-science trained, staff. Also, the *in vivo* variant of tissue engineering will demand that materials that are robust enough to be handled in surgical procedures are available. Processing is also a key issue. Ideally, hydrogels that can be formed into fibres and foams are required.

In this area imaginative reactive processing routes will undoubtedly play a role. The structural role of collagen can be mimicked in several ways. For example, we are attempting to do this by producing hydrogel materials that also contain hydrophobic domains, which act as physical cross-links.[48] Alternatively, the use of hybrid materials in which natural collagens are combined with synthetic polymers is a useful strategy that is already used. The advantage of this procedure is that many of the important biochemical signalling sequences that are present within the collagen assemblies can be retained. For example collagen I contains sequences that bind fibronectins: the important cell adhesion glycoprotein. A biomimetic ECM must also be biodegradable but it must degrade at the required rate and give non-toxic degradation products. The issue of non-toxic degradation products is frequently equated to naturally occurring products but this is of course an oversimplification since many natural products are toxic and many synthetic polymers/oligomers are benign. The ideal situation however is one in which the degradation products do not effect the composition of the naturally produced ECM and where they can be used to provide cellular nutrition. An increasingly important aspect of new scaffolds to be considered is the inclusion of recognition and signalling elements. Usually these are peptidic but they need not be. They may in fact be important advantages in using non-peptidic, biofunctional groups attached to the scaffold. Not least of these advantages would be increased resistance to steam sterilization.

Most of the effort in scaffold design centres is on the need to use biodegradable scaffolds and the necessity of controlling that degradation. However, not all applications would be best served by a scaffold that is eventually absorbed. Applications that involve load bearing by soft tissue might benefit from the continued presence of the scaffold. Such scaffolds will need to be biostable rather than degradable. In these materials very critical balances between initiating the immune response and allowing useful cells such as endothelial cells to colonise the implant need to be established. The challenge lies in the fact that several of the important cell adhesion/growth molecules are common to more than one cell type. For example, the RGD sequence is currently the most important cell adhesion sequence known, so that RGD functional polymers are often found to be superior scaffolds to non-functional polymers. However, RGD and fibronectin are also implicated in platelet adhesion[47–50] so adhesion promoting species specific for endothelial cells will have to be used in blood contacting applications.

4.2.2 Fabrication. Porosity and surface roughness are important physical parameters for any useful scaffold. The scaffold can be classified into two basic forms: either fibrous or foam-like. In order to form woven or non-woven meshes the scaffold must be amenable to spinning and this limits the number of materials that can be produced in this form. The most important fibre forming scaffold are the α-hydroxy acid based materials.[51–53] Other fibre forming materials that had suitable biological properties would be useful. In particular materials that could be spun to give, eventually, hydrogel fibres might be a major advance. Several methods of preparing porous polymer structure are available including freeze drying,[54] particulate leaching[55] or solvent leaching.[56] A very recent example uses the carbon dioxide released from the decomposition of ammonium bicarbonate as a poreogen.[57] Much is now known of the physics of foams and macroporous polymers and this knowledge will clearly continue to impact on the area. For example, Howdle *et al.* have recently shown the utility of supercritical carbon dioxide for the formation of porous scaffolds for regeneration of bone.[58] Another emerging technology in the area is the technique of three dimensional printing.[59,60] The process involves the successive fusing of polymer particulates, in layers, until a three dimensional shape is formed. The technique can be used to prepare a wide variety of asymmetrical shapes that cells can be seeded into. Clearly, this is a very powerful technology for the manufacture of a whole range of scaffold shapes.

5 Conclusions

New scaffolds will need to meet several important criteria. Scaffolds based on modelling the ECM seem to be attractive but it will be necessary to process them into fibres, foams or three dimensional printed shapes, *etc.* and they will need to withstand sterilization techniques. These processing/manufacturing aspects appear to favour the use of synthetic polymers rather than natural polymers, which are often difficult to process and would probably not withstand three dimensional printing. However, the advantage of natural materials is that they and their degradation products are probably benign and they may contain important signalling sequences. Therefore, as with many aspects of materials science, compromises in materials properties will be necessary and as is often the case individual applications will require different property balances. Therefore, hybrid and formulated materials will undoubtedly become the materials of choice as the field matures.

The principles required for the molecular design of polymers that can be spun or processed into foams and other porous structures are already known but the biochemical structure/property relationships of synthetics are not well understood. Currently, new materials are appearing in parallel with important discoveries in cell and molecular biology and it is probably true that discoveries in one field are feeding off work in the other. This is an unusual but exciting and challenging situation. In other more mature interdisciplinary fields it is generally possible to identify one way traffic from theory through to practice but in tissue engineering this is not yet possible, despite the enormous advances in our understanding of cell biology in recent years.

Finally, the ethical and public perception issues should not be underestimated. Currently tissue engineering is in its infancy and few outside the field are aware of the impact that technology derived from our endeavours will have on society.

References

1 *e.g.* P.B. van Wachem, L.A., Brouwer, R. Zeeman, P.J. Dijkstra, J. Feijen, M. Hendriks, P.T. Cahalan, M.J.A. van Luyn, *J. Biomed. Mater. Res.*, 2000, **53**, 18.
2 *e.g.* S.C. Vasudev, T. Chandy *J. Biomed. Mater. Res.*, 1997, **35,** 357.
3 *e.g.* K.C. Sung, E.M. Topp, *J. Memb. Sci.*, 1994, **92**, 157.
4 *e.g.* M. Martson, J. Viljanto, T. Hurme, P. Saukko, *Euro. Surg. Res.*, (1998), **30**, 426.
5 *e.g.* S.V. Madihally, H.W.T. Matthew, *Biomater.*, 1999, **20,** 1133.
6 J. Pasch, A. Schiefer, I. Heschel, G. Rau, *Cryobiology*, 1999, **39**, 158.
7 W.M. Rhee, R.A. Berg, US Patent 5,510,418 (1996).
8 A.G. Mikos, M.G. Papadaki, S. Louvroukoglou, S.L. Ishang, R.C. Thompson, *Biotechnol. Bioeng.*, 1994, **43,** 673.
9 O. Bostman, E. Hirvensalo, J. Makinen, P. Rokkanen, *J. Bone Joint Surg.*, 1999, **72B**, 52.
10 J. Suganama, H. Alexander, J. Traub, J.K.L. Ricci, in *Tissue Inducing Biomaterials*, 1992, 339.
11 C. Martin, H. Winet, J.Y. Bao, *Biomater.*, 1996, **17**, 2373.
12 *e.g.* A.G.M. van Dorp, M.C.H. Verhoeven, H.K. Koerten, C.A. van Blitterswijk, M. Ponec, *J. Biomed. Mater. Res.,* 1999, **47**, 292.
13 R.J.B. Sakkers, R.A.J. Dalmeyer, J.R. de Wijn, C.A. van Blitterswijk, *J. Biomed. Mater. Res.,* 2000, **49**, 312.
14 Y.L. Xiao, J. Riesle, C.A. Van Blitterswijk, *J. Mater. Sci.-Mater. Med.,* 1999, **10**, 773.
15 A.G.M. van Dorp, M.C.H. Verhoeven, H.K. Koerten, C.A. van Blitterswijk, M. Ponec, *J. Biomed. Mater. Res.,* 1999, **47**, 292.
16 H.R. Allcock, S.R. Pulcher, A.G. Scopelianos, *Macromolecules*, 1994, **13**, 1338.
17 *e.g.* D.K. Han, J.A. Hubbell, *Macromolecules*, 1997, **30**, 6077.
18 *e.g.* L.J. Suggs, J.L. West, A.G. Mikos, *Biomater.*, 1999, **20**, 683.
19 *e.g.* D.K. Han, K.D. Park, J.A. Hubbell, Y.H. Kim, *J. Biomater. Sci.-Polym. Ed.*, 1998, **9**, 667.
20 *e.g.* A.S. Sawhney, C.P. Pathak, J.A. Hubell, *Macromolecules*, 1993, **26,** 581.
21 T. Tokiwa, M. Kodama, *Mater. Sci. For.*, 1997, **250**, 97.
22 A. Martinsen, G. Skjak-Braek, O. Smidsrod, *Biotechnol. Bioeng.*, 1989, **33,** 70.
23 F. Lim, A.F. Sun, *Science*, 1980, **210**, 908.
24 M.A. LeRoux, F. Guilak, L.A. Setton, *J. Biomed. Mater. Res.,* 1999, **47**,46.
25 J.A. Rowley, G. Madlambayan, D.J. Mooney, *Biomater.,* 1999, **20**, 45.
26 M. Yamamoto, Y. Tabata, Y. Ikada, *J. Bioact. Comp. Polym.*, 1999, **14,** 474.
27 H.W. Kang, Y. Tabata, Y Ikada, *Biomater.,* 1999, **20**, 1339.
28 T. de Chalain, J.H. Phillips, A. Hinek, *J. Biomed. Mater. Res.,* 1999, **44,** 280.
29 V.F. Sechriest, Y.J. Miao, C. Niyibizi, A. WesterhausenLarson, H.W. Matthew, C.H. Evans, F.H. Fu, J.K. Suh. *J. Biomed. Mater. Res.,* 1999, **49,** 534).
30 E. Ruoslahti, M.D. Pierschbacher, *Science,* 1987, **238,** 491.
31 A.D. Cook, J.S. Hrkack, N.N. Gao, I.M. Johnson, U.B. Pajvani, S.M. Canizzaro, R. Langer, *J. Biomed. Mater. Res.*, 1997, **35**, 513.
32 J.S. Tija, B.J. Aneskievich, P.V. Moghe, *Biomater.* 1999, **20,** 2223.
33 K.C. Dee, T.T. Anderson, R. Bizios, *Mater. Res. Soc. Symp.*, 1994, **331**, 115.
34 B.K. Mann, A.T. Tsai, T. Scott-Baden, J.L. West, *Biomater.*, 1999, **20**, 2281.
35 E.M. Lieze, E.J. Leize, M.C. Leize, J-C. Vogel, A. van Dorsselaer, *Anal. Biochem.*, 1999, **272**, 19.

36 S.L. McArthur, P. Kingshott, H.A.W. St. John, K.M. McLeam, H.J. Griesser, *Inv. Opth. Vis. Sci.*, 1999, **40**, No. 4 abstract number 3198-B56, page 608.
37 K.M. McLean, S.L. McArthur, R.C. Chatelier, P. Kingshott, H.J. Griesser, *Coll. Surf. B-Biointer.*, 2000, **17**, 23.
38 P. Kingshott, H.A.W. St John, R.C. Chatelier, H.J.J. Griesser, *Biomed. Mater. Res.*, 2000, **49**.
39 P. Kingshott, H.A.W. St John, H.J. Griesser, *Anal. Biochem.*, 1999, **273**, 156.
40 A.S. Breitbart, J.M. Mason, C. Urmacher, M. Barcia, R.T. Grant, R.G. Pergolizzi, D.A. Grande, *Annal Plast. Surg.*, 1999, **43**, 632.
41 S. Brocchini, K. James, V. Tangpasuthadol, J. Kohn, *J. Biomed. Mater. Res.*, 1998, **42**, 66.
42 E. Tziampazis, J. Kohn, P.V. Moghe, *Biomater.*, 2000, **21**, 511.
43 K. James, H. Levene, J.R. Parsons, J. Kohn, *Biomater.*, 1999, **20**, 2203.
44 S. Brocchini, K. James, V. Tangpasuthadol, J. Kohn, *J. Biomed. Mater. Res.*, 1998, **42**, 66.
45 S. Brocchini, K. James, V. Tangpasuthadol, J. Kohn, *J. Am. Chem. Soc.*, 1997, **119**, 4553.
46 S. Rimmer, P. Tattersall, J.R. Ebdon, N. Fullwood, *React. Polym.* 1999, **41**, 177.
47 S.A. Mousa, J.M. Bozarth, M. Forsythe, G. Cain, A. Slee, *Blood,* 1999, **94**, 979.
48 M.M. Rooney, D.H. Farrell, B.M. van Hemel, P.G. deGroot, S.T. Lord, *Blood*, 1998, **92**, 2374.
49 N.P. Murphy, D. Pratico, D.J. Fitzgerald, *J. Pharmacol. Exp. Therap.*, 1998, **286,** 945.
50 S. Nakatani, T. Hato, Y. Minamoto, S. Fujita, *Thromb. Haem.,* 1996, **76**, 1030.
51 H. Steuer, R. Fadale, E. Muller, H.W. Muller, H. Planck, B. Schlosshauer, *Neuro. Lett.,* 1999, **277,** 165.
52 L. Fambri, A. Pegoretti, R. Fenner, S.D. Incardona, C. Migliaresi, *Polymer*, 1997, **38**, 79.
53 C. Migliaresi, L. Fambri, *Macromol. Symp.*, 1997, **123**, 155.
54 *e.g.* R.Y. Zhang, P.X. Ma, *J. Biomed. Mater. Res.*, 1999, **44**, 446.
55 *e.g.* J.M. Pachence, *J. Biomed. Mater. Res.*, 1996, **33**, 35.
56 *e.g.* A.K. Nyhus, S. Hagen, A. Berge *J. Polym. Sci. A-Polym. Chem.*, 1999, **37,** 3973.
57 Y.S. Nam, J.J. Yoon, T.G. Park, *J. Biomed. Mater. Res.*, 2000, **53,** 1.
58 A.I. Cooper, S.M. Howdle, *Materials World* 2000, **8**, 10.
59 B.M. Wu, S.W. Borland, R.A. Giordano, L.G. Cima, E.M. Sachs, M.J. Cima, *J. Control. Rel.*, 1996, **40**, 77.
60 B.M. Wu, M.J. Cima, *Polym. Engin. Sci.,* 1999, **39**, 249.

9

Biomaterials: Recent Trends and Future Possibilities

Neil R. Cameron

DEPARTMENT OF CHEMISTRY, UNIVERSITY OF DURHAM, SOUTH ROAD, DURHAM DH1 3LE, UK

1 Introduction

Currently, one of the most rapidly expanding areas of polymer science is at the interface with biology. Polymer materials play a huge role in cutting edge research in such areas as tissue engineering, drug delivery, gene therapy and biosensors. In addition, polymer scientists have begun to use nature's methods of synthesis to make novel, highly complex and functional materials. This 'cross-fertilisation' between two traditionally disparate disciplines is likely to continue to expand at an increasing rate in the foreseeable future. This chapter will highlight some of the current themes of biomaterials research in the fields of polymer-based therapeutics, biomimetics, biopolymers, biocatalysis and bioassays; however, it is not intended as a comprehensive review. A brief overview of recent developments in each area will be given, followed by some speculative predictions of future developments.

2 Polymers for the Delivery of Therapeutic Agents

2.1 Controlled Release

Controlled release devices based on polymers have been the subject of much research activity for many years. Original investigations were into biodegradable hydrogels as depots for the release of agents; however, this has largely been superseded by the development of injectable microparticulate formulations that do not require surgical implantation. Many polymers used in controlled release have been based on aliphatic polyesters (see Figure 1). Kissel's group in Marburg have examined poly(lactide) (PLA), poly(lactide-co-glycolide) (PLGA) and linear and star PLA block copolymers with poly(ethylene oxide) (PEO), chiefly for the delivery of plasma proteins[1] and genetically

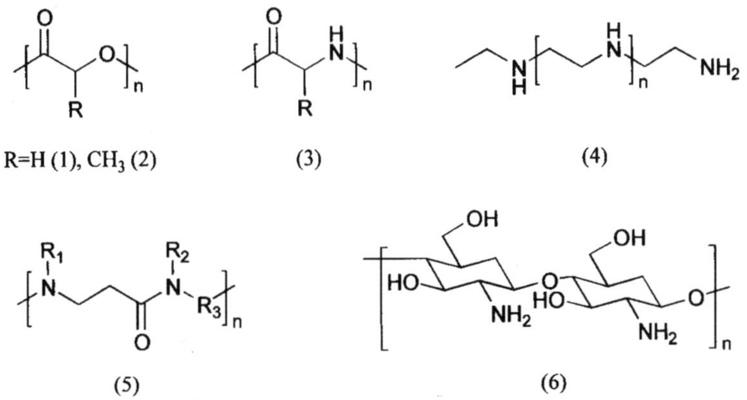

R=H (1), CH_3 (2) (3) (4)

(5) (6)

Figure 1 *Structures of polymers commonly used in biomaterials applications: PGA (1), PLA (2), PAAs (3), linear PEI (4), PAMAMs (5) and chitosan (6) (see text for abbreviations)*

engineered human erythropoietin (a glycoprotein hormone used in red blood cell proliferation).[2] The loaded particles are prepared in one step from linear polymer either by an emulsion processing method or by spray drying. These processes yield particle sizes in the range from 1 to 100 μm.

Further work from Kissel's group has looked at developing poly(acrylic acid) dispersions by inverse emulsion polymerisation. These 'bioadhesive' particles are intended to have increased residence time in the intestine, giving enhanced hydrophilic drug delivery. Sizes in this case were much smaller (*ca.* 100 nm) than the particles described above.[3] Korean workers have used a 'polymer therapeutics' (*vide infra*) approach in which a model drug, Fmoc-Trp(Boc), is linked *via* an ester group to PLGA. The conjugate is then formulated into microspheres.[4]

Poly(amino acid)s (PAAs) have also been used in drug delivery; PEO-(L-aspartic acid) block copolymer 'nano-associates', formed by dialysis from a dimethyl acetamide solution against water, could be loaded with vasopressin.[5] PLA-(L-lysine) block copolymer microcapsules loaded with fluorescently labelled (FITC) dextran showed release profiles dependent on amino acid content.[6] In a nice study, poly(glutamate(OMe)-sarcosine) block copolymer particles were surface-grafted with poly(*N*-isopropyl acrylamide) (PNIPAAm) to produce a thermally responsive delivery system; FITC-dextran release was faster below the lower critical solution temperature (LCST) than above it.[7] PAAs are prepared by ring-opening polymerisation of *N*-carboxyanhydride amino acid derivatives, as shown in Scheme 1.

Despite advances in particulate-based systems, work on hydrogels continues. Hoffman *et al.* have shown that the release of drugs from side-chain modified poly(acrylic acid) hydrogels depends on the hydrophobicity of the side groups.[8] Novel hydrogel drug delivery systems based on PNIPAAm have also been described.[9] These show different swelling behaviour in the presence of the drugs ephedrine and ibuprofen, which is dependent on the hydrophilicity of

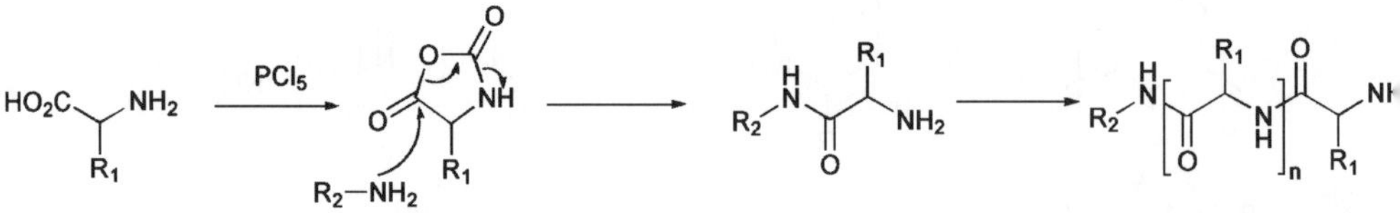

Scheme 1

the gel. Other responsive highly porous hydrogels have been prepared from sucrose-based monomers.[10] Incorporation of carboxyl groups leads to pH-sensitive materials with fast swelling–deswelling kinetics.

2.2 Targeted Delivery

A more specific approach to delivering agents to biological hosts is to incorporate some selectivity mechanism – this is known as targeted delivery. Selective delivery can be achieved by controlling the size of the polymer particle (vector), or through some chemical functionality on the particle surface that is recognised by the host.

The use of polymeric micelles to deliver drugs to the central nervous system (CNS) has been reviewed recently.[11] One strategy employs polymers with specific ligands to enhance transport across the blood-brain barrier (BBB), *via* interaction with receptor proteins on the BBB. Many of the polymers employed are derivatives of PEO–PPO–PEO block copolymers (Pluronics).

Targeting of antibiotics is achieved using entrapment in poly(cyano acrylate) nanoparticles, prepared by emulsion polymerisation in the presence of the drug.[12] The particles are taken up in the liver by phagocytes and degraded by esterases, and targeting is presumably based on size. Similarly, sub-micron (down to 90 nm) PLGA particles have been developed by a two-stage processing procedure, for targeted delivery.[13]

Polysaccharides play an important role in many biological functions, in particular in the recognition of species by proteins (lectins) on cell surfaces. Consequently, polymers containing sugar residues (glycopolymers) are being developed for targeted delivery.[14] A representative example of a glycopolymer precursor monomer, (7), is shown. A different approach involves the use of 'sugar balls', which are prepared by anchoring disulfide-containing poly(sugar acrylate)s to colloidal silver particles.[15] These were found to aggregate in

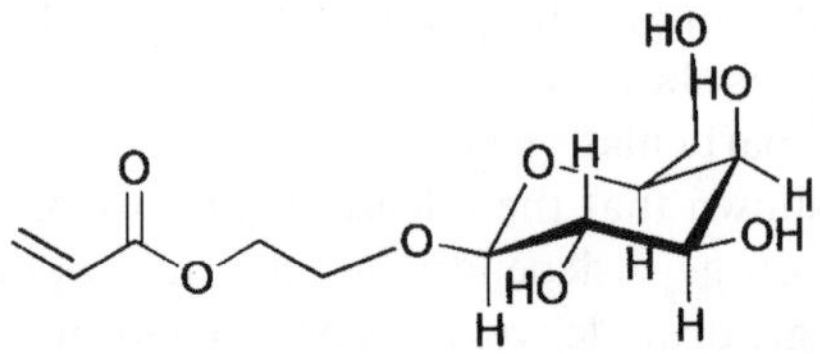

(7)

solutions containing the lectin concanavalin A (Con A). An alternative therapeutic strategy involves preparing polymers that inhibit the enzyme β-glucuronidase.[16] This species causes the release of potentially toxic xenobiotics in the intestine, *via* hydrolysis of their parent glycoconjugates. Inhibition of the enzyme results in the foreign entity being safely excreted. Glycoconjugates have been prepared by ring-opening polymerisation of *N*-carboxyanhydrides by sugars (see Scheme 1), and glycosylated PAMAM dendrimers have also been described.[17] The general use of dendrimers and hyperbranched polymers for drug delivery has been reviewed in the recent past.[18] Antibodies have also recently been employed as a method of targeting. Polymerisable Fab' (a fragment of IgG_1) antibody fragments were prepared and copolymerised with a polymerisable drug derivative.[19] The resulting polymers showed adsorption to specific cells *in vitro*.

2.3 Polymer Therapeutics

One of the main difficulties in drug delivery is ensuring a high loading of the agent in the carrier. Covalent attachment is one method by which maximum loading can be guaranteed, and the resulting drug–polymer conjugate is known as a polymer therapeutic. Once the carrier reaches the desired site, the drug is release by cleavage (usually by hydrolysis) of the linkage to the polymer.

Ruth Duncan's group has described conjugates between the anti-tumour agent emetine and poly(*N*-(2-hydroxypropyl) methacrylamide (HPMA).[20] The drug was released through hydrolysis of tetrapeptide links and some influence on survival times of tumour-possessing animal models was observed. In addition, toxicity of the conjugate was significantly lower than that of the free drug. Other workers have linked the antitumour drug daunomycin to a branched polypeptide.[21] Again, long-term survival of models was found, and toxicity was much reduced, relative to the free drug. Doxorubicin hydrochloride (Adriamycin) is another popular anti-cancer agent to be delivered by polymer conjugates[22,23] (see (8)). In one study, delivery was found to be enhanced as conjugate hydrophobicity increased.

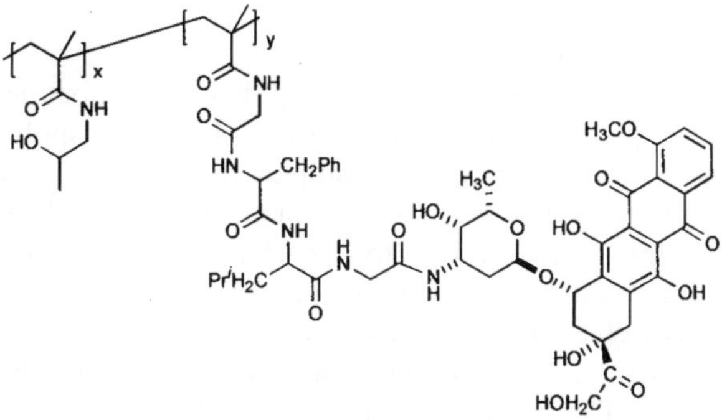

(8)

Uptake by and concentration in tumour tissue is reckoned to occur *via* the enhanced permeability and retention (EPR) mechanism, which is due to differences in permeability of tumour and healthy tissue. Following uptake by the cell, the conjugates are located in the endosomes that mature to become lysosomes; the latter have a lower pH and possess enzymes that hydrolyse the peptide spacers selectively, as shown in Figure 2. PEG-based Dox conjugates bearing acid sensitive hydrazone linkers that are cleaved in either the endosome or the lysosome have also been described.[24] Other work has described the preparation of peptide hormone anti-cancer analogues conjugated to poly(*N*-vinylpyrrolidone-co-maleic anhydride) copolymers. Significant enhancements in antitumour activity compared to the free agent were observed in both *in vitro* and *in vivo* tests.[25]

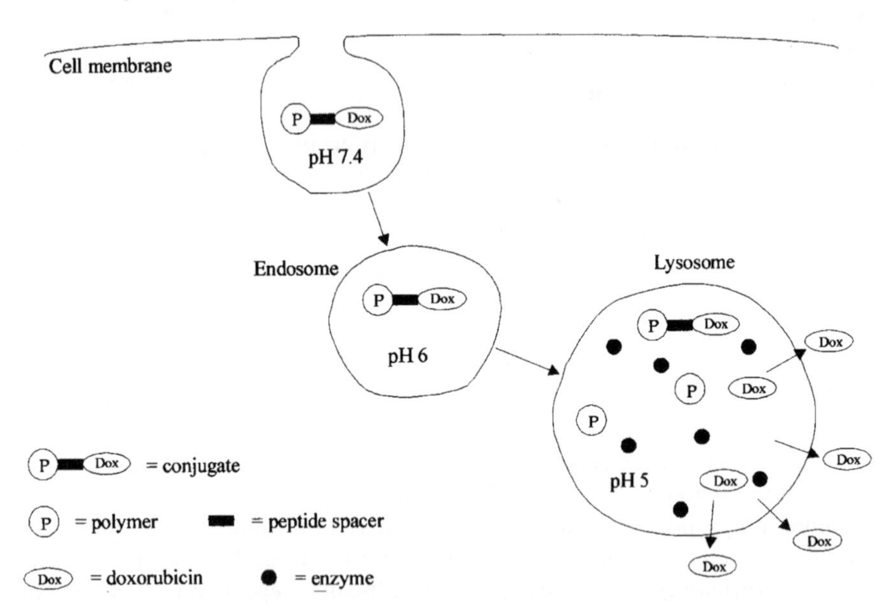

Figure 2 *Schematic of the process of intracellular release of drug from a polymer conjugate following uptake and enzymatic hydrolysis of the peptide spacer in the lysosome (redrawn from ref. 23, © the American Association of Cancer Research, 1999)*

Duncan's group has also investigated a polymer therapeutic approach to platinate (*e.g.* cisplatin) anti-tumour agents. HMPA copolymers bearing peptide side chains were prepared and complexed to Pt *via* ethylene diamine or carboxylate termini, to give polymeric cisplatin analogues.[26] Again, activity was similar to free cisplatin but toxicity was greatly reduced. Similarly, PAMAM–platinate conjugates, including dendrimers, have been described.[27]

The same researchers have described an alternative approach based on polymer–enzyme conjugates, and known as PDEPT (polymer-directed enzyme prodrug therapy).[28] This involves attaching both the drug and an enzyme that will cleave the drug-polymer linker, to the polymer. *In vivo* model studies indicated a 3.6 times increase in blood retention time, and a two-fold increase in accumulation in the tumour, of the conjugate, relative to when the free enzyme was used. A novel, responsive-release conjugate for the delivery of antibiotics has recently been described.[29] The antibiotic gentamicin was linked to poly(vinylalcohol) (PVA) *via* a peptide spacer that is cleaved enzymatically in the presence of thrombin. Previous studies had shown that thrombin activity was high in wounds infected with *Staphylococcus aureus.* This approach lowers the overall amount of antibiotic needed to be administered, thereby reducing side effects and leading to less resistance. Antibacterial activity was also the target application for recently developed glycopolymers.[30] Terpolymers of sugar-containing monomers were prepared by radical polymerisation and modified enzymatically to produce chains bearing antigens and ligands for bacterial cells. The binding of the polymers to bacteria was demonstrated.

Polymer–drug conjugates for the delivery of other species have also been described in the literature. Carriers of vitamin E have been prepared from α-tocopherol methacrylate monomers.[31] These would have potential use as antioxidants to protect against cellular damage. Poly(acrylate)s that release non-steroidal anti-inflammatory drugs (NSAIDs) through hydrolysis of links to the backbone have also been described.[32]

Recently, a drug for multiple sclerosis, in which the active ingredient is itself a polymer, has been described. This is a copoly(amino acid) of lysine, glutamic acid, tyrosine and alanine, and is known as Cop 1 or Copaxone. It is reckoned to be the first entirely polymeric drug for the treatment of a specific disease.[33]

2.4 Gene Therapy

The delivery of oligonucleotides (ONs) using polymeric vectors is an alternative approach to viral or liposomal delivery methods used traditionally in gene therapy. Generally, ONs are conjugated *via* their anionic sugar phosphate moieties to cationic polymers,[34] and popular carriers are poly(ethylene imine) (PEI), poly(amidoamines) (PAMAMs) and PAAs (see Figure 1). PEIs are prepared by cationic ring opening polymerisation of aziridine, giving branched material, or 2-alkyl-2-oxazolines, which produce linear polymers.[35] Targeting of PEI vectors can be achieved by functionalisation with suitable ligands. Molecular weight of branched PEI was shown to influence transfection ability, increasing as molar mass increased to 70 k. However, variation of pH did not influence delivery efficiency.[36] In a separate study, PEI of very high molecular weight (M_w~1.6 M) was shown to be two orders of magnitude less efficient than that of *ca.* 12 k. In addition, the latter showed lower cytotoxicity.[37] An influence of the extent of branching (as inferred from the polydispersity) was invoked as a possible explanation, this being lower for the low molar mass sample. Another popular class of carrier used in polymer-based gene delivery is the poly(amido amine)s (PAMAMs).[38] These are prepared by polycondensation between, *e.g.*, bisacryloyl piperazine, methyl piperazine and *N,N'*-bis(2-hydroxyethyl)ethylene diamine (see Scheme 2). The PAMAM products were found to be relatively non-toxic, and were suggested to be capable of transfer

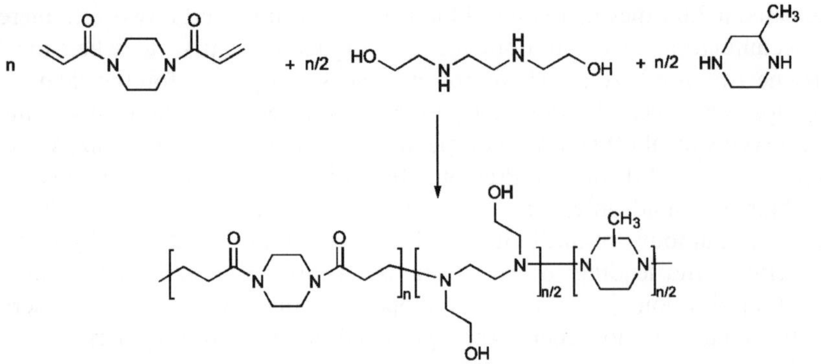

Scheme 2

to the cytoplasm after intravenous administration. PEG-PAMAM block copolymers have also been described;[39] the PEG blocks aid solubilisation.

Poly(L-lysine) (PLL) has also been used in gene delivery, however it is quite toxic. Employing similar chemistry to that used to prepare PAMAMs, Ferrick's group have modified the lysine primary amino side groups by reaction with *N,N*'-disubstituted acrylamides.[40] This was shown to decrease toxicity relative to unmodified PLL. Nanocomplexes (100 nm diameter) between DNA and PLL-hydrophilic polymer block copolymers have been shown to exhibit enhanced transfection and reduced toxicity.[41] Block copolymers of PEG and poly(aspartic acid) form micelles with anionic cores in aqueous media.[42] These can be used for the delivery of DNA and other nucleotides; the PEG corona gives steric stabilisation and ensures long circulation times *in vivo*. HPMA copolymers with a quaternary ammonium methacrylate monomer have also been used to complex DNA.[43] The stability of the complex to salt treatment was enhanced by high levels of HPMA. A recent review of glycotargeting in gene therapy, employing glycosylated PLL, has highlighted the influence of the sugar moiety on nucleic acid delivery.[44] Paradoxically, several instances of an inverse relationship between ON-polymer complex uptake and gene expression were reported; this was ascribed to the complexity of the biological processes occurring during uptake and expression, and the fact that these depend on many factors other than the sugar type. The natural polymer chitosan has also been investigated as a potential delivery vector.[45] It was found to be non-toxic, able to bind DNA and able to be administered intravenously without accumulation in the liver.

One difficulty associated with polymer-based gene therapy is that delivery efficiency is generally low, due to redirection of the polymer complexes to and enzymatic hydrolysis of the DNA in the lysosome, rather than release of DNA from the endosome (see Figure 2). A strategy to circumvent this is to include polymers that disrupt the endosomal membrane. This is achieved if they are active (*i.e.* protonated) at pH 6.5 or below, but inactive (deprotonated) at pH 7.4, since the endosomal pH is around one unit lower than the cytoplasm. Hoffman, Tirrell and co-workers[46] have developed synthetic polymers such as poly(alkyl acrylic acid)s and poly(acrylate-co-acrylic acid)s that show enhanced membrane disrupting activity between pH 6.3 and 5.0, and no activity at 7.4. In this manner, DNA can be released before hydrolysis in the lysosome occurs.

3 Biomimetics

Polymers are widely used in nature to regulate many processes involving materials growth, such as the development of bone tissue and the production of shells by sea-creatures. The design of synthetic polymers to control processes such as crystallisation and deposition of (in)organic materials is known as biomimetics, and is a rapidly expanding area of polymer science.

Materials with potential for bone replacement were prepared by controlled crystallisation of apatite (calcium phosphate) in a porous polyglycolide matrix.[47] However, control over crystal morphology inside the matrix was difficult. Hydrophilic block copolymers of PEO and alkylated methacrylic acid were used to control calcium phosphate precipitation to form complex organic/inorganic hybrid materials.[48] Tirrell's group elegantly demonstrated that the poly(amino acid) poly(aspartate) caused deposition of calcium carbonate films, similar to the aragonite layers found in nacre. In addition, in some cases, intriguing helical vaterite crystal structures were obtained (see Figure 3).[49] In another study, deposition of calcite films was produced using an amphiphilic porphyrin in the presence of poly(acrylic acid) as a crystal growth inhibitor.[50]

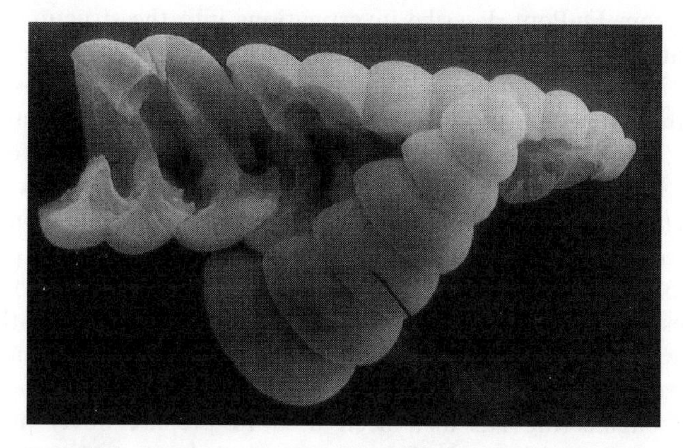

Figure 3 *Helical vaterite crystals grown in the presence of poly(aspartic acid) (reprinted from ref. 49, with permission from Elsevier Science)*

Biomimetic processes other than inorganic crystallisation have also been described. Wright and co-workers templated the production of the polymer hemozoin, a crystalline heme aggregate used by malarial parasites in a detoxification process. Understanding and controlling this could lead to new antimalarial drugs. Peptide dendrimers containing multiple repeats of the Ala–His–His sequence known to mediate hemozoin formation were prepared and indeed were shown to induce its production.[51]

A potentially important development for the future of biomimetic research was published recently. Using a procedure developed by Whitesides, nucleophiles micropatterned on a gold surface were used as initiators to prepare poly(γ-benzyl-L-glutamate). Thus, an array of poly(amino acid) layers covalently attached to the surface was obtained. These would potentially act as mediators of biomimetic reactions.[52] German workers have described the production of protein-DNA double and triple layers adsorbed on surfaces.[53] Adhesion between layers was provided by biotin-streptavadin interactions and binding was detected by a quartz crystal microbalance. These studies may have some relevance to the understanding of protein DNA interactions at interfaces in biological systems.

4 Biopolymers

The natural world is replete with polymeric materials: proteins, polysaccharides and nucleotides being some examples. In this section, we shall limit ourselves to two topics of particular interest to the polymer/materials scientist: silk as an engineering material, and the development of artificial proteins. Viney's group has been investigating the structure and mechanical properties of spider silk for some time. Spider silk is processed from a liquid crystalline solution in the spinneret with fibroin molecules aligning under shear to form oriented fibres.[54] Several groups are involved in bioengineering artificial silk proteins that will have potentially improved mechanical properties.[55] Researchers from DuPont describe investigations into the structure of dragline silk and attempts to prepare synthetic analogues. Knowing the amino acid sequences in the silk component proteins, spidroin 1 and 2, it was possible to describe their structure as resembling block copolymers with hard and soft segments. The soft segment block conformation was speculated to undergo a transition from a three-fold helical to an extended β-sheet conformation on extension (Figure 4). This could explain the property of high elongation to a structure of higher mechanical strength. Subsequently, genes expressing key elements of the protein sequences were established in foreign hosts and silk was harvested and spun. However, the mechanical properties of the artificial material were inferior to the natural analogue, due to differences in hard domain size and intersheet spacing.[56]

Genetic engineering has also been used to prepare novel protein-based polymers for drug delivery.[57] The precise control over sequence, molecular weight and stereochemistry allows the design of polymers with altered recognition, degradation and adhesion properties. Tirrell and co-workers have reported the production of artificial analogues of the extracellular matrix proteins elastin and fibronectin as potential vascular reconstruction materials.[58] Adhesion of endothelial cells to the proteins coated on glass slides was demonstrated. The same group has also demonstrated that non-natural amino acids, such as homoallylglycine, can be introduced into proteins using bacterial hosts.[59] These units provide the materials with a wider range of functional groups, increasing the scope of protein engineering. Responsive protein hydro-

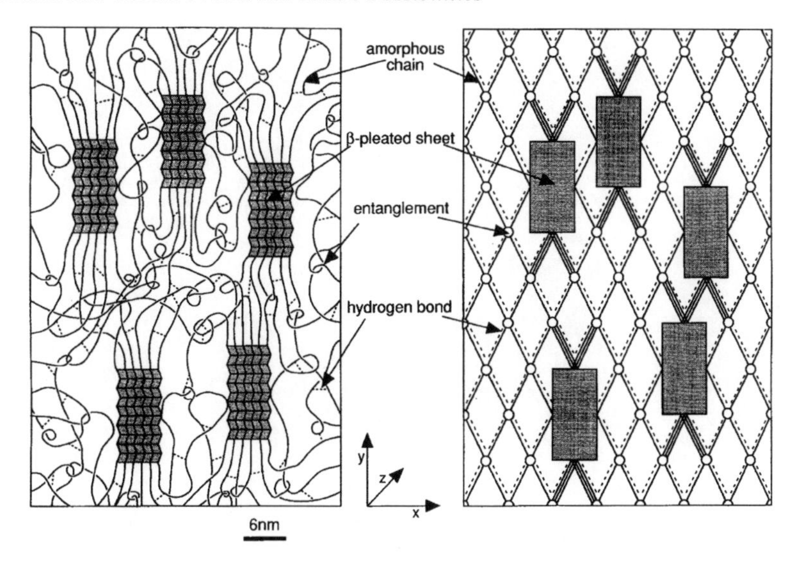

Figure 4 (a) *Model and* (b) *schematic representations of a spider dragline (reprinted from ref. 56, with permission from Wiley-VCH)*

gels with potential as biomaterials have also been described.[60] "Triblock" proteins, prepared through expression of artificial genes in bacterial hosts, consisting of helical domains flanking a random polyelectrolyte coil, were shown to undergo a gel–sol transition on heating, the temperature of the change decreasing with increasing pH from above 100 °C at pH 5 to around 30 °C at pH 11.

5 Materials for Bioassays and Biocatalysis

Polymer materials are also being used in the preparation of devices to detect, regulate or induce biologically important reactions. Many of these polymers are stimuli-responsive ('smart').[61] Immobilisation of glucose oxidase on a poly(acrylic acid) hydrogel led to the production of a chemical valve (Figure 5). In the presence of glucose, the pH dropped and the hydrogel deswelled, allowing transport of insulin across the membrane. Similar, totally synthetic, devices based on glucose binding to boronic acid residues were developed by Japanese workers.[62] Copolymerisation of a monomeric antibody and corresponding antigen derivative to form a semi-IPN resulted in the production of a responsive hydrogel.[63] In the presence of the free antigen, the polymeric antibody-antigen interactions were broken and swelling occurred. This could be employed to deliver some therapeutic species in the presence of the specific antigen.

A synthetic device for detecting galactose transfer activity, implied in the development of cancer and other diseases, has been described. A polyacrylamide containing *N*-acetyl glucosamine moieties was incubated with galactosyl transferase (GalT), leading to polymer bound galactose units. These were

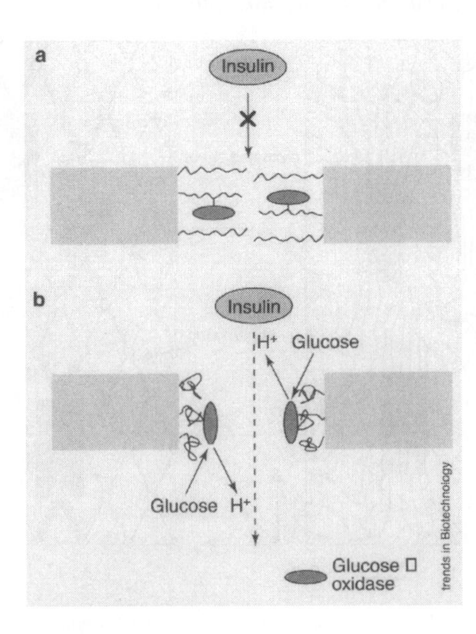

Figure 5 *Schematic representation of a chemical valve.* (a) *swollen poly(acrylic acid) blocks insulin transport.* (b) *glucose oxidation causes a decrease in pH that results in deswelling and insulin transport across the membrane (reprinted from ref. 61, with permission from Elsevier Science)*

specifically assayed using a lectin-peroxidase labelled antibody combination.[64] Thus, the device could be used to detect and quantify GalT activity. The use of electroactive polymers in bioaffinity sensor devices has been reviewed recently.[65] To make a bioaffinity sensor, a biological sensing element (*e.g.* antibody, receptor, protein, *etc.*) is incorporated into a conducting polymer. Binding of the analyte causes changes in the redox potential of the polymer. This forms the basis of the bioaffinity signal. A 'reagentless' controlled release biosensor for the nucleotides ATP, ADP and AMP based on the release of luciferin has also been described. Commercial acrylate/methacrylate copolymers (Eudragit®) were formulated into microspheres, from solutions containing luciferin. These particles were then cast in a UV-curable PVA film, which was cured. The composite was used in a trienzymatic (adenylate kinase, creatine kinase and luciferase)-based sensor device; zero-order release of luciferin (a cosubstrate for the above enzymatic system) was observed in the presence of any one of the nucleotides, allowing three hours of reproducible detections. However, the sensitivity was limited to 80% of the control (luciferin in solution) for ADP and AMP and only 30% for ATP.[66] Biomimetic glucose sensors based on molecularly-imprinted polymers (MIPs) have also been prepared, by electropolymerisation of *o*-phenylenediamine in the presence of glucose as a template.[67] Amperometric glucose devices were also obtained by photopolymerisation of vinyl ferrocene and PEG-diacrylate in the presence of glucose oxidase.[68]

A multidisciplinary team involving chemists, biologists and chemical and biomedical engineers has recently reported an exciting and novel method of

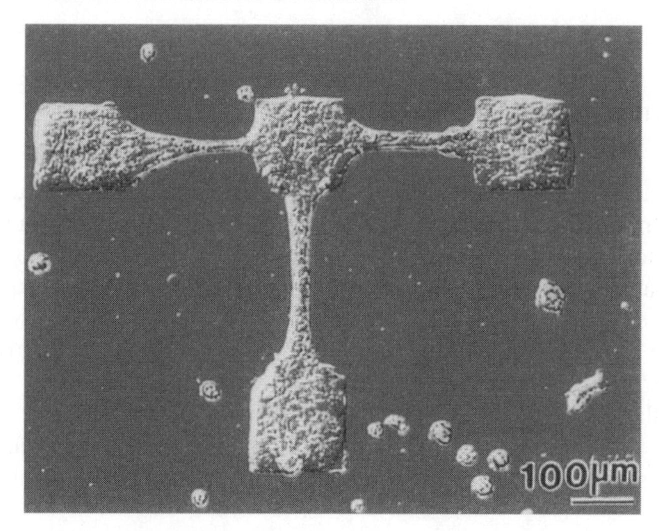

Figure 6 *Human epidermal carcinoma cells growing on squares of oligopeptides possessing a cell-adhesion motif, arranged in a T-shaped pattern (reprinted from ref. 69, with permission from Elsevier Science)*

cell manipulation that will facilitate the future study of cells and their molecular processes.[69] Using a micro-printing process, a pattern of PEG oligomers was attached to a gold surface *via* terminal -SH groups. The surface was then immersed in a solution of an oligopeptide, which contained a specific cell adhesion sequence of amino acids. These became attached to the unmasked areas of the gold surface, *via* thiols on a cysteine residue at the C-terminus. The resulting surfaces were able to align cells in specific arrays and patterns, including lines and linearly connected squares, by interaction with the immobilised oligopeptides (Figure 6).

'Smart' hydrogels have found use in bioseparation.[70] Materials responsive to temperature were used in liquid chromatography and electrophoretic applications, to separate proteins, steroids and DNA fragments. Antibody-containing materials for the concentration of antigens were also described.

Spanish workers have recently disclosed the preparation of lipase-silicone biocatalysts. Lipase is adsorbed onto poly(hydroxymethyl siloxane); to this, a mixture of polysiloxane oligomers and crosslinkers were added to form, on curing, a solid biocomposite. Various morphologies (solid monoliths, particulates, foams, membranes, *etc.*) could be formulated to give materials with high enzymatic activity and stability in aqueous and organic media. Extending the technique to other enzymes was a real possibility.[71] Polymer-supported enzymatic catalysts have been prepared and used by other groups.[72] Hydrogels obtained by copolymerising a sucrose acrylate with acrylated enzymes were used to catalyse transesterifications; the supported species had lower activities in aqueous media but higher in organic solvents than the native enzymes.[73] In addition, much higher (in some cases two orders of magnitude) thermal stabilities were observed.

6 Future Directions

6.1 Polymer-based Therapeutics

There are several factors that are likely to influence how this area develops. Possibly the most important is toxicity; obviously, materials to be used *in vivo* should be non-toxic and entirely biocompatible. This is probably an unrealistic goal for most potential biomaterials; indeed, the biocompatibility of materials currently approved such as PLGA (*vide supra*) has recently been questioned. These are designed to degrade to metabolites and ultimately be excreted, but even these seemingly innocuous byproducts (lactic acid from PLA or PLGA) can cause problems when generated in the wrong location or at the wrong time. Similarly, one might imagine that poly(amino acid)s would be acceptable due to the proliferation of proteins in the body, but these have been shown in certain cases to be toxic, immunogenic and otherwise to cause adverse responses.

The major stumbling block to the implementation of new biomaterials for use *in vivo* is, and always will be, regulation. Potentially marketable products, either from novel materials or where approved materials have been covalently modified, will need to demonstrate very significant advantages over existing materials, or new properties currently unavailable, in order to justify the huge investment required for biological testing.

If we accept that new materials will need to be tested rigorously and that biodegradability may in actual fact be disadvantageous, another direction may be more fruitful for polymer based therapy. This is the use of biologically inert, non-degradable materials that will not concentrate in the body and will be easily excreted. Such materials (*e.g.* HPMA) are used as carriers in polymer therapeutic approaches. The difficulties, other than toxicity issues, will be in designing materials with sufficient residence times *in vivo* to be effective, and some strategies towards achieving this have been described above. Also, developments in controlling interactions of materials with cells, proteins, *etc.*, will help in generating desired biological responses and suppressing undesired ones. The base materials of such polymers are likely to be hydrophilic to prevent accumulation in adipose tissue.

Despite the above-mentioned hurdles, the benefits to society of polymer-based therapeutics are potentially huge and the investment required is likely to be made in the most promising cases. This will undoubtedly continue to inspire exciting and novel fundamental research in this area.

6.2 Biomimetics

Significant advances in controlling the morphology of crystalline inorganic materials with polymers have been achieved, but the greatest challenge remains; how to influence the association and organisation of these into complex composite materials. Progress could be made following close examination of the methods used in nature to prepare composite materials such as bone and sea shells. These processes are generally mediated by peptides, which have an amazing range of options for structural diversity: 20 natural amino acids; complete control over molecular weight; monodispersity; absolute stereochemistry; complete control over monomer sequence down to each individual unit; secondary and tertiary structures. In contrast, man-made polymers are limited to rather more crude architectures (block, graft and alternating copolymers; dendrimers). However, recent advances in controlled polymer synthesis, in particular those tolerant of aqueous media and wide ranges of functional groups, are extending the range of possibilities for the synthetic polymer chemist. Recombinant DNA technology to engineer designer proteins (*vide infra*) is a concurrent advance that can aid in achieving these long-term goals. In addition, advances in enzymatic polymerisation may allow the preparation of a wider range of controlled materials. It is safe to assume that astonishing progress will be made; after all, the suggestion ten years ago that living transition metal mediated polymerisations could be carried out in aqueous solution would have been met with widespread derision.

6.3 Biopolymers

Advances in genetic engineering to make designer materials will be the key to future progress in this and related areas of biomaterials science. Two directions are important: controlling protein structure using existing monomers (amino acids); and introducing non-natural amino acids to induce novel properties. In conjunction, increased understanding of the biosynthesis of other biopolymers could lead indirectly to the production of new materials, through genetic manipulation of the proteins used to make them. Protein modification by chemical means may also bring this about. These modified enzymes could potentially lead to the synthesis of non-natural polymers in aqueous media, allowing the synthesis of designer man-made materials by environmentally friendly enzymatic methods.

6.4 Bioassay and Biocatalysis

Increased understanding of biological processes together with improved materials synthesis will be the drivers in the advancement of this wide collection of topics. The development of novel sensors that detect wider ranges of biologically relevant species is one area of advance. In addition, the range of detection methods will be extended; advances in electrochemical sensing are currently generating considerable excitement. Composite materials will be important in the development of self-contained sensors. There be will be a requirement for materials with built-in filters, pre-concentration layers and selective adsorbents, for use in 'real' applications where contaminants (soluble and insoluble) are present. Another direction of progress will be in miniaturisation, towards entirely self-contained sensor/detector devices ('Lab-on-a-Chip').

Since many of the advances in this area will require materials with specific shapes, surface areas, porosities, swelling/deswelling kinetics, *etc.*, controlled methods of polymer synthesis will be essential. Recent advances in this area have brought us to the pleasant situation where functional materials of controlled molar mass and distribution, with specific end-groups and complex architectures such as block copolymers and dendrimers, can be made in aqueous solution. In addition, novel templating procedures allow the preparation of porous materials with a specific pore size and narrow size distribution. Further advances in this area can only be of benefit to this aspect of biomaterials science.

7 Conclusions

The future for synthetic chemists working at the interface between materials and biological science is extremely bright. The rapidly expanding toolkit of novel materials and architectures, with diverse properties, together with major advances in the understanding of biological science, can only mean that this area will become increasingly fruitful for exciting new interdisciplinary science.

References

1 M. Lück, K.-F. Pistel, Y.-X. Li, T. Blunk, R. H. Müller and T. Kissel, *J. Controlled Release*, 1998, **55**, 107.

2 B. Bittner, M. Morlock, H. Koll, G. Winter and T. Kissel, *Eur. J. Pharm. Biopharm.*, 1998, **45**, 295; M. Morlock, T. Kissel, Y.-X. Li, H. Koll and G. Winter, *J. Controlled Release*, 1998, **56**, 105; K.-F. Pistel, B. Bittner, H. Koll, G. Winter and T. Kissel, *J. Controlled Release*, 1999, **59**, 309.

3 B. Kriwet, E. Walter and T. Kissel, *J. Controlled Release*, 1998, **56**, 149.

4 J. E. Oh, Y. S. Nam, K. H. Lee and T. G. Park, *J. Controlled Release*, 1999, **57**, 269.

5 T. Aoyagi, K. Sugi, Y. Sakurai, T. Okano and K. Kataoka, *Coll. Surf. B: Biointerf.*, 1999, **16**, 237.

6 T. Kidchob, S. Kimura and Y. Imanishi, *J. Controlled Release*, 1998, **54**, 283.

7 T. Kidchob, S. Kimura and Y. Imanishi, *J. Controlled Release*, 1998, **50**, 205.

8 T. Inoue, G. Chen, A. S. Hoffman and K. Nakamae, *J. Bioact. Biocompat. Polym.*, 1998, **13**, 50.

9 T. L. Lowe, J. Virtanen and H. Tenhu, *Polymer*, 1999, **40**, 2595.

10 J. Chen and K. Park, *Carbohydrate Polym.*, 2000, **41**, 259.

11 D. W. Miller and A. V. Kabanov, *Coll. Surf. B: Biointerf.*, 1999, **16**, 321.

12 H. Pinto-Alphandery, A. Andremont and P. Couvreau, *Int. J. Antimicrob. Agents*, 2000, **13**, 155.

13 P. D. Scholer, A. G. A. Coombes, L. Illum, S. S. Davis, M. Vert and M. C. Davies, *J. Controlled Release*, 1993, **25**, 145.

14 T. Miyata and K. Nakamae, *TRIP*, 1997, **5**, 198 and references therein; B. G. Davis, *J. Chem. Soc., Perkin Trans. 1*, 1999, 3215 and references therein.

15 A. Yoshizumi, N. Kanayama, Y. Maehara, M. Ide and H. Kitano, *Langmuir*, 1999, **15**, 482.

16 K. Hashimoto, R. Ohsawa, N. Imaiand M. Okada, *J. Polym. Sci. Part A: Polym. Chem.*, 1999, **37**, 303; K. Hashimoto, R. Ohsawa and H. Saito, *ibid.*, 1999, **37**, 2773.

17 M. Okada, K. Aoi and K. Tsutsumiuchi, *Proc. Japan Acad. Ser. B*, 1997, **73**, 205.

18 K. Uhrich, *TRIP*, 1997, **5**, 388.
19 Z.-R. Lu, P. Kopeckova and J. Kopecek, *Nat. Biotech.*, 1999, **17**, 1101.
20 S. Dimitrijevic and R. Duncan, *J. Bioact. Biocompat. Polym.*, 1998, **13**, 165.
21 D. Gaál and F. Hudecz, *Eur. J. Cancer*, 1998, **34**, 155.
22 K. Abdellaoui, M. Boustta, M. Vert, H. Morjani and M. Manfait, *Eur. J. Pharm. Sci.*, 1998, **6**, 61.
23 P. A. Vasey, S. B. Kaye, R. Morrison, C. Twelves, P. Wilson, R. Duncan, A. H. Thomson, L. S. Murray, T. E. Hilditch, T. Murray, S. Burtles, D. Fraier, E. Frigerio and J. Cassidy, *Clin. Cancer Res.*, 1999, **5**, 83.
24 P. C. A. Rodrigues, U. Beyer, P. Schumacher, T. Roth, H. H. Fiebig, C. Unger, L. Messori, P. Orioli, D. H. Paper, R. Mülhaupt and F. Kratz, *Bioorg. Med. Chem.*, 1999, **7**, 2517.
25 J. Pató, M. Móra, I. Mezö, J. Seprödi, I. Teplán, B. Vincze, A. Kálnay and I. Pályi, *J. Bioact. Biocompat. Polym.*, 1999, **14**, 304.
26 E. Gianasi, M. Wasil, E. G. Evagorou, A. Keddle, G. Wilson and R. Duncan, *Eur. J. Cancer*, 1999, **35**, 994.
27 P. Ferruti, E. Ranucci, F. Trotta, E. Gianasi, F. G. Evagorou, M. Wasil, G. Wilson and R. Duncan, *Macromol. Chem. Phys.*, 1999, **200**, 1644; N. Malik, E. G. Evagorou, R. Duncan, *Anti-Cancer Drugs*, 1999, **10**, 767.
28 R. Satchi and R. Duncan, *Brit. J. Cancer*, 1998, **78**, 149.
29 M. Tanihara, Y. Suzuki, Y. Nishimura, K. Suzuki, Y. Kakimaru and Y. Fukunishi, *J. Pharm. Sci.*, 1999, **88**, 510.
30 J. Li, J. Zacharek, X. Chen, J. Wang, W. Zhang, A. Janczuk and P. G. Wang, *Bioorg. Med. Chem.*, 1999, **7**, 1549.
31 C. Ortiz, B. Vázguez and J. San Román, *Polymer*, 1998, **39**, 4107; C. Ortiz, B. Vázquez and J. San Román, *J. Biomed. Mater. Res.*, 1999, **45**, 184.
32 C. Parejo, A. Gallardo and J. San Román, *J. Mater. Sci. Mater. Med.*, 1998, **9**, 803.
33 M. Sela, *Acta Polym.*, 1998, **49**, 523.
34 Z. H. Israel and A. J. Domb, *Polym. Adv. Tech.*, 1998, **9**, 799.
35 W. T. Godbey, K. K. Wu and A. G. Mikos, *J. Controlled Release*, 1999, **60**, 149 and references therein.
36 W. T. Godbey, K. K. Wu and A. G. Mikos, *J. Biomed. Mater. Res.*, 1999, **45**, 268.
37 D. Fischer, T. Bieber, Y. Li, H.-P. Elsässer and T. Kissel, *Pharm. Res.*, 1999, **16**, 1273.
38 S. Richardson, P. Ferruti and R. Duncan, *J. Drug Targeting*, 1999, **6**, 391.
39 M. C. Garnett, F. C. Maclaughlin, S. S. Davis, F. Bignotti and P. Ferruti, *WO 97/ 25067*, 1997.
40 P. Ferruti, S. Knobloch, E. Ranucci, R. Duncan and E. Gianasi, *Macromol. Chem. Phys.*, 1998, **199**, 2565.
41 V. Toncheva, M. A. Wolfert, P. R. Dash, D. Oupicky, K. Ulbrich, L. W. Seymour and E. H. Schacht, *Biochim. Biophys. Acta*, 1998, **1380**, 354.
42 A. Harada and K. Kazunori, *Macromolecules*, 1998, **31**, 288.
43 D. Oupicky, C. Konák and K. Ulbrich, *J. Biomater. Sci. Polym. Edn.*, 1999, **10**, 573.
44 M. Monsigny, P. Modoux, R. Mayer and A.-C. Roche, *Biosci. Rep.*, 1999, **19**, 125.
45 S. C. W. Richardson, H. V. J. Kolbe and R. Duncan, *Int. J. Pharm.*, 1999, **178**, 231.
46 N. Murthy, J. R. Robichaud, D. A. Tirrell, P. S. Stayton and S. Hoffman, *J. Controlled Release*, 1999, **61**, 137; J. Kim and D. A. Tirrell, *Macromolecules*, 1999, **32** , 945.
47 K. Schwarz and M. Epple, *Chem. Eur. J.*, 1998, **4**, 1898.

48 M. Antonietti, M. Bruelmann, C. G. Göltner, H. Cölfen, K. K. W. Wong, D. Walsh and S. Mann, *Chem. Eur. J.*, 1998, **4**, 2493.
49 L. A. Gower and D. A. Tirrell, *J. Cryst. Growth*, 1998, **191**, 153.
50 G. Xu, N. Yao, I. A. Aksay and J. T. Groves, *J. Am. Chem. Soc.*, 1998, **120**, 11977.
51 J. Ziegler, R. T. Chang and D. W. Wright, *J. Am. Chem. Soc.*, 1999, **121**, 2395.
52 T. Kratzmüller, D. Appelhans and H.-G. Braun, *Adv. Mater.*, 1999, **11**, 555.
53 K. Ijiro and H. Ringsdorf, *Langmuir*, 1998, **14**, 2796.
54 C. Viney, *Supramol. Sci.*, 1997, **4**, 75.
55 A. E. Barron and R. N Zuckermann, *Curr. Opin. Chem. Biol.*, 1999, **3**, 681.
56 J. P. O'Brien, S. R. Fahnestock, Y. Termonia and K. H. Gardner, *Adv. Mater.*, 1998, **10**, 1185.
57 H. Ghandehari and J. Cappello, *Pharm. Res.*, 1998, **15**, 813.
58 A. Panitch, T. Yamaoka, M. J. Fournier, T. L. Mason and D. A. Tirrell, *Macromolecules*, 1999, **32**, 1701.
59 J. C. M. van Hest and D. A. Tirrell, *FEBS Lett.*, 1998, **428**, 68.
60 W. A. Petka, J. L. Harden, K. P. McGrath, D. Wirtz and D. A. Tirrell, *Science*, 1998, **281**, 389.
61 I. Y. Galaev and B. Mattiasson, *TIBTECH*, 1999, **17**, 335 and references therein.
62 K. Kataoka, H. Miyazaki, M. Bunya, T. Okano and Y. Sakurai, *J. Am. Chem. Soc.*, 1998, **120**, 12694.
63 T. Miyata, N. Asami and T. Uragami, *Nature*, 1999, **399**, 766.
64 M. Oubiki, K. Kitajima, K. Kobayashi, T. Adachi, N. Aoki and T. Matsuda, *Anal. Biochem.*, 1998, **257**, 169.
65 O. A. Sadik, *Electroanal.*, 1999, **11**, 839.
66 P. E. Michel, S. M. Gautier-Sauvigné and L. J. Blum, *Talanta*, 1998, **47**, 169.
67 C. Malitesta, I. Losito and P. G. Zambonin, *Anal. Chem.*, 1999, **71**, 1366.
68 K. Sirkar and M. V. Pishko, *Anal. Chem.*, 1998, **70**, 2888.
69 S. Zhang, L. Yan, M. Altman, M. Lässle, H. Nugent, F. Frankel, D. A. Lauffenburger, G. M. Whitesides and A. Rich, *Biomater.*, 1999, **20**, 1213.
70 J. J. Kim and K. Park, *Biosep.*, 1999, **7**, 177.
71 I. Gill, E. Pastor and A. Ballesteros, *J. Am. Chem. Soc.*, 1999, **121**, 9487.
72 J. S. Dordick, S. J. Novick and M. V. Sergeeva, *Chem. Ind.*, 1998, **(1)**, 17 and references therein.
73 S. J. Novick and J. S. Dordick, *Chem. Mater.*, 1998, **10**, 955.

Surfaces and their Modification

10

Plasma Polymerisation: Improved Mechanistic Understanding of Deposition, New Materials and Applications

Robert D. Short and A. Goruppa

DEPARTMENT OF ENGINEERING MATERIALS,
UNIVERSITY OF SHEFFIELD, SIR ROBERT HADFIELD BUILDING,
MAPPIN STREET, SHEFFIELD S1 3JD, UK

1 Introduction

Plasma polymerisation is a process whereby thin organic coatings (typically 10–100 nm) may be deposited from the glow-discharges of volatile organic compounds. Although widely investigated since the 1960s, the current level of interest in plasma polymerisation is perhaps greater than at any time in the preceding 40 years. There are two principal and related reasons for the current level of interest in this research topic: (1) Plasma parameters have been established that allow the deposition of new classes of coatings, previously not accessible by plasma techniques. These materials are highly functionalised and may 'retain' much of the original starting compound's structure and chemical functionality. (2) The emergence of new technologies that depend upon coatings which actively direct processes through their surface chemistry.

Given the commercial pressures that now drive much academic and most industrial research, mechanistic understanding will, as a research topic, lag behind the effort given to materials and applications, and for this reason the author gives prominence to mechanistic understanding in this paper. By choosing to highlight the need for improved mechanistic understanding, the author's aim is to draw attention to the topic, which has been much neglected to date. This important topic may hold the key to significant advances in the fabrication of new materials and in the design of materials to meet new or specific application needs.

2. Plasma Polymerisation

Radio-frequency (RF) induced plasmas of volatile organic compounds have been the subject of intensive research efforts for many years.[1] From these plasmas, polymeric materials (plasma polymers) may be obtained. Experimental conditions may be varied to control the physical and chemical nature of the plasma polymer. For many applications, the plasma polymer may be most usefully obtained in the form of a thin film coating ($\ll 1$ μm) deposited on a substrate. Early studies (pre-1980s) generally concentrated on the plasma deposition of coatings from higher power plasmas, with applications described where the coating functioned through its physical presence (*e.g.* as a barrier),[1,2] the exception being work undertaken with fluorocarbon compounds.[2] In the early 1980s papers began to appear describing the deposition of coatings, which through their surface chemistries participated in the control of specific processes. For example, as early as 1982 Heider *et al.* describe the modification of graphite electrodes using plasma polymerised acrylic acid coatings.[3] Over the decade 1980–1990 researchers started to appreciate that by using lower plasma powers, functionalised films could be obtained, and that in many cases the starting compound's original functional group(s) could be incorporated into the coating. The lower the power employed the greater the level of incorporation. Ward[4] termed this incorporation 'retention' on the grounds that the functional group was being incorporated from the plasma without fragmentation. The pace picked up substantially in the early 1990s, with numerous reports appearing on the plasma deposition of highly functionalised films containing alcohols,[5] carbonyls,[6] carboxyls,[7] amines[8] *etc.*

In the author's own laboratory and elsewhere, efforts have been made to sustain continuous wave (CW) plasmas using the lowest power possible. Notable amongst the recent low power CW studies is that of Alexander and Duc.[7] These researchers have shown that in the CW plasma polymerisation of acrylic acid, coatings may be deposited that retain nearly 70 % of the carboxyl group of the starting compound (with a further 10 % being converted to ester). The calculation of retention is based on the starting compound ('monomer'), which contains one acid for every three carbons. If this ratio of carboxyl/carbon were achieved in the plasma polymer deposit it would be described as 100 % retention. In the coatings of Alexander and Duc there was one acid functional group for every ~4.5 carbons. Using secondary ion mass spectrometry (SIMS) Alexander and Duc have shown that these coatings contained segments of at least five acrylic acid units bound in a linear fashion. This study corroborates earlier work from the author's own laboratory with alkyl methacrylates.[9,10] Taken together, these studies (and others) strongly suggest that at low plasma power, a significant fraction of the coating forms by processes akin to those in conventional chain growth polymerisation.

By pulsing of the plasma, extraordinarily low (average) powers can be achieved.[11] This approach to plasma deposition is particularly exciting as it offers a way to power regimes ($\ll$ 1W) that simply can not be accessed by CW. These studies have demonstrated with a wide range of compounds that have very high levels of functional group and structural retention can be obtained.[12–16] Reports describe films prepared from perfluoroallylbenzene[12] (ring retention), hydroxyl[15] (functional group retention) and 3-methyl-1-vinylpyrazole and 1-allylimidazole (heterocyclic ring retention).[16] This list is by no means exhaustive. Not only do films obtained from the pulsed plasmas of these precursors contain higher levels of structural retention than might be anticipated from low power CW plasmas, but also superior film quality is often described.[12,16]

3 Mechanistic Understanding

3.1 Plasma

A plasma consists of electrons, ions (mainly positive, but in certain cases negative), neutral atoms and / or molecules. There may be a wide spectrum of electromagnetic radiation from infrared through to vacuum UV. The plasma state is sometimes referred to as the fourth state of matter since it demonstrates quite different properties from the gaseous, liquid or solid state. The term plasma was first used by Langmuir in 1927 to describe an ionised gas. The term plasma comes from the Greek word 'plasso' (to shape or mould). The name conveys that the various plasma components (electrons, ions, *etc.*) are 'moulded' together, determining the whole plasma environment.

Plasmas are generally divided into two groups according to the temperature of the plasma species, which can reach 10^6–10^8 K in 'high temperature' plasmas (stars, thermonuclear reactors), but are well below 10^6 K in 'low temperature' plasmas. The latter class of plasmas can be further divided into 'hot plasmas' and 'cold plasmas'. The former have gas temperatures (T_g) of above 1000 K, normally of the order of 10^4–10^5 K and occur in lightning, electrical arcs and other high-power discharges and are not relevant to this article. The latter have a T_g of less than 1000 K, normally of the order of 300–400 K, including various glow discharges at low pressures. In the low temperature, low pressure plasmas which are relevant, the two charged components, the electrons and ions, are not in thermodynamic equilibrium. The electrons can have temperatures of 10^4–10^5 K (1–10 eV), while the ion temperature can be as low as room temperature. This temperature difference arises from the difference in mass between an electron and an ion. It is the cause of a very rich chemical environment. For a plasma of a two component gas mixture several dozen reaction equations are required to describe the processes of ionisation, excitation, recombination, charge transfer *etc.*

Plasmas can be created and sustained using electrical, radiant and thermal energy sources. The most common means of injecting energy into the system is to use an electrical discharge. In order to ignite a plasma, an initial 'ionisation event' is essential, *i.e.* a molecule or atom must lose an electron to create an ion. Some free electrons are always present in a gas, as a result of ionisation by naturally occurring radioactivity or cosmic rays The application of an electric field accelerates these free electrons, which acquire sufficient energy to ionise atoms or molecules on collision. Apart from ionisation, electrons may cause an increase in the rotational and vibrational temperatures of molecules and electronic excitation in atoms and molecules.

Ion (n_i) and electron (n_e) densities are important parameters which characterise any plasma. These are assumed to be equal to each other ($n_e = n_i = n$ in units of particles.cm^{-3}) and the plasma is described as 'quasi-neutral', where n is the 'plasma density'. For the low temperature, low-pressure plasmas, which are used to deposit polymeric materials, the plasma density is much less than the density of the neutrals (n_n). In a low temperature, low-pressure plasma of 10^{-2} mbar n is between 10^9 and 10^{12} cm^{-3}. Even the higher of these values is four to five orders of magnitude below n_n.

The electron energy distribution function $f(W)$ for a plasma is a critically important parameter because it governs the reactions which can take place. The rate at which new species are created within the plasma depends on the $f(W)$ and n_e. The $f(W)$ is known to approximate to a Maxwellian energy distribution for non-equilibrium plasmas. An important feature of the energy distribution is the high-energy tail. Although the average electron energy in plasmas is rarely of the magnitude sufficient to break chemical bonds (*ca.* 5 eV) or cause ionisation (8–25 eV), electrons of appropriate energy are available within the tail of the distribution. These electrons control the rate at which ionisation and molecular fragmentation occur in plasmas.

Any surface which is in contact with a plasma will be bombarded by electrons and ions. Since electrons escape much faster from the plasma to the surface, a 'sheath' (a positively-charged) region is formed. The electrons are described as being more mobile. In this way a potential difference between a grounded surface and the plasma volume is maintained. This is referred to as the 'plasma potential' (V_p). If the surface is allowed to float, *i.e.*, it is not electrically earthed, the floating surface also acquires a negative potential with respect to the plasma. Ions are accelerated across this potential and impart energy to the surface.

3.2 Polymer Deposition from Plasma

Polymeric materials can be deposited from low temperature plasmas in a vessel containing the vapour of an organic compound under low pressure. These plasmas are sometimes referred to as 'cold plasmas'. The power supplies that can be used to ignite and sustain the plasma are direct current (DC), radio frequency (RF) or microwave (MW). Power may be delivered by means of an electric field (using electrodes) or electromagnetic (using a coil or MW). The electrodes may be placed either within the plasma vessel (direct coupling) or external to it (indirect coupling). With direct coupling, higher ion energies are encountered. This may or may not be an advantage, depending upon the type of deposit required. The disadvantage of direct coupling is deposition of polymer onto the electrodes. Indirect coupling eliminates the problem of polymer deposition because the electrodes are not in contact with the plasma. Indirect coupling by means of external electrodes or a coil is more common in the deposition of polymers. This leads to the majority of surfaces being at floating potential. The potential difference between a plasma and a floating surface is less than that for a plasma and a grounded surface, and this helps reduce ion energies.

RF discharges are classified in accordance to their electrode or electrodeless nature into capacitively or inductively coupled. In the case of the inductively coupled discharge, a helical or flat copper coil is utilised. Inductive coupling is generally used at AC frequencies greater than 1 MHz as it becomes less efficient at lower frequencies. Although systems are often described as inductively coupled this may not be the case as the 'skin depth' (the distance over which the electromagnetic radiation is absorbed) is often in excess of the reactor dimensions. This is certainly the case for the 13.56 MHz systems widely employed for polymer deposition.

3.3 Mechanistic Understanding: The Orthodox View

Given the complexity of even a simple gas plasma, it is not surprising that the mechanisms that lead to the formation of plasma polymer films are poorly understood, in either CW or pulsed mode. Yasuda[1] has described plasma polymerisation as proceeding through a rapid step-growth polymerisation (RSGP) mechanism. The scheme put forward by Yasuda is a very general one, and it was probably not intended to cover the entire range of conditions under which plasma polymer deposits form. However, in the absence of other schemes, it is widely quoted.[17] There are two intersecting cycles to the scheme. The first describes the reactions of monofunctional activated species (M*). The nature of the reactive functionality is not made clear, but it is presumed to be a radical (M$^\bullet$) based on the greater abundance of radicals in the plasma (*cf.* ions). This radical can combine with a neutral species, to produce a further (larger) radical ($M_x^\bullet + M_y \rightarrow M_{x+y}^\bullet$), which reacts on, or with another radical to produce a neutral compound ($M_x^\bullet + M_y^\bullet \rightarrow M_{x+y}$). The neutral compound can be reactivated in the plasma. The second cycle describes the reactions of difunctional species (*M*). This species can react with a neutral species to yield a larger difunctional species (${}^\bullet M_x^\bullet + M_y \rightarrow {}^\bullet M_{x+y}^\bullet$), or with another difunctional species (${}^\bullet M_x^\bullet + {}^\bullet M_y^\bullet \rightarrow {}^\bullet M_{x+y}^\bullet$). The cycles cross-over where a difunctional species reacts with a monofunctional species (${}^\bullet M_x^\bullet + M_y^\bullet \rightarrow M_{x+y}^\bullet$). As written, this scheme implies that plasma-phase reactions are non-specific and that a wide range of new compounds may form in the plasma. For many (high power) systems this description is probably valid. However, its applicability to low power (CW or pulsed) RF plasmas is questionable in the light of the functional group/structural retention reported in these systems. Further, the scheme provides no insight as to where the polymer forms. Two possibilities may be considered.

3.3.1 Growth at polymer/plasma interface. Deposition occurs by grafting of gas-phase material onto an 'activated' surface. This could be particularly favoured if the grafted compound contained unsaturation. It is usually argued that grafting is *via* a free radical mechanism; surface free radicals may be created by:

1. The plasma VUV/UV component;
2. Ion impact; or
3. Reactive (metastable) or fast neutrals.

The weightings given to each process would probably depend upon the plasma reactor, frequency, power, pressure, monomer chemistry, *etc.*

3.3.2 Polymer chain growth in the plasma gas-phase. This could proceed by:

1 Radical chemistry (radical-radical and radical-neutral combinations);
2 On-molecule reactions (including radical cations).

A less widely quoted scheme is the activated growth mechanism (AGM) of d'Agostino and coworkers[2] developed to describe fluorocarbon deposition. In the AGM $CF^{\bullet}_x$ radicals stick onto activated sites on the depositing polymer surface. The activation is triggered by low energy positive-ion bombardment. The deposition rate is described as first-order with respect to the $CF^{\bullet}_x$ species and a complex function, $f(I^+)$, which represents both the ion flux and energy in positive-ion bombardment. This model may have more relevance in lower power plasmas than the RSGP.

3.4 Mechanistic Understanding: The Controversial View

The RSGP model of Yasuda was not developed to explain the deposition of the highly functionalised films that are the current focus of intensive research effort. The high levels of functional group and structural retention reported in these films are not consistent with Yasuda's scheme. For this reason, in 1994 the author initiated an investigation of plasma deposition at lower plasma powers. Mass spectrometry has been used to probe the glow region of the plasma-gases of alkyl methacrylates,[18] organic acids,[19] alcohols[20] and amines.[21] The mass spectrometer is shown Figure 1. Despite some considerable effort, the author and co-workers have been unable to find any mass spectral evidence, with a single exception,[18] for the types of neutral species predicted by the RSGP scheme. Indeed, no neutral species larger than the monomer were detected in any of these studies (with the exception of ref. 18). Save for simple $R^{\bullet} + H^{\bullet}/CH_3^{\bullet}$ (re-) combinations, no evidence has yet been found for radical processes in the plasmas of these compounds. Of course, our inability to detect these species by one specific technique does not constitute unambiguous proof that larger neutral species are present. Neutral radicals may be very difficult to detect by electron impact mass spectrometry. Even using low electron energies of 20 eV, such species may fragment in a manner to yield only low mass-charged species.

However, substantial evidence has been found in the positive ion mass

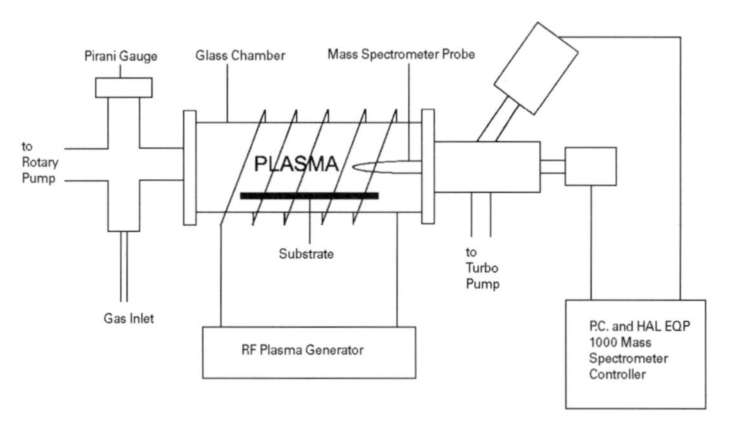

Figure 1 *Schematic of the plasma apparatus and mass spectrometer. The mass spectrometer may be used in one of three modes. 1. To obtain the electron impact mass spectrum of the compound under investigation (no plasma). 2. To probe the plasma-gas phase neutrals, by means of electron impact. 3. To probe the charged species present within the plasma-gas phase. In this mode, the electron impact is switched off and the lenses within the spectrometer are tuned for transmission of either positively- or negatively-charged ions. [Adapted from J. Chem. Soc., Faraday Trans., 1995,* **9***, 1363]*

spectra for 'oligomer' growth in the plasma-phase, which may be accounted for by associative ion–molecule reactions,[18–21] see Figure 2 and Scheme 1. These associative ion-molecule reactions appear to be an important part of the plasma-gas phase chemistry and the ions thus formed may contribute more significantly to the plasma deposit (at low plasma power) than has been previously suggested. For some compounds a correlation between the abundance of the dimeric and trimeric cations and functional group retention [as measured by XPS] has been reported.[18–21] Both appear to be inversely correlated with power. For example, in the plasma polymerisation of acrylic acid a direct correlation between the degree of dimer and trimer formation (plasma-gas phase) and functional group retention (product) with power has been established.[19] Plasma-phase reaction schemes have been proposed for acrylic acid[19] and allyl alcohol,[20] and for the latter compound the feasibility of this scheme was tested by means of a selected ion flow tube, see Schemes 1 and 2.[22] Scheme 1 was first suggested based on the mass spectral data in Figure 2. Part of the scheme, describing the formation of the *m/z* 81 ion was revised based on SIFT data subsequently acquired. Alternative pathways to the *m/z* 97 and 67 ions have also been proposed. In attempting to model plasma-phase reactions it is important to recall that the ions and neutrals are essentially at room temperature, and therefore, irrespective of how the ions are created they must subsequently be thermalised prior to reaction with neutrals. For this reason the use of a selected ion flow tube is most suited for this type of investigation. However, it should be noted that there is no unique pathway to deposit formation and in carrying out such investigations only one of several

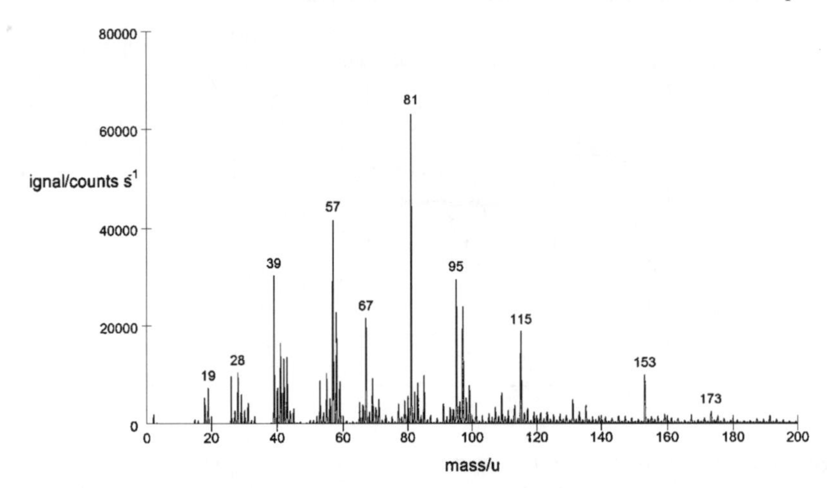

Figure 2 *Positive-ions mass spectrum of allyl alcohol plasma, obtained in external ion mode with a plasma power of 1 W. Figure shows positive ions that correspond to the monomer m/z 57 $(M\text{-}H)^+$, dimer m/z 115 $(2M\text{-}H)^+$, and trimer m/z 173 $(3M\text{-}H)^+$ of allyl alcohol. [Adapted from J. Chem. Soc., Faraday Trans., 1997,* ***93****, 1961]*

deposition pathways is being modelled. The importance of this pathway will then require weighting.

The idea that extensive plasma-gas phase oligomerisation is occurring by means of ion-molecule reactions is neither new nor particularly controversial; however, attempting to link the cations thus formed to the deposit growth has proven controversial, and is by no means 'mainstream' thinking.

3.5 Mechanistic Understanding: The Way Forward

A number of authors describe significant improvements in the polymer coating quality obtained by pulsed plasma polymerisation (*cf.* CW). If these claims are substantiated, it would seem likely that research effort will focus increasingly on this mode of plasma deposition. The question arises as to whether the pulsed environment is simply an extrapolation of the low power CW environment, or whether it represents an entirely different plasma environment altogether.

As discussed, our knowledge of the low power CW environment is fragmentary. Important intrinsic plasma parameters such as electron temperature, the electron energy distribution function, plasma density and plasma potentials have yet to be systematically investigated in these plasmas. It is for this reason that researchers are restricted to describing plasma power in terms of the external input power rather than some intrinsic property of the plasma. Measurements of the intrinsic plasma parameters are made (in non-depositing systems) by means of electrostatic probes, but in a depositing environment the application of these probes is severely restricted by the growth of an

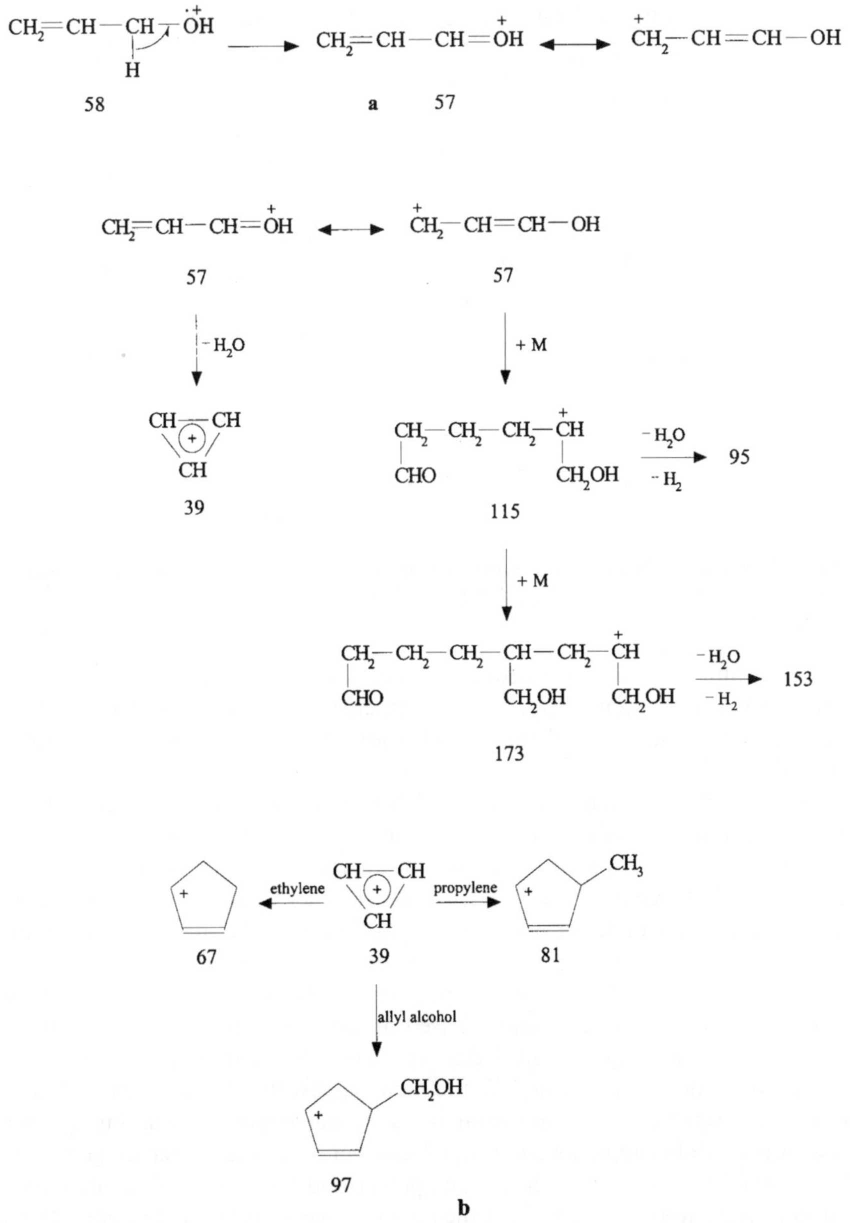

$CH_2{=}CH{-}CH(H){-}\dot{O}H^{+}$ → $CH_2{=}CH{-}CH{=}\overset{+}{O}H$ ↔ $\overset{+}{C}H_2{-}CH{=}CH{-}OH$

58 **a** 57

$CH_2{=}CH{-}CH{=}\overset{+}{O}H$ ↔ $\overset{+}{C}H_2{-}CH{=}CH{-}OH$

57 57

$-H_2O$ +M

39 115

$-H_2O$, $-H_2$ → 95

+M

173

$-H_2O$, $-H_2$ → 153

ethylene propylene

67 39 81

allyl alcohol

97

b

Scheme 1 *Reaction scheme based purely upon mass spectral data displayed in Figure 2.* a *describes the formation of m/z 57 ion,* b *the possible plasma-phase reactions of this ion. [Adapted from J. Chem. Soc., Faraday Trans*, 1997, **93**, *1961]*

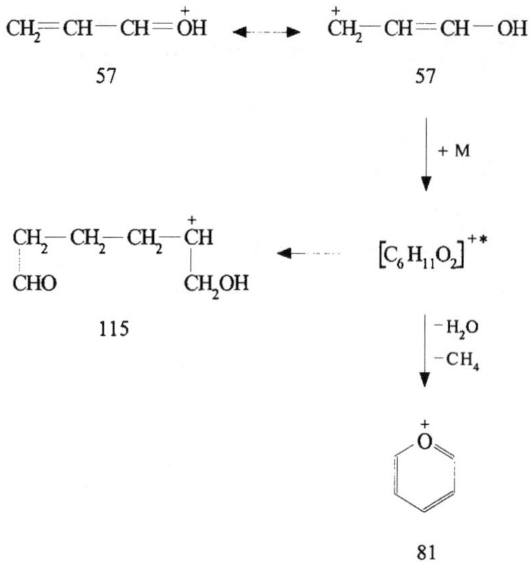

$CH_2{=}CH{-}CH{=}\overset{+}{O}H$ ⟷ $\overset{+}{C}H_2{-}CH{=}CH{-}OH$

57 57

+ M

$[C_6H_{11}O_2]^{+*}$

$CH_2{-}CH_2{-}CH_2{-}\overset{+}{C}H$ (CHO, CH_2OH)

115

$-H_2O$

$-CH_4$

81

Scheme 2 *Revised Scheme for the formation of m/z 81 based upon SIFT data. [Adapted from J. Chem. Soc., Faraday Trans. 1997,* ***93****, 1961]*

insulating film on the active probe surface. The latest generation of electrostatic probes has been designed to compensate for this problem using a pulsed bias RF method and their application in these plasmas should afford new insights.[23]

Limited, but promising progress has been made on this front. Recently, by means of a pulsed bias RF electrostatic probe,[23] ion flux measurements have been made in low power CW plasmas of acrylic acid.[24] In conjunction with quartz mass balance and mass spectral measurements, it has been possible to make a crude estimate of the total mass being delivered to a collecting substrate by the ion flux (assuming an ion sticking probability of 1). In a 1 W plasma, it is estimated that the ion flux could account for up to 50 % of the total deposit mass. This estimate is based on an extrapolation of the available data and it should only be regarded as approximate. More importantly, as the plasma power decreases below 1 W the data suggest that the ion flux makes an increasing contribution to the total mass of the deposit (assuming the ion sticking probability remains constant). However, over the power range 0.5–15 W, the relationship between the plasma power and the (potential) contribution that ions make to the total mass of the deposit is very complex. The complexity of this relationship probably indicates that the sticking probability changes as a function of the plasma input power.

The sticking probability and the general role of ions on arrival to a surface will depend strongly upon their energy at the collecting substrate surface. This energy will be determined by the potential difference between the plasma and the substrate surface. The plasma will be at a positive potential with respect to the collecting substrate surface (due the very mobile nature of the 'free' plasma electrons). The potential of the collecting substrate will depend upon whether it is grounded, or allowed to 'float', and in radio-frequency (RF)-driven plasma, it depends upon the RF potentials that develops at the surface. Very low energy ions, <5 eV, may be 'soft landed' at a surface (without fragmentation in the ion or surface).[25] Such low energy ions would contribute substantially to the molecular structure of the deposit, if arriving in sufficient quantity. Above 5 eV some fragmentation in either the incoming ion or target surface must be anticipated. Ions with energies in excess of 100 eV would only contribute to the deposit surface in an 'atomic sense', that is, we cannot expect a contribution from the ion to the molecular structure of the deposit. High-energy ions (>100 eV) would also create free radical sites in the surface for radical grafting, and may also remove (ablate) surface material.

Our knowledge of the pulsed plasma environment is even more restricted than our knowledge of the same CW environment. Researchers[12,16] have explored the effects of plasma (duty cycle) 'on-' and 'off-' periods and the ratio of 'on/off' time on product chemistry and structure. In general, it is observed that as the off-time increases (fixed on-time) functional group and structural retention increase. Conversely, as on-time increases (fixed off-time) functional group and structural retention decreases. The requirement of a sufficient off-time is thought to be needed for polymerisation to take place. This is shown when the duty cycle ratio is fixed: at long off-time (and correspondingly long on-time) retention is high, but at the same ratio with a short off-time retention is poor. However, to the best of the author's knowledge, no investigation has been undertaken of the plasma-gases of these plasmas (MS, electrostatic probe, *etc.*). This in the author's opinion represents a major deficiency in the field and demands immediate attention.

4 New Materials, New Applications

Despite the complexity of the plasma deposition process, and the multitude of possible plasma-phase and plasma-surface reactions, it is possible from low power plasmas to obtain highly functionalised films that closely resemble their conventional analogue. Sometimes there is not a readily available conventional poly-analogue, for example, in the case of allyl compounds. Even when structural or functional group retention is far from 100 %, as in the case of deposition of amine compounds,[26] highly functionalised deposits are obtained and applications for these materials are being described.[27] Highly functionalised plasma polymer coatings have realised or potential applications in both existing and emerging technologies. These applications cut across the fields of adhesion,[28] biomaterials,[26] electronics,[29] fibres and textiles,[30] *etc.*

4.1 'Biology Will Define Scientific Progress in the 21st Century'[31]

The author has chosen to draw attention to some possible applications for plasma polymers in the Life Sciences. Whilst considerable attention is given (quite rightly) to recent significant advances in our understanding of biological systems, *e.g.* the genome, the supporting roles played by materials, chemistry and physics in making these advances possible are often overlooked, to the peril of these disciplines.

For several years, we have been working upon the plasma synthesis of coatings with specific hydration properties, to promote cell attachment, spreading, growth and proliferation.[32,33] Most cells are anchorage-dependant, *i.e.* they must attach to a solid surface before they can function 'normally' and attachment is a pre-requisite in cell culture. Hydration may allow the delivery of basic nutrients, growth factors *etc.* to the basal surface of the cell. The attachment of a wide range of cell-types (*e.g.* fibroblasts, osteoblasts, keratinocytes) has been correlated with the carboxyl functionality.[32,33] Coatings containing carboxyl functional groups can be readily deposited from low power plasmas of acrylic acid. However, the resultant plasma polymers are of low molecular weight, and they are partially soluble in water. The solubility problem can be eliminated by the addition of a cross-linking compound to the plasma. In our studies, we have been using 1,7-octadiene and hexane as the cross-linker. By controlling the ratio of acrylic acid:cross-linker in the plasma-gas feed we have been able to deposit plasma (co)polymers and in these coatings we can (crudely) control the extent of hydration. Although the surface chemistries of these deposits are not dissimilar to tissue culture plastics (TCPS, treated polystyrenes), cell cultures on these plasma-deposited surfaces often outperform cultures on TCPS. This we have tentatively credited to plasma polymer hydration. As yet, we know nothing about how the functionalised compounds and cross-linkers interact (if at all) in the plasma-gas phase, or why indeed this plasma copolymerisation 'trick' works! This clearly makes a case for more mechanistic work.

Cellular attachment is equally important in the engineering of whole organs, or of specific tissues, where biodegradable scaffolds are to be used to provide the 3-dimensional shape of the organ or tissues. Typically, the scaffold materials are polyglycolic acid or polylactic acid, but cellular attachment to these materials can be poor. A number of strategies are currently being investigated to improve cellular attachment, and amongst these are plasma treatment and plasma polymerisation.

Unstable coatings are not always undesirable and in specific applications advantage can be taken of the coating's instability. For example, coatings that dissolve *in vivo* could find applications in the delivery of cells to a wound bed. Here, the trick is to control the rate of dissolution in the coating. In the design of surfaces for (the eventual) delivery of keratinocytes (skin cells) to a non-healing wound, *e.g.* a chronic leg ulcer, we have demonstrated that suitable coatings may be fabricated from the CW plasmas of acrylic acid. These coatings support the culture of keratinocytes in the laboratory for 5–10 days, but on placing in an artificial wound bed model release healthy, proliferative cells into the wound.[34] In this application, control over the cross-link density in the coating is critical, and this has been achieved through a rudimentary understanding of how to manipulate the plasma input power and the ratio of acrylic acid: cross-linking compound (a crosslinker is also added to the plasma gas).

Many other applications for plasma polymers in the Life Sciences have been cited, often in relation to implantable medical devices or materials, with the goal of 'concealing' the device from the bodies defence mechanisms, or improving cell colonisation of the material, *e.g.* endothelial cell growth into vascular grafts.[30] A number of excellent studies from the group of Hans Griesser (CSIRO, Australia) describe the use of plasma polymers as substrates to which biomolecules can be immobilised.[26,27] These immobilisations have been demonstrated to enhance the medium-term 'acceptability' of contact lens materials[27] and may prove relevant to implantable devices.

Extending upon the ideas of Griesser *et al.*, it may be possible through the plasma copolymerisation of two or more compounds to achieve complex surface immobilisations, whereby, several different molecules or biomolecules could be immobilised to a plasma polymerised surface coating. Surfaces containing two or more different functional groups could be deposited by plasma copolymerisation of two or more functionalised compounds. It should be possible to control their respective surface densities by their ratio in the plasma-gas feed. These different surface functional groups could then be used to immobilise at least two different molecules. Such coatings would be valuable, for example, in biosensor applications. Immobilization one might, for example, be of a PEO-like molecule which would confer general anti-protein fouling properties, whilst immobilization(s) two (three *etc.*) would be of a specific biomolecule(s) which would bind with high affinity to the target compound(s) to be 'sensed'.

5 Conclusions

Plasma polymerisation is an effective route to new surface chemistries. Plasma polymerisation is widely employed as a coating technology, whereby the value of a 'commodity' material is substantially enhanced by the provision of a specific (new) surface property. By employing low plasma powers, highly functionalised surfaces can be fabricated. Such surfaces may be used to direct specific surface processes, *e.g.* cell adhesion.

It appears that the lower the plasma power that is employed, the less fragmentation of the starting compound occurs in the plasma. By pulsing of the plasma, stable plasmas of extraordinarily low (average) powers can be sustained. Hence, pulsed plasma is the subject of considerable research and commercial interest. The author makes a case for more effort to be put into developing our understanding of the pulsed and low power CW plasma environments. He argues that such understanding is important in new materials research, and that it will aid significantly in the engineering of coatings to meet specific application needs.

Acknowledgments

Thanks are given to Mr. Jason Whittle and Mr. Stuart Fraser for reading this manuscript and for helpful comments.

References

1 H. Yasuda, Plasma Polymerisation', Academic Press, London, 1985.
2 R. d'Agostino, V. Colaprico, P. Favia, F. Fracassi and R. Lamendola, *Plasmas and Polymers*, 1998, **3**, 177.
3 G. H. Heider, M. B. Gerbert and A. M. Yacynch, *Anal. Chem.*, 1982, **54,** 324.
4 R. J. Ward, Ph.D. Thesis, 'An Investigation of the Plasma Polymerisation of Organic Compounds and its Relationship with Surface Photopolymerisation' University of Durham, UK, 1989.
5 A. P. Ameen, R. J. Ward, R. D. Short, G. Beamson and D. Briggs, *Polymer*, 1993, **34**, 1795; A. P. Ameen, R. D. Short and R. J. Ward, *Polymer*, 1994, **35**, 4382.
6 A. Chilkoti, B. D. Ratner and D. Briggs, *Anal. Chem.*, 1991, **63**, 1612.
7 T. Duc and M. R. Alexander, *J. Mater. Chem.*, 1998, **8**, 937.
8 H. J. Griesser, R. C. Chatelier, T. R. Gengenbach, G. Johnson and J. G. Steele, *J. Biomedical Sci. Polym. Ed.*, 1994, **5**, 531.
9 A. J. Ward and R. D. Short, *Polymer*, 1993, **34,** 4179.
10 A. J. Ward and R. D. Short, *Polymer*, 1994, **36,** 343.
11 C. R. Savage, R. B. Timmons and J. W. Lin, 'Structure Property Relations in Polymers' Advances in Chemistry Series 236; American Chemical Society; Washington, DC, 1993, 745-768.
12 A. M. Hynes and J. P. S. Badyal, *Chem. Mater.*, 1998, **10,** 2177.
13 Q C. R. Savage, R. B. Timmons and J. W. Lin, *Chem. Mater.*, 1991, **3**, 575.
14 Q S. J. Limb, K. K. Gleeson, D. J. Edell and E. F. Gleason, *J. Vac. Sci. Technol.*, 1997, **A15,** 1814.
15 C. L. Rinsch, X. L. Chen, V. Panchalingam, R. C. Eberhart, J. H. Wang, and R. B. Timmons, *Langmuir*, 1996, **12,** 2995.
16 L. M. Han, R. B. Timmons, D. Bogdal and J. Pielichowski, *Chem. Mater.*, 1998, **10,** 1422.
17 A. Grill, 'Cold Plasmas in Materials Fabrication', IEEE Press, Piscataway, 1993.
18 A. P. Ameen, L. O'Toole, R. D. Short and F. R. Jones, *J. Chem. Soc., Faraday Trans.*, 1995, **91,** 1363.
19 A. J. Beck, L. O'Toole, R. D. Short, A. P. Ameen and F. R. Jones, *J. Chem. Soc., Faraday Trans.*, 1995, **91**, 3907.
20 L. O'Toole and R. D. Short, *J. Chem. Soc., Faraday Trans.*, 1997, **93,** 114.
21 A. J. Beck, A. Candan, R. M. France, F. R. Jones, L. O'Toole and R. D. Short, *Plasmas and Polymers*, 1998, **3**, 97.
22 R. D. Short, L. O'Toole and C. A. Mayhew, *J. Chem. Soc., Faraday Trans.*, 1997, **93,** 1961.
23 N. St. J. Braithwaite, J. P. Booth and G. Cunge, *Plasma Sources Sci. Technol.*, 1996, **5,** 677.
24 S. Candan, A. J. Beck, L. O'Toole, R. D. Short, A. Goodyear and N. St. J. Braithwaite, *Phys. Chem. Chem. Phys.*, 1999, **1,** 3117.
25 E. T. Ada, O. Kornienko, L. Hanley, *J. Phys. Chem.*, 1998, **102**, 3959.
26 L. Dai, H. A. W. St. John, J. Bi, P. Zientek, R. C. Chatelier and H. J. Griesser, *Surf. Interface Anal.*, 2000, **29,** 46.
27 H. J. Griesser, Surface Characterization of Biomolecules and Protein Layers, Invited lecture at the 46th International Symposium of the American Vacuum Society, Seattle, Washington, 1999.
28 M. R. Alexander and T. M. Duc, Polymer, 1999, **40,** 5479.
29 L. C. Han, R. B. Timmons and W. W. Lee, *J. Vac. Sci. Technol.*, 2000, **18,** 799.

30 H. Biederman and D. Slavínská, *Surf. Coating Technol.*, 2000, **125,** 371.
31 *Business Week*, March 10, 1997.
32 R. Daw, A. J. Beck, R. D. Short, A. J. Devlin and I. M. Brook, *Biomaterials*, 1998, **19,** 1717.
33 D. B. Haddow, R. M. France, R. A. Dawson, S. McNeil and R. D. Short, *J. Biomed. Mater. Res.*, 1999, **47,** 379.
34 UK Patent Application No. 9914616.9 'Detachment Surfaces'; R. M. France and D. B. Haddow, *Chemistry in Britain*, 2000, **36, (6),** 29.

11

The Self Assembly of Polymer Films

Mark Geoghegan[a] and Georg Krausch[b]

[a] DEPARTMENT OF PHYSICS AND ASTRONOMY, UNIVERSITY OF SHEFFIELD, HOUNSFIELD ROAD, SHEFFIED S3 7RH, UK
[b] LEHRSTUHL FÜR PHYSIKALISCHE CHEMIE II AND BAYREUTHER ZENTRUM FÜR KOLLOIDE UND GRENZFLÄCHEN (BZKG), UNIVERSITÄT BAYREUTH, D-95440 BAYREUTH, GERMANY

1 Introduction

The importance of polymer surfaces and interfaces has been well documented. Polymers can be used to tailor surfaces with specific properties, to strengthen interfaces and to structure substrates. As well as these practical uses, polymers are excellent model systems to test and to extend our understanding of statistical physics. Interest in polymer surfaces and interfaces has exploded in the past ten years and it is probably true to say that this field is some way off maturity. The conformation of polymers at interfaces is of enduring relevance but even early on in the development of the field the idea of self-assembly took root. After all, the properties of polymers at surfaces and interfaces represent only one side of the coin. The other side, that which interests us here, is how the polymers get to where they are. In contemplating how and why a particular structure develops, one should consider two issues. The first is simply a question of kinetics, encompassing entanglements and reptation. The second response to the question of the origin of a particular structure would be to ask why it is there at all. Here one would consider the nature of the surface or interface, the surface and interfacial energies of the polymers and confinement effects. An understanding of the importance of such parameters is a prerequisite for understanding the kinetics.

Rather than consider the development of self-assembly chronologically, we shall break it up into the constituent parts that we believe will be of importance in future years, as well as to propose some likely possible experiments. We begin with surface segregation, however, because this provides an historical foundation for the theme of self-assembly.

2 Surface Segregation

Probably the simplest example of self-assembly is that of surface segregation. A binary polymer blend film is cast from a common solvent. The two components have different surface energies and so the component with the lower surface energy preferentially segregates to the surface. The segregation may occur during the spin casting process and/or during subsequent annealing. The first quantitative experiments demonstrating surface segregation were ion beam experiments on a miscible isotopic polystyrene blend.[1] On account of the slightly different polarizabilities of the D–C and H–C bonds, deuterated polystyrene has a lower surface energy than its non-deuterated analogue. Simple mean field theory[2] was used to quantify the segregated amount and compare this with that measured. The agreement between experiment and simple mean field theory was good. There have been improvements, in terms of both experiment and theory, but the simple experiment and result remains a valid template for future experiments. The importance of surface segregation is that it provides a means to create a surface layer without actually having to place a layer at the surface. An extremely good example of how this might be useful has been provided by Mayes and co-workers.[3] In this work a comb polymer consisting of a poly(methyl methacrylate) (PMMA) backbone and poly(ethylene oxide) side-chains is mixed with poly(vinylidene fluoride). The comb polymer preferentially segregates to the surface on annealing, providing a coating resistant to proteins from a water solution. Such resistance is necessary when polymer membranes are being used for the filtration of oil and protein solutions. A great benefit of such an arrangement is that, if the surface layer is eroded, it may be further regenerated on subsequent annealing prolonging the lifetime of the layer.

We have given just one example of how surface segregation might be important, but it does not stretch the imagination too much to see that this might be extended to other systems and situations. In what follows, we shall discuss the strengthening of interfaces. It has long been known that diblock copolymers can strengthen the interface between two immiscible components.[4] A layer of A–B diblock copolymer will strengthen an interface between homopolymers A and B by aligning itself at the interface with the A block protruding into the A homopolymer matrix and the B block protruding into the B homopolymer matrix (figure 1). More recently it has been demonstrated that such an interface may be strengthened by an A–B random copolymer (figure 1). This is because, over a given length of a random copolymer, there will be an excess of the A component with respect to the B component.[5] This part of the chain will preferentially find itself in the A matrix. Similarly, the parts of the chain that contain more B elements will lie mostly on the B side of the interface. In fact, strengthening interfaces with random copolymers is noticeably more effective than with diblock copolymers because a random copolymer is expected to cross the interface many times. However, to demonstrate such an effect, a trilayer system has been necessary. That is a layer of A, on which the copolymer is placed, and finally a layer of B on top of

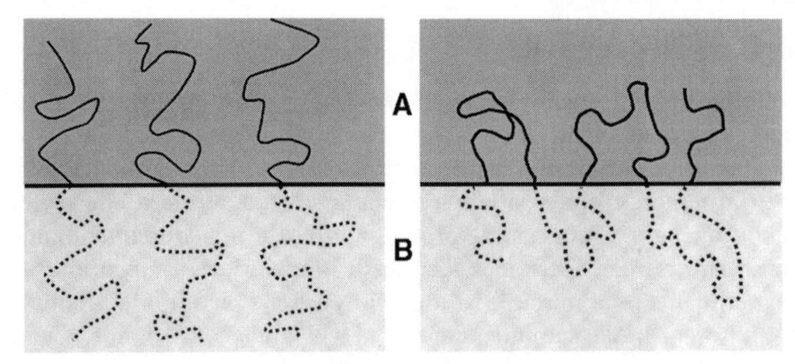

Figure 1 *A–B diblock copolymers provide adhesion at an A–B interface because they cross the interface once (left). An A–B random copolymer can provide even better adhesion because it crosses the interface several times, depending on its length (right). How can we get these copolymers to the interface in the first place, without having to deposit a separate layer of copolymer between the A and B layers?*

that. On annealing, the copolymer must align itself. Ideally, following the theme of surface segregation, one would mix the copolymer in either the A or B matrices (or both) and anneal. The copolymer would then segregate to the interface simplifying our system. In practice, even such high-energy interfaces do not prove attractive to the copolymers, which would rather segregate to the air (vacuum) interface or to an inorganic substrate. Because of the importance of adhesion, and the desirability of achieving the above simple solution, it is likely that work will be directed in this area. If one requires homogeneous A and B matrices, it is important that the copolymer be limited in concentration to ensure that neither micellisation nor phase separation occurs. Perhaps an easier solution to this problem would be to mix the copolymer with either the A or B homopolymer films before forming the bilayer (at a concentration lower than the critical micelle concentration). In this case, the layer can be annealed allowing the copolymer to segregate to the surface before forming the bilayer. Once the two layers are connected, the copolymer can arrange itself as necessary, strengthening the interface. By varying the annealing time and concentration of copolymer, one can vary the amount of segregated material. This would be particularly important in the case of the random copolymer because the effect relies on the copolymer in a matrix of A and B. If too much copolymer segregates to the interface, the effect is diminished because separate matrices of A and B are less visible to the copolymer. We return to the issue of random copolymers at interfaces in Section 5 below.

We have seen therefore that the application of surface segregation to create coatings is well acknowledged. We have also proposed that surface segregation may be useful as a precursor to an interface. In moving on, we shall be much more general, and more speculative. The natural extension of surface segregation is the related phenomenon of wetting. We focus on this next.

3 Wetting and Dewetting

We have mentioned above that one component of a miscible polymer blend will preferentially segregate to the surface. Most polymer systems are not miscible however, so one asks what would happen in a partially miscible or immiscible binary system. In an immiscible bulk polymer mixture, phase separation proceeds by spinodal decomposition.[6] Phase separation is isotropic and dominated by a characteristic length scale (the spinodal wavelength). In a polymer film the phase separation is triggered by the surface and a layered structure penetrates from the surface, towards the bulk of the film. This structure is broken up by thermal noise deeper in the film, whereupon the bulk structure dominates. Such "surface-directed spinodal decomposition" has indeed been observed in different polymer systems[7] and is a very good way of creating a lamellar system by self-assembly. The phase separation can evolve through annealing and during solvent casting. The length scale during surface driven phase separation is expected to be the same as that in the bulk. However, in thinner films confinement alters this wavelength,[8,9] and in very thin films a lateral structure will dominate.[10]

In metastable mixtures, the surface nucleates a wetting layer. Such a wetting layer would grow indefinitely except for finite size effects curtailing the growth due to lack of material in the bulk. Wetting might also be stopped by nucleation in the bulk starving the wetting layer of material. The approach to wetting has been studied in an isotopic polystyrene blend by neutron reflectometry,[11] but phase separating systems are perhaps more useful for fundamental studies of wetting. Indeed, there have been several useful experimental[8,9,12–16] and theoretical[17] studies of wetting and it is probably true to say that wetting in polymer mixtures is well understood. Future work is expected to concentrate on wetting in different geometries.

Dewetting has also been the subject of much work. The two major forms of dewetting are spinodal dewetting and nucleation and growth.[18] The names reveal the analogy with spinodal decomposition and wetting, but it is probably true to say that dewetting is a more challenging subject. In the case of dewetting in thin films, the surface energies and the Hamaker constants of the different materials are the important parameters. In thicker films, gravity might also be important but it can usually be neglected. For a film of a polymer A on a B substrate, dewetting can occur when the spreading coefficient, S is negative. S is given by

$$S = \gamma_B - (\gamma_{AB} + \gamma_A) \tag{1}$$

where γ is the surface energy and the subscripts B, A, and AB refer to the B–air, A–air and A–B interfaces respectively. Dewetting can also occur when the Hamaker constant is negative. In the former case, dewetting proceeds by nucleation and growth, whilst a system with a negative Hamaker constant will spontaneously dewet (spinodal dewetting). There will also be systems that will have both a negative spreading coefficient and a negative Hamaker constant. The dominance of one form of dewetting over the other will clearly be very complicated. As such, it is desirable to study films whereby either one, or the other, but not both applies. In practice, this can be awkward because the γ_{AB} term can mean that nucleation can exist when a layer of polymer A is on a layer of polymer B or *vice versa.* A good example of such a situation is that of PMMA and polystyrene (PS). With PMMA as the upper layer, both nucleation and growth and spinodal dewetting[19] can occur. When a PS film is placed on a PMMA layer dewetting can proceed only by nucleation and growth.[20] Such behaviour has been well studied but there is still more to do. After all, in a bilayer with a negative spreading coefficient, but a positive Hamaker constant, a point should come when these two terms cancel each other out and the film, instead of dewetting, becomes stable. In the PS on PMMA system, this occurs when the PS layer is very thin. Such work is in progress. Another area of interest is when one considers the effect of the substrate. This is particularly important for spinodal dewetting in very thin bilayers. In the bilayer system that we use for our example, PMMA on PS on silicon, we have three different Hamaker constants. A PMMA layer on silicon should be stable because PMMA is a polar material. Inserting a thin layer of polystyrene between the PMMA and silicon will only make the system unstable if the PS is thick enough (typically thicker than about 50 nm). Work here is also in progress. It is likely that substrate effects will dominate research on dewetting in the foreseeable future.

In the above the substrate is planar. In real life this is not generally the case. It therefore is desirable to learn more about wetting on rough surfaces. With this in mind, a particularly interesting area of research in the future will be when the substrate is patterned, or when there is curvature involved. As an example of how this could be interesting, we consider recent work on corrugated silicon substrates. One can take a single crystal of silicon, for example, cut at a slight angle (1–2 °C) off the <113> crystal plane and remove its oxide layer. On careful annealing of the silicon at elevated temperatures (about 850 °C), a facetted surface can form.[21] The end result will be a substrate with a striped surface structure, having a sawtooth structure in cross-section. The height of the steps is typically about 3 or 4 nm and the repeat length is of the order of 100 nm. It might be expected that such a shallow depth with a lateral length scale of the order of 100 nm should not cause significant changes to the morphology of a film. However, this is not so, as we show in Figure 2.[22]

The polystyrene aligns itself along the troughs to lower its free energy. The reason why the polystyrene prefers to be in the troughs is contentious. It may be due to chain confinement effects, or due to the spin casting process producing a film in which the polymer chains are quenched into a non-equilibrium state, creating internal tension in the film. We observe that the critical thickness, above which the film is stable, scales with chain length (the radius of gyration).[22] The polystyrene film, which was spin coated onto the substrate, is thicker above the troughs, than above the peaks of the corrugations. It therefore dewets more rapidly from the peaks, forcing the polymer into the troughs of the corrugations. This is a very subtle effect with only the

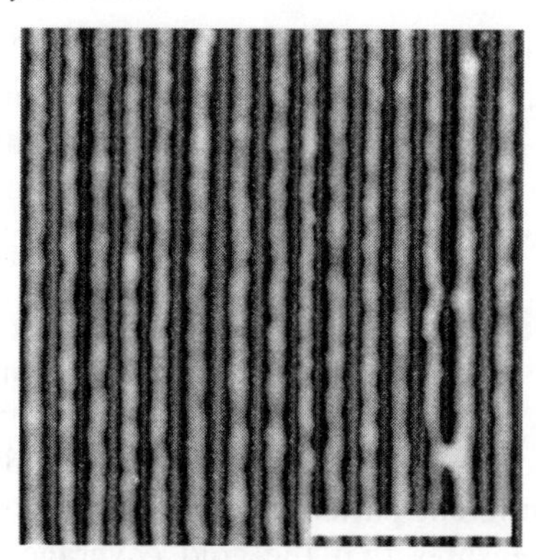

Figure 2 *AFM topography image of a film of polystyrene (nominally 4.5 nm thick) on stepped silicon substrates. The height of the steps before polystyrene deposition was about 6 nm. The film was annealed for three hours at 150 °C. The polystyrene dewets the elevated regions of the structure, forming channels in the troughs. The scale bar is 2 μm and the bright regions are 10 nm higher than the dark regions*

small topographic feature of the structured substrate forcing the ordering along one axis.

The work on the stability of polystyrene films on corrugated substrates[22] has shown to us that the wetting and stability of polymer films thinner than the chain size is a very poorly understood subject. There have also been similar studies on flat substrates,[23] but a key issue is likely to be the conformation of polymer chains close to surfaces. Scattering experiments have not yet lead to a clear answer to this question, despite several experiments being performed.[24–27] Theoretically, the issue is also open.[28–30] We expect that in the next few years this issue will be resolved, probably by neutron or X-ray scattering experiments.

Returning to the wetting of structured substrates, one can consider what phenomena would be different to planar substrates. It has been theoretically demonstrated that the critical behaviour on non-planar substrates is different, with new transitions (unbending and unbinding, for example) that cannot be observed on flat substrates.[31] Whether polymeric systems are ideally suited to elucidating such phenomena remains to be seen. However, a substrate corrugation could also have interesting effects close to the wetting transition for a polymer blend. The wetting transition exists when both complete wetting and partial wetting have the same energies[32] and has recently been located for two different polymer blends.[15,16] One could envisage that this first order transition would also exist on a structured substrate but that its location would move. If we consider two identical systems with a planar and structured substrate, it would be interesting to see if partial wetting dominates on the structured substrate when the blend is close to a wetting transition on a planar substrate. The domination of partial wetting might be expected because topographic constraints would stop a wetting layer from forming on the peaks of the structure. It is also conceivable that wetting on structured substrates could give rise to dynamical instabilities such as that described by Mullins and Sekerka.[33] Such an instability arises when there is a small perturbation in the shape of a growing wetting layer. These perturbations can amplify giving rise to non-uniformity in the growth of a wetting layer. The question arises as to whether how such instabilities would exist at substrates where perturbations in shape are already present.

Another way of advancing such research is to tailor the properties of the substrate. In Bayreuth, we are approaching this problem by evaporating gold onto the silicon. Gold provides a very useful precursor layer because it can form the foundation of further grafting processes allowing many further possibilities to create different patterns with a variety of chemical properties. The stepped structure ensures that the gold coverage is not complete. This gives us a structure alternating with silicon oxide and gold stripes. Immediately, we have changed the characteristics of the film. Since the gold is evaporated onto the elevated part of the stepped structure, and since polystyrene prefers gold, we immediately have a competition between possible confinement effects and the enthalpic preference for gold. Other experiments have been performed on a similar system.[34] In this work, preferential dewetting of a thin layer of polystyrene along the stripes was observed. When a phase separating film of PS and PMMA was annealed on such a substrate, the morphology was very different from that on a homogenous surface, with the gold stripes clearly attracting the polystyrene. In the same paper[34] it was also seen that block copolymers of PS and PMMA aligned themselves with the stripes when the size of the blocks was similar to the length scale of the structured substrate. First theoretical work on patterned substrates has predicted that, for strong enough surface interactions, phase separating polymer mixtures can follow the patterned behaviour.[35,36] Possible experiments still remain when the surface field is comparable to the enthalpic driving force for phase separation. These experiments would be concerned with the competition between the lateral length scale of the phase separation and the pattern size. Another effect due to a chemical pattern was demonstrated for hydrophobic substrates patterned with hydrophilic stripes. Water wets the hydrophilic stripes but when the amount of water is too large to lie only along the hydrophilic lines, a shape instability forms *via* a bulge with a cigar shape, ordered along the hydrophilic lines.[37,38]

4 Block Copolymers

In a large part of what we have discussed above, we considered binary polymer mixtures. However, the situation is somewhat different, if instead of polymer blends, thin films of block copolymers are investigated. Due to the molecular connectivity of the different blocks, the inherent length scale is now determined by the size of the molecules. Early experiments focussed on the thin film morphology in symmetric diblock copolymers,[39] where surface interactions tend to orient the block copolymer lamellae parallel to the boundary surfaces. In contrast to most bulk specimens, the planar interfaces lead to the formation of laterally extended monodomains. The competition between the inherent length scale (the lamellar period) and the film thickness (given simply by the amount of material deposited onto the substrate) has also been the subject of intensive study.[40,41] In ultra-thin films, regular lateral patterns are formed.[42] It is probably fair to say that the equilibrium structure of symmetric diblock copolymer thin films is well understood by now. However, other aspects emerged from these early experiments that deserve further investigation in the near future.

As a first example, aligning effects of a planar substrate tends to facilitate the morphological analysis. While bulk specimens often exhibit a complex multidomain structure, thin films of the same material may be far easier to analyse experimentally. Powerful experimental techniques (for example, high resolution depth profiling, scanning probe microscopies) become applicable and a conventional "bulk technique" like transmission electron microscopy becomes even more powerful, when large-scale alignment is present and difficult sectioning procedures can be omitted. Therefore it can be expected that thin film studies of more complex block copolymer systems (multiblock copolymers, graft and star copolymers, copolymer blends) will become attractive alternatives to the well-established bulk investigations. The same holds for the role of different solvents for the resulting block copolymer structures. Since thin films exhibit a large surface to volume ratio, they can easily be swollen in solvent vapours and even investigated *in situ*. Current work in this direction involves the study of ABC triblock copolymer thin films exposed to different solvent vapours.[43] Along the same lines, non-equilibrium phenomena will become increasingly important as well, since the achievement of the overall lowest free energy configuration is only rarely possible in controlled experiments, not to mention in industrial processes.

In recent years, much work was devoted to the use of block copolymer thin films for the controlled creation of regular lateral patterns. Different routes have been taken to utilise the nanoscopic length scales inherent in block copolymer microdomains in a bottom-up lithography approach. As an example, we quote the extensive work by Möller and co-workers[42] who developed a scheme for the deposition of regular arrays of monodisperse metal particles of different chemical nature with the use of block copolymer micelles.[44] Based on poly(styrene-*b*-vinylpyridine) diblock copolymers dissolved in a selective solvent for polystyrene (toluene), the authors were able to "load" the poly(vinylpyridine) cores quantitatively with suitable metal salts, which were turned into metallic nanoparticles by reduction. After the deposition of a micellar monolayer onto a suitable substrate and the removal of the organic components in a plasma, nanopatterns of metal particles (with sizes of between 1 and 10 nm) can be achieved. To quote two possible applications; these particle arrays can be used as masks in further processing steps (lithography)[45] or as starting points for the selective binding of biomolecules. First steps are presently taken along both directions.

An alternative route to laterally patterned surfaces involves self-organisation of more complex block copolymer structures, where different blocks carry different functions (selective binding to substrate, hydrophobicity, hydrophilicity, selective binding sites, to name four). First experiments along these lines recently demonstrated that a thin layer of ABC triblock copolymer with a selectively adsorbing B middle block can be used to compatibilize an A/C homopolymer blend thin film.[46] Here, the strong binding of the B block to the surface ensures a stable, self-organised coating, which withstands repeated washing in good solvents for the A and C homopolymers. Other experiments may involve the use of similar block copolymers to create surfaces with laterally patterned wettability. Since the lateral scale of the wettability pattern is of molecular size, studies of the adsorption of homopolymers on these structures (as a function of the size of the homopolymer, for example) become feasible.

5 Random Copolymers

The idea of random copolymers being capable of self-assembly may well seem counter-intuitive and we discuss briefly here the reasons why research in this field is necessary in the coming years. It has been known for some time, for example, that certain proteins can form micelles in solution. An example that we shall consider is that of a particular protein, β-casein. The ability of this protein to form micelles had been thought to be due to hydrophobic head groups, but the detailed structure of the protein made this seem difficult, as the hydrophobic and hydrophilic groups are distributed randomly. In terms of the underlying physics of soft condensed matter, micelles are formed by surfactants and block copolymers, and not random biomacromolecules. To this end recent neutron scattering experiments have been performed by Leclerc and co-workers in order to obtain a better understanding of the micellisation of a particular protein, β-casein,[47] and were able to propose a possible structure.[48] Above the critical micelle concentration, the micelles are monodisperse, coexisting in solution with single protein molecules. To unravel the behaviour of such proteins, the early theoretical treatment of this problem considered multiblock copolymers in solution.[49] As surfaces are particularly efficient at instigating ordering, these ideas of a multiblock copolymer of alternating components A and B were also applied to interfaces between pure phases of A and B.[49,50]

Although multiblock copolymers are not random, later theoretical work on the adsorption of random copolymers of monomers A and B at a pure A–B interface has shown that the random copolymer will not only be localised at the interface, but can also order at this interface. At very low bulk concentrations of random copolymers, the copolymer is flattened at the interface. At larger bulk concentrations, excluded volume interactions force the chains into the bulk with a length scale of the order of the chain size. However, at even larger bulk concentrations, the chains are squeezed together – there is little energy cost in doing this – and form brushes at the interface.[51] That random copolymers can organise at interfaces in such a way is a very surprising result that has only been predicted theoretically and experimental work on this theme is much needed. This result for random copolymers is the same as the results for multiblock copolymers.[50]

We started this section by considering the difficulties in understanding how proteins can form micelles, and linked it to the issue of random copolymers at interfaces. This issue of interface selectivity will be very important in the years to come. We have already described in Section 2 how random copolymers can increase interfacial strength, and we have suggested here how such research could be important in biological systems, particularly as it may have implications for the understanding of the mechanisms of such complex phenomena as protein folding; indeed, it may help us understand how proteins were able to develop to begin with. Another property of the biological systems mentioned is that the self-organisation is more often than not reversible, and this brings us to the final part of the article.

6 Reversible Self Assembly

So far we have been motivated by the goal obtaining structures with a view to their final properties either as a means to an end, or for the evaluation of theory. However, it is also likely that the future will bring situations when reversible behaviour is desirable. In the equilibrium structures characteristic of much of the research thus far studied, reversibility can be obtained by changing the temperature. For example, if we have a polymer blend film displaying surface-directed spinodal decomposition, we can, by changing the temperature, move the mixture into the one-phase region and have only a small amount of phase separation (due to surface segregation). Then, by changing the temperature again, we can again observe surface-directed spinodal decomposition. However, most polymer systems do not have a large enough temperature window for such a treatment to be possible. It might be that, by increasing the temperature to enter the one phase regime, one or both of the polymers begins to degrade. Alternatively, miscible polymers will vitrify on lowering the temperature before passing through a phase boundary. It would also be undesirable for nucleation and growth to occur on passing through the metastable region of the phase diagram. Reversible self-assembly is more readily observed in block copolymers because of the presence of the order to disorder transition. Here, quite elaborate effects can be reproduced by a temperature variation.[52]

A perhaps more general way of forcing reversible self-assembly would be to change the environment of the system. A very simple recent example of this behaviour has been performed on grafted polymers.[53] Brushes of polystyrene and poly(2-vinylpyridine) (PVP) are attached to silicon by chemically grafting the two components. By subsequent exposure to different solvent atmospheres, the surface structure of the binary brush layer is altered. Exposure to toluene forces PS to the upper part of the layer and exposure to HCl brings PVP to the surface. Such behaviour was demonstrated to be reversible. A similar effect was also demonstrated in triblock copolymer films, whereby the surface topography could be reversibly altered by briefly exposing thin films of poly(styrene-*b*-2-vinylpyridine-*b*-*tert*-butyl methacrylate) triblock copolymer to different solvent vapours.[43] The component swollen at the surface depended on the selectivity of the solvent.

We have seen therefore that varying the temperature or the use of different solvent atmospheres can be used to control reversible systems. Another possibility is to study such systems in a liquid environment. By varying the quality of the solvent one can vary the behaviour of the material. For example, one can consider a polyelectrolyte brush (a polyelectrolyte grafted by one end to a substrate) in water. By changing the salt concentration of the water, one can change the quality of the solvent. As an example, sulfonated polystyrene will be soluble in water providing that the sulfonation is above about 30%. Such a brush will be swollen in water. However, on addition of salt, the brush will collapse, expelling water. A similar behaviour can be expected from a polymer network. A film of a polyelectrolyte network will expand when in good solvent, allowing material through, but collapse when the solvent quality changes (rather like a valve). The question that we then ask ourselves is what would happen when we put this system together. The penetration of polystyrene brushes into polystyrene networks in the molten state has already been studied.[54]

If films of the network and brush are made from the same polymer, could we control the assembly of the material by changing the solvent quality? We consider a brush and network of the same sulfonated polystyrene in water. Some brush penetrates into the network because both are swollen by the solvent. On decreasing the solvent quality the brush and network will expel water and collapse, locking each other into place. If the solvent quality is improved, it will be possible to separate the two again. We can see now that we have a reversible lock, with possible applications in adhesion. The lock is open in a good solvent and closed in poor solvent. Such behaviour is easily controlled if we use flowing water because changing the solvent quality is easily reversible. Now we consider the situation when the grafted component is more sulfonated than the network. In a good solvent there will still be the penetration of one by the other, but as the salt concentration is increased, the network will collapse first. It could well expel the brush before collapsing. A decent lock may therefore not be obtained. It might be that one would have to adjust the relative amount of sulfonation in the network and brush in order to optimise the reversible penetration and exclusion. Other problems may well concern the degree to which the network and brush actually collapse. An investigation of this behaviour is being planned. Neutron reflectometry measurements on sulfonated polystyrene brushes show significant chain shrinkage, but the brush layer was not observed to collapse.[55]

This is just one example of how the assembly of polymers at interfaces can be controlled reversibly. There are likely to be many others. Imagine the sawtooth silicon substrates with gold aligned along the ridge of the stripes. One could combine this with a grafted block-copolymer such as that described above.[53] One would expect the grafting to take place either on the exposed silicon or to the gold, but not both. The wetting properties of the substrate can then be adjusted by changing the solvent atmosphere. One could then envisage that a polymer overlayer would wet along one of the stripes. A change of the solvent atmosphere could then cause complete wetting of the substrate. It would be very interesting if one could make such systems reversible. A major difficulty is that one would need the solvent to be effective on a brush layer that is already covered by a wetting layer.

7 Summary and Conclusions

The quantitative study of polymer surfaces and interfaces is about ten years old and so is a relatively recent subject. It is therefore not surprising that there are many possibilities for future research. Many of the phenomena that occur at polymer interfaces are related to self-assembly. Even the simplest example, that of surface segregation of one component of a polymer blend to an interface is self-assembly. This is because where we initially had a homogeneous film, we now have one with two layers, a bulk layer and a surface segregated layer. We have shown how even such a well-studied phenomenon as surface segregation can have a future because with it one can generate a surface which could provide an excellent precursor to an interface.

Although we suggest that fundamental studies of wetting on planar substrates will no longer be critically important, related studies on corrugated substrates are only just beginning. These studies will be important because they will provide a route to understanding the behaviour of roughness in real situations. Patterned substrates also provide a means to compare wetting and dewetting. In the past these two related topics have largely been studied separately. Now the interrelation between the two should be stronger.

Copolymers also have a role to play in future work on self-assembly. It is expected that in the next few years, the advantages of lateral nanostructures based on block copolymers will be revealed in real applications, be they industrial or scientific. Aside from that, the thin film behaviour of more complex block copolymers (*i.e.* other than symmetric diblock copolymers) has hardly been studied, let alone understood. Here, first experiments indicate a complex competition between different phenomena (affinity of the different blocks to the confining interfaces, spatial confinement, inherent interactions between the different blocks). Many of the resulting structures are not yet understood; although they appear to be highly ordered over large spatial scales. As the number of blocks in copolymers increase, ordering processes remain, although not necessarily on smaller length scales. We have described how multiblock and random copolymers can behave similarly, as well as the consequences for biological macromolecular systems.

We have also speculated that reversible self-assembly will also play a significant role in future research. As an example, we discussed how a grafted polymer layer and a polymer network could contribute to research on adhesion by together generated a lockable system. There are a wide variety of reversible systems that might well emerge in areas of nonotechnology, which should play a major role in the development of soft condensed matter.

Acknowledgement

We thank Dr Robert Magerle for useful discussions during the preparation of the manuscript. The authors are grateful for financial support through the Deutsche Forschungsgemeinschaft (SFB 481).

References

1. R. A. L Jones, E. J. Kramer, M. H. Rafailovich, J. Sokolov, and S. A. Schwarz, *Phys. Rev. Lett.* 1989, **62**, 280.
2 I. Schmidt and K. Binder, *J. Physique*, 1985, **46**, 1631.
3 J. F. Hester, P. Banerjee, and A. M. Mayes, *Macromolecules*, 1999, **32**, 1643.
4 H. R. Brown, V. R. Deline, and P. F. Green, *Nature*, 1989, **341**, 221.
5 C.-A. Dai, B. J. Dair, K. H. Dai, C. K Ober, E. J. Kramer, C.-Y. Hui, and L. W. Jelinski, *Phys. Rev. Lett.*, 1994, **73**, 2472.
6 For a review, see T. Hashimoto, *Phase Trans.*, 1988, **12**, 47.
7 For a review, see G. Krausch, *Mater. Sci. Eng. Rep. R*, 1995, **14**, 1.
8 G. Krausch, C.-A. Dai, E. J. Kramer, J. F. Marko, and F. S. Bates, *Macromolecules*, 1993, **26**, 5566.
9 M. Geoghegan, R. A. L. Jones, and A. S. Clough, *J. Chem. Phys.*, 1995, **103**, 2719.
10 S. Walheim, M. Böltau, U. Steiner, and G. Krausch, in 'Polymer Surfaces and Interfaces III', edited by R. W. Richards and S. K. Peace, Wiley, Chichester, 1999, pp 75-99.
11 M. Geoghegan, T. Nicolai, J. Penfold, and R. A. L. Jones, *Macromolecules*, 1997, **30**, 4220.
12 P. Wiltzius and A. Cumming, *Phys. Rev. Lett.*, 1991, **66**, 3000.
13 G. Krausch, C.-A. Dai, E. J. Kramer, and F. S. Bates, *Phys. Rev. Lett.*, 1993, **71**, 3669.
14 U. Steiner and J. Klein, *Phys. Rev. Lett.*, 1996, **77**, 2526.
15 J. Rysz, A. Budkowski, A. Bernasik, J. Klein, K. Kowalski, J. Jedliński, and L. J. Fetters, *Europhys. Lett.*, 2000, **50**, 35.
16 M. Geoghegan, H. Ermer, G. Jüngst, G. Krausch, and R. Brenn, *Phys. Rev. E*, 2000, **62**, 940.
17 For a review, see S. Puri and H. L. Frisch, *J. Phys.: Condens. Matter*, 1997, **9**, 2109.
18 F. Brochard Wyart and J. Daillant, *Can. J. Phys.*, 1990, **68**, 1084.
19 M. Sferrazza, M. Heppenstall-Butler, R. Cubitt, and D. Bucknall, J. Webster, and R. A. L. Jones, *Phys. Rev. Lett.*, 1998, **81**, 5173.
20 P. Lambooy, K. C. Phelan, O. Haugg, and G. Krausch, *Phys. Rev. Lett.*, 1996, **76**, 1110.
21 S. G. J. Mochrie, S. Song, M. Yoon, D. L. Abernathy, and G. B. Stephenson, *Physica B*, 1996, **221**, 105.
22 N. Rehse, C. Wang, M. Hund, M. Geoghegan, R. Magerle, and G. Krausch, *Eur. Phys. J. E*, in press.
23 W. Zhao, M. H. Rafailovich, J. Sokolov, L. J. Fetters, R. Plano, M. K. Sanyal, S. K. Sinha, and B. B. Sauer, *Phys. Rev. Lett.*, 1993, **70**, 1453.
24 A. Brûlet, F. Boué, A. Menelle, and J. P. Cotton, *Macromolecules*, 2000, **33**, 997.
25 R. L. Jones, S. K. Kumar, D. L. Ho, R. M. Briber, and T. P. Russell, *Nature*, 1999, **400**, 146.
26 K. Shuto, Y. Oishi, and T. Kajiama, *Polymer*, 1995, **36**, 549.

27 J. Kraus, P. Müller-Buschbaum, T. Kuhlmann, D. W. Schubert, and M. Stamm, *Europhys. Lett.*, 2000, **49**, 210.
28 T. Pakula, *J. Chem. Phys.*, 1991, **95**, 4685.
29 A. Silberberg, *J. Colloid Interface Sci.*, 1982, **90**, 86.
30 A. N. Semenov, *J. Physique II*, 1996, **6**, 1759.
31 C. Rascón, A. O. Parry, and A. Sartori, *Phys. Rev. E*, 1999, **59**, 5697.
32 J. W. Cahn, *J. Chem. Phys.*, 1977, **66**, 3667.
33 W. W. Mullins and R. F. Sekerka, *J. Appl. Phys.*, 1964, **35**, 444.
34 L. Rockford, Y. Liu, P. Mansky, T. P. Russell, M. Yoon, and S. G. J. Mochrie, *Phys. Rev. Lett.*, 1999, **82**, 2602.
35 D. Petera and M. Muthukumar, *J. Chem. Phys.*, 1997, **107**, 9640.
36 C. Seok, K. F. Freed, and I. Szleifer, *J. Chem. Phys.*, 2000, **112**, 6452.
37 H. Gau, S. Herminghaus, P. Lenz, and R. Lipowsky, *Science*, 1999, **283**, 47.
38 P. Lenz and R. Lipowsky, *Phys. Rev. Lett.*, 1998, **80**, 1920.
39 S. H. Anastasiadis, T. P. Russell, S. K. Satija, and C. F. Majkrzak, *J. Chem. Phys.*, 1990, **92**, 5677.
40 G. Coulon, B. Collin, D. Ausserre, D. Chatenay, and T. P. Russell, *J. Physique*, 1990, **51**, 2801.
41 P. Lambooy, T. P. Russell, G. J. Kellogg, A. M. Mayes, P. D. Gallagher, and S. K. Satija, *Phys. Rev. Lett.*, 1994, **72**, 2899.
42 J. P. Spatz, S. Sheiko, and M. Möller, *Adv. Mater.*, 1996, **8**, 513.
43 H. Elbs, K. Fukunaga, R. Stadler, G. Sauer, R. Magerle, and G. Krausch, *Macromolecules*, 1999, **32**, 1204.
44 J. P. Spatz, A. Roescher, and M. Möller, *Adv. Mater.*, 1996, **8**, 337.
45 J. P. Spatz, T. Herzog, S. Mößmer, P. Ziemann, and M. Möller, *Adv. Mater.*, 1999, **11**, 149.
46 K. Fukunaga, H. Elbs, and G. Krausch, *Langmuir*, 2000, **16**, 3474.
47 E. Leclerc and P. Calmettes, *Phys. Rev. Lett.*, 1997, **78**, 150.
48 E. Leclerc and P. Calmettes, *Physica B*, 1998, **241–3**, 1141.
49 E. Leclerc, Doctoral thesis, Université de Paris VI, Paris, 1996.
50 E. Leclerc and M. Daoud, *Macromolecules*, 1997, **30**, 293.
51 G. Peng, J.-U. Sommer, and A. Blumen, *Eur. Phys. J. B*, 1999, **8**, 73.
52 J. Ruokolainen, R. Mäkinen, M. Torkkeli, T. Mäkelä, R. Serimaa, G. ten Brinke, and O. Ikkala, *Science*, 1998, **280**, 557.
53 A. Sidorenko, S. Minko, K. Schenk-Meuser, H. Duschner, and M. Stamm, *Langmuir*, 1999, **15**, 8349.
54 M. Geoghegan, C. J. Clarke, F. Boué, A. Menelle, T. Russ, and D. G. Bucknall, *Macromolecules*, 1999, **32**, 5106.
55 Y. Tran, P. Auroy, and L.-T. Lee, *Macromolecules*, 1999, **32**, 8952.

Biopolymers

12

Polymers and the Living World

Athene M. Donald

CAVENDISH LABORATORY, UNIVERSITY OF CAMBRIDGE'
MADINGLEY ROAD, CAMBRIDGE CB3 0HE, UK

1 Introduction

Whereas biology tends to talk in terms of macromolecules, in the synthetic world similar long chain molecules are known as polymers. Whatever the name – and this field is littered with words that have already been appropriated for something other than the interface that forms the heart of this paper – the basic physical principles are the same. Hence, all the ideas that have been derived from the synthetic world can equally well be applied to *in vivo* rather than *in vitro* situations. This provides the basis for developing rigorous physics and chemistry to apply to the complex world of biology. Furthermore, since so much of the biological world is comprised of macromolecules, the polymer scientist is well placed to make a significant contribution to the 'Life Sciences Interface', if a fear of complexity can be overcome.

It is first desirable to define the potential area. Biomaterials and biophysics both have well-defined meanings and active research communities. Biomaterials tends to mean materials for prosthetics – which may indeed be polymeric; this is traditional materials science, where properties such as strength and stiffness matter, but the material is hard and it is largely the macroscopic lengthscale which matters. Biophysics deals with membranes and structure/function relationships of individual molecules, typically proteins. The whole field of analysing the structure of proteins is a vital and growing one, and will continue to grow in importance as gene sequences are completed and we enter the 'post-genomic' era of proteomics, where the nature and roles of the proteins which the gene is known to produce must be identified. Although neither of these two comparatively mature fields are clearly beyond the remit of polymer scientists, it is not these where I believe they have the most potential to make a contribution to.

Polymer scientists are used to dealing with ensembles of molecules, frequently involving a solvent, and these two factors are often literally vital to life. Specific interactions are likely to be crucial, perhaps mediated by water – since this is nearly always going to be the appropriate solvent. In many cases the ensembles may self-assemble, and there may be different lengthscales involved which are each important, from nms to microns, possibly in a hierarchical structure.

Polymer science can be expected to interact with biology in a variety of different ways, each of which will require a different skill base but possibly the same set of tools from the battery the polymer scientist is familiar with. The following list, which is probably far from comprehensive, indicates some possibilities, which will be expanded upon in the remaining paper.

Firstly, polymer science should be able to make a significant contribution to unravelling *in vivo* biological systems in several different arenas:

- biopolymers at (and in) wet interfaces;
- interactions involving macromolecules and adhesion;
- supramolecular organisation in, *e.g.*, plant cell walls secretions and whole tissue;
- the impact of genetic modification on all of the above properties.

Secondly, there is the knowledge polymer science can contribute to manipulating biopolymers outside living systems. This could cover:

- food;
- drug delivery and controlled release;
- biosensors and the so-called 'laboratory on a chip'.

The third area in which biology and polymers may profit each other is in learning from a natural biopolymer to design a molecule or a system which may have novel functionality. This implies approaches such as the utilisation of molecular biology methods of expression in organisms such as *E Coli*, and the ability to synthesise oligopeptides using protein sequencers. What may be sought here is the design of a molecule which may spontaneously self-assemble under certain conditions (*e.g.* pH, temperature) and disaggregate otherwise, or to open and shut a switch. Many possibilities can be envisaged. It should be noted that these are the same principles – of self-assembly and conformational change – which lie at the heart of the article by Richard Jones in this volume on 'Polymers and Soft Nanotechnology'.

Finally, there are many examples where the well-controlled architecture of biopolymers, especially proteins, means they are ideal model systems for unravelling the behaviour of polymers of all kinds.

2 Recent History in the Field

To demonstrate both the breadth and the potential power of the field a few illustrative examples will be given of where polymer science has cast light on a biopolymer system. These will be divided up into the same four divisions as listed above.

2.1 *In vivo* Systems: Liquid Crystallinity

It is well recognized that many polymers of biological origin are inherently stiff – often because of intrachain hydrogen bonding – and hence have the potential to form main chain liquid crystalline phases. Examples include DNA[1] and cellulose derivatives,[2] as well as the synthetic polypeptide poly (γ-benzyl-L-glutamate) (PBLG) which formed the basis for so much of the early work on lyotropic phase diagrams.[3] Recently examples have started to appear of side-chain liquid crystalline biopolymers for molecules of complex shape. Mucin, which is the main polymer in mucus secretions such as those that line the stomach wall or are left in a slug's wake, is a glycoprotein. It has a linear protein backbone with alternately rather stiff segments which are highly glycoslyated (that is, there are side-chain branches of sugar chains up to around 100 repeats long), and more flexible segments. Its shape has been said to resemble a bottlebrush, with the side-chain branches being both stiff and so densely packed they are forced to stick out essentially straight from the backbone. This means that over a certain range of conditions of concentration and temperature – the details of which are probably species dependent – an anisotropic phase forms, as revealed by optical microscopy.[4] Since it is well known that the viscosity of liquid crystalline solutions is lower than somewhat more dilute isotropic solutions, there may be a biological reason for the existence of the anisotropic phase: the secretions flow more easily to line the stomach wall or the slug's path.

A second instance of side-chain liquid crystallinity recently discovered is the case of the polysaccharide amylopectin,[5] one of the two principal ingredients in the starch granule. Amylopectin is also a highly non-linear molecule, containing a complex hierarchy of branches. Within the native granule it has long been recognized that it is the amylopectin component – and not the linear amylose – which crystallises. The crystallisation occurs *via* the formation of double helices between neighbouring branches which then line up. But these double helices can themselves be regarded as mesogens which line up into smectic layers. In the native granule this does indeed occur, and there is a well-defined periodicity at around 9 nm.[5] If the granule is dehydrated from its normal water content of around 12%, however, the antagonistic effects of backbone and side-chain[6] act in opposition and pull the mesogens out of register, effectively forming a nematic phase, and the 9 nm peak (reversibly) disappears (Figure 1).[7] In this case, viscosity is clearly not an issue, and the role of liquid crystallinity may simply be because it is a convenient way to pack the highly branched molecule as it is synthesised within the granule, although there are many features of the synthesis which remain a mystery.

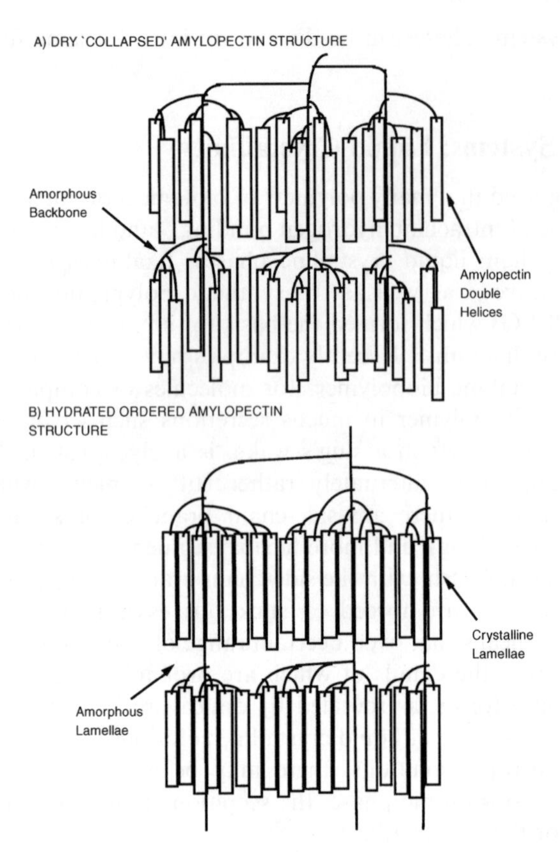

Figure 1 *Schematic view of the amylopectin molecule in the dry (nematic) and hydrated (smectic) states; the double helices are represented as the block like mesogens*

2.2 *In vivo* Systems: Plant Cell Walls

Plant cell walls contain predominantly, but not exclusively, cellulose. This is the stiff backboned polymer which enables trees to stand proud without collapsing under their own weight. The cellulose is not entirely oriented along the cell axis, but rather there is a helicoidal packing, closely resembling plywood, and for the same reason – to provide good mechanical properties isotropically. However, there are also other biopolymers present, and the interactions between cellulose and these polymers may have crucial roles to play in some of the more subtle responses of the plant. These interactions have formed the focus of recent studies aimed at understanding the behaviour of cell walls in more detail, although a model system of cellulose produced by a bacterium *acetobacter acetii* has been used.[8] This has been done since it is easier to produce the cellulose into a culture with the desired second polymer added in this way. Information has been sought on the structure of the complex fibres formed, and the degree of 'crosslinking' introduced into the network of fibres. One of the interesting aspects of many of these cell wall biopolymers is that they are polyelectrolytes, although their charge density as in the case of mucin, may be very unevenly distributed. The importance of such charges in determining the response of the cell wall to changing conditions during growth, or osmotic stress is beginning to receive attention.

2.3 Manipulating Biopolymers: Food and Drugs

Many uses of biopolymers involve substantial processing. Here we are not talking about *in vivo* systems, but control of their structure to achieve a particular end-use. This may relate to the consumer's perception of crunchiness or smoothness in the mouth, or the ability for a drug to diffuse through a matrix to give controlled release over extended time periods.

In both these areas we are dealing with aspects very familiar to polymer scientists. The methodologies required for unravelling processing – structure-property relationships are essentially the same for a polyethylene gas pipe as for an encapsulation medium for a drug, the questions asked broadly similar. What is the interplay between crystallinity and diffusion? How can one control processing to achieve the requisite macroscopic shape without compromising the optimum crystallinity? For foods the system may be complicated by the presence of water. So often the systems are 'soft', although not necessarily fluid, with water being a key element of many foods. Furthermore, the choice of processing conditions may be limited by the additional constraints of ensuring food safety and maintaining nutritional quality.

As one illustrative example in the arena of food, the case of low-fat spreads will be considered. In fact these may have elements in common with photographic film, involving protein (such as gelatin) – polysaccharide (*e.g.* dextran) mixtures in water. The key feature from the consumer's point of view, is that these mixtures must have the same textural attributes as butter, but without the fat content. Gelatin – the main constituent of jelly – of course can gel. Thus for such aqueous biopolymer mixtures a competition between phase separation and gelation ensues, and equilibrium is unlikely to be achieved. Nevertheless, despite the fact that the constituent polymers may be ill-defined (gelatin, although a protein, is actually denatured collagen and can vary tremendously depending on its source), standard tools from the polymer world can be fruitfully applied. Two types of study are worth mentioning. Firstly, in the food community it has traditionally been assumed that in such tertiary systems each phase contains only one polymer. FTIR methods were developed for the case when gelation did not intervene, to permit the complete evaluation of the composition of each phase. For the case of gelatin mixed with highly branched amylopectin, the amylopectin-rich phase can accommodate more of the linear gelatin than conversely, as one would expect.[9] Light scattering methods were applied to the study of the phase separation itself.[10] Spinodal decomposition was found to occur, and the same scaling laws were found to be obeyed as for the traditional binary synthetic polymer mixtures studied by Wiltzius and Bates,[11] despite the heterogeneity of the constitutent polymers. Furthermore, a combination of microscopy and rheological measurements showed how processing history (or correspondingly depth of quench into the two phase region) substantially affected the relative rates of phase separation and gelation, leading to alterations in the connectivity of the phases and consequently to the rheological response.[12] Such studies provide a rationale for processing strategies to ensure the correct spreadability and mouth feel, although derived from rigorous physical approaches well familiar to the synthetic polymer community.

2.4 Designing New Molecules

There are two broad classes of biologically derived methods which are making an impact on the design and synthesis of novel amino acid-based polymers. These involve recombinant DNA methods ('genetic engineering') and peptide synthesis machines. Either of these routes may prove fertile, if there is a clear idea of what molecule is required. The first stage is thus to identify some particular sequence which has an attractive property, and then to build it into a synthetic protein. The group of Tirrell *et al.* have been exploring the recombinant DNA strategy in a variety of contexts. The basic strategy is to create an artificial gene encoding the polymer of interest, which is expressed in bacteria, typically *E Coli.* One strand of their work explored the possibility of making homologues of polymers previously prepared by conventional synthetic techniques, but now with the added advantage of a monodisperse preparation. In the case of PBLG, well known to form cholesteric phases, when the monodisperse form was prepared it was found to form smectic phases:[13] since all chains were the same length such a layered structure is geometrically possible, whereas it is not in the conventional monodisperse material. This system proves ideal for directly comparing monodisperse and polydisperse behaviour. In practice this route for biosynthesis may be difficult to achieve successfully, one of the common problems being the low yield of the polymer of interest. To get round this, for a different system (poly(L-alanylglycine)), chosen because of interest in its chain folding, the reactor in which the bacteria sat was deliberately fed additional oxygen, with a consequent increase in yield of over an order of magnitude.[14]

As an illustration of the second approach to biosynthesis, the work of Boden *et al.* will be taken as an example. Again the first step is to identify an interesting motif to be incorporated in a molecule. The motif that they have focussed on is the β-sheet.[15] By incorporating a sequence which facilitates the formation of this structure it becomes possible to form a peptide which spontaneously self-assembles into β-sheet tapes. In appropriate solvents these peptides can form gels whose rheological properties can then be controlled by external parameters such as pH. Thus, what they have termed 'rational design' coupled with the peptide synthesiser may yield macromolecules with a specific set of controlled properties.

2.5 Biopolymers as Model Systems

Because biopolymers may have properties uncommon in synthetic systems, they can be very attractive as model systems to test specific ideas. An early example of this can be seen in the work on PBLG, a synthetic polypeptide. Although the motivation for its original synthesis failed, it provided a firm basis for many of the early studies on lyotropic liquid crystalline polymers. It was one of the first systems to have its phase diagram characterised,[3] for comparison with Flory's predictions, and a study of its viscosity demonstrated that there is a non-monotonic increase in viscosity[16] with concentration as the liquid crystalline phase is entered.

Much more recently DNA has become an attractive model polymer, both because it can be obtained as an extremely long molecule, with chain lengths up to μms or even mms in length, but also because it can be conveniently fluorescently tagged. Labelling a small proportion of chains in a sea of other DNA chains, means that the motion of individual chains can be directly imaged by fluorescence microscopy. In this way some of the basic tenets of reptation theory have been explicitly visualised, as contorted chains were followed as they relaxed.[17] Using optical tweezer techniques, Chu *et al.* have studied in great detail the different modes by which DNA can relax,[18] and related these to familiar theories such as those due to Zimm.

In both these cases, the fact that the polymers are of biological origin (or related) is not relevant, but the polymers can provide excellent models and new insights. In this way the synergy between polymers and biology is working in the opposite direction from some of the earlier examples, and perhaps in a direction that feels more comfortable to the polymer community at large.

3 Looking to the Future

The list of topics covered above are almost exclusively drawn from work in the last few years. This area has been expanding rapidly, and in the new century we should expect this expansion to continue apace. There will be three main drivers, taking polymer scientists further into the biological arena. Firstly there are a wealth of new techniques – and improvements of older ones too – which make the wet and complex world of living systems much more accessible than in the past. Secondly, as biologists take better control of producing polymers *via* genetic manipulation, there will be two knock-on effects: they will want to understand the impact of the GM approach on the new polymers produced, for which they will need to look to the polymer community, and they will potentially produce polymers with exactly those effects polymer scientists have been looking for. Thus in the post-genomic era it will be vital for polymer scientists to work closely with the molecular biologists in order to produce breakthroughs in both directions. The third motivation will, inevitably, be money. With the Life Sciences Interface of the EPSRC now up and running, and with money on the table, there is every reason for each member of the polymer community to find a niche close to the living world, and to develop further some of the ideas presented here into new directions. This will be healthy for the advance of science as well as the individual.

3.1 The Impact of Novel and Improved Experimental Techniques

In both real space and reciprocal space there have been tremendous instrumental advances in the last decade, whose benefits are still being appreciated in the polymer field. A wealth of new microscopies have been developed, which seem particularly suited to this area of biological polymeric systems. With the exception of the scanning probe microscopies, which do seem fairly widespread, many of these are almost certainly not yet playing as important a role that they have the potential to do. Thus the general field of new microscopies seems likely to be one of rapid expansion in its application to the sort of systems of interest here, and some specific (rather than exhaustive) examples will be mentioned.

Scanning Probe Microscopies, and particularly Atomic Force Microscopy (AFM), are making a major impact, and not just in imaging. They have also been shown to be tremendously powerful in manipulating and measuring the response of single molecules. Although microscopy is often regarded with some suspicion as a non-quantitative technique, the ability to measure the mechanical properties of single molecules directly is proving a great boon. The elastic properties of individual polymers of titin have been studied by several groups. In 1997 two different groups published, essentially simultaneously, observations on how at low strain rates, forces of around 20–30 pN may unfold individual immunoglobulin and fibronectin domains within the molecule; under other conditions the molecule behaves as a non-linear entropic spring.[19,20] At the same time the same properties of the same molecule were studied *via* the technique of optical tweezers,[21] another new generation technique already alluded to above. This explosion of activity concentrated on titin because it is one of the largest macromolecules known, and is therefore more easily handled than many others. However, it is indicative of directions for the future, when a wide range of different biologically important molecules – proteins and polysaccharides – can be expected to be examined at the single chain level.

Scanning probe microscopies are inherently surface techniques, and are frequently used for topographic studies alone. They have the potential for atomic resolution on the right kind of surface, but the molecules may require some kind of interaction with the surface to pin them down sufficiently. However, in certain cases it is possible to study the very shape of the individual molecules,[22] maintaining the natural hydration (and pH) state of the *in vivo* environment during observation. This ability to maintain an appropriate environment around the molecules is something that conventional electron microscopies cannot achieve, due to the requirements of maintaining a high vacuum throughout the column. Although biologists have become very skilled at sample preparation, the danger of artefacts for delicate samples which have been through a lengthy preparation route involving some are all of freezing, dehydration, embedding and sectioning cannot be ignored. Thus any new microscopy, with resolution better than a conventional optical method and which permits maintenance of an appropriate aqueous environment offers tremendous possibilities for the study of biopolymers in their natural state, either singly or as ensembles in tissue, bacteria *etc.*

AFM is clearly not the only new methodology to offer this possibility, although as already mentioned it is the microscopy which is the most well established in the laboratory. Two other obvious candidates can also be mentioned. Scanning near field optical microscopy is rapidly making inroads into this area. The technique is much newer, and less well developed, than the scanning probe microscopies, although it is often used in conjunction with such a system (specifically the scanning force microscope) and may use a similar probe. This technique gets round the diffraction limit of a conventional light microscope by working in the 'near field', as its name implies, and so utilising the nonpropagating evanescent parts of the light wave. Work under water is only just beginning,[23] but has been demonstrated to work, with topographic detail at the ~1 nm level being recorded, although its spatial resolution is less excellent. Additionally it can be used with real living systems, and dynamic processes have been recorded, as in the beating of cardiac myocytes in culture with nm height resolution.[24] There are additionally a wide range of new techniques involving fluorescence which are certainly appropriate for wet systems; their only drawback is the requirement of tagging some suitable fluorophore (frequently, for biological tissues, green fluorescent protein) onto the molecules of interest. Revealing though such fluorescent images may be, they do not overcome the basic problem of minimal invasion or specimen preparation, and there is always the danger – which can of course be checked – that artefacts are introduced either into the image itself or more particularly into the response of the system due to the tagging process.

Another microscopy which offers the possibility of maintaining biological systems in their native state is Environmental Scanning Electron Microscopy (ESEM). Although this technique has been commercial for a number of years, it has not yet been widely applied to wet systems. As with the case of SNOM, the knowhow making it possible to image such systems reliably is still not yet widespread, although ESEM's application to biopolymers has been amply demonstrated. As with any scanning EM technique, it will never reach molecular resolution, and it is still a surface technique, but for ensembles of molecules such as gels,[25] or for whole tissue studies[26,27] it opens up new possibilities without the dangers of artefacts due to preparation route (although beam damage will still be a potential source of artefact). An example of its utilisation is shown below, for the case of the 'network' structure of gluten. The idea of quite a coarse network structure for this important protein found in wheat has come from SEM. However, as the ESEM images (Figure 2) clearly show, if the gluten is never dehydrated the structure is featureless at the micron length scale. If, however, an inappropriate pumpdown sequence is used, mimicking the conditions used to dehydrate the sample for conventional SEM imaging, the coarse fibrillar structure is immediately recovered, illus-

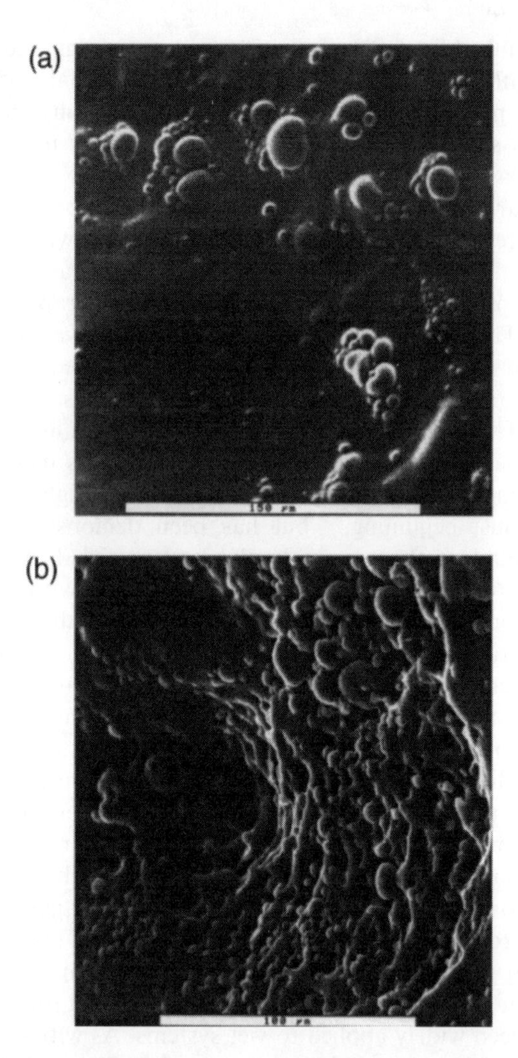

Figure 2 *ESEM images of gluten structure.* (a) *A featureless surface is seen if the sample is always maintained wet;* (b) *following a protocol akin to that used for conventional SEM imaging, a coarse network structure becomes visible due to the harsh dehydration conditions used (after*[25]*)*

trating its source is due to an artefact arising from the specific sample preparation route used.

International facilities for both X-ray and neutron scattering, are going to play a key role in future advances at the polymer/biology interface. As sources continue to develop, with accompanying improvements in detectors, plus spatial and temporal resolution, their applications to the complexity of biological systems will develop in parallel. For the case of neutrons, both small angle and reflectivity measurements are currently being used to study complex proteins, although the resolution of the technique may only be coarse-grained.[28] Reflectivity methods are being developed so that the structure of proteins *within* membranes can be examined,[29] or the penetration of a protein into a liquid at an air/liquid interface revealed.[30] As of now, however, many reactors provide insufficient flux of neutrons to follow dynamic effects in real time, and increased flux is likely to be a strong driving force for advances in this area in the near future; nevertheless, even now it is quite possible to carry out 'before and after' experiments to examine the effect of drugs or virulence factors on, *e.g.*, a membrane.[31]

Flux is much less of a problem at X-ray sources; indeed sometimes it needs to be reduced to avoid detector saturation. The whole field of protein crystallography is heavily dependent on the use of synchrotron sources, an enormous area of application which has made a spectacular impact – and no doubt will continue to do so. The high flux additionally means that the application of real time studies is much further advanced for both the wide and small angle regimes.[32,33] Additionally, X-rays are much more easily focussed, and can therefore be used at high spatial resolution to produce microfocus data. For biological systems this has already led to some interesting studies, *e.g.* on starch[34] and flax.[35]

There are of course a large number of other emerging techniques which do not fit into either of these broad categories, and for which there is no room for discussion here. Suffice to say that the wealth of new methodologies that are currently under development can only be beneficial for this field. As new techniques become available, so do new opportunities.

But technique development is only part of the story. The other reason that this field will be expanding in the years ahead relates to the impact of molecular biology – loosely defined – on the topic. Genetic modification, that emotive term, has tremendous potential to alter the polymers available to us from plant sources. In its simplest manifestation, the most benign and probably also the most relevant to this forum, is the possibility of 'classical breeding speeded up', when wild-type cultivars are effectively crossed by genetic engineering methodologies. This may be done blind, simply to see what may result. Increasingly however, it will be important to have predictive power both at the plant biosynthetic level, so that known changes may be effected, but also at the level of knowing what these changes do to any desired end polymer functionality. The end use will then determine what changes must be sought in the biosynthesis. Currently there are few researchers working in this middle region linking changes in the biochemistry to end-use functionality. This will be a vital community to build.

My own personal work in the starch area, has indicated the challenges – and the satisfaction – that may be obtained from working in this interdisciplinary field. The expertise in manipulating the different enzymes involved in starch synthesis, now mean that particular ones can be knocked out reasonably straightforwardly; there is even some control over the degree of inactivation of each enzyme. However, the end results of these changes are far from obvious at both the molecular and supramolecular levels.[36] In part this is because the interplay between the various enzymes involved in the synthesis of both the branched amylopectin and linear amylose are really not understood. Changing the activities of these enzymes does not only mean more or less of one of these two polysaccharides are produced, but that the whole pattern of branching and molecular weight distribution is modified, producing molecules that are not akin to the wild-type variant of either amylopectin or amylose. These may or may not be useful, and much work is being done in this area to see which mutations can usefully be harnessed. The changes at the molecular level then spill over into the assembly of the whole granule itself; certain changes lead to the granule structure being almost completely destroyed, although the reasons for this are still far from clear. Since the hierarchical structure of the granule is intimately involved in the way the granule is processed,[37] modifying it – albeit uncontrollably – will clearly also have implications for subsequent processing: it may make it easier or harder or slower or faster and so on. As starch is such a vital commodity both for foodstuffs and for industrial uses, any of these sorts of alterations at the molecular or larger lengthscale could have a significant impact on local economies. This example highlights the importance of bridging between the genetics of production and the bulk use of a polymer. Other plants, other biopolymers, are likely to have distinct but similar issues connected with them. The polymer scientists need to get in there.

The example of starch deals with a biopolymer which is already heavily used in a broad range of applications. Plants may of course produce other polymers for which no use has yet been identified, either because it has been uneconomic or because the wild form is indeed not useful. However, the pressure is on, for environmental reasons, to make better use of natural resources and this provides another motivation for the polymer scientist to explore the potential harvest. Thus by understanding both what is naturally produced and what would ultimately be desirable, the community can play a vital role in driving the biology in a useful direction.

The issues are of course not restricted to the production of bulk industrial materials. By looking at the production of polymers in the cells themselves, we can help move on from gene sequence to understand the operation of whole organisms. This is not to duplicate the work of biochemists, who are concerned with individual molecules, but much more to do with ensembles of macromolecules and how structures may be built up, disassembled, migrate and/or stick. The experimental tools and mind set required are the same as for *in vitro* experiments; the handling of samples may be very different and need close collaboration with life scientists.

The final driver is of course financial, and not one to be underrated! The EPSRC has made its commitment to this 'Life Sciences Interface' with many millions of pounds already promised. The message is clear: interdisciplinarity is timely and potentially will be rewarded. The requirement on our community is to make sure that we learn enough of the biologist's language to be able to forge working collaborations beneficial in both directions. The problem of language is not a trivial one and cuts both ways since every discipline is buried in its own jargon. However, as long as both partners are committed to bridging the divide such collaborations can be tremendously stimulating. Now the pot of money is clearly identified, with mechanisms being put in place to permit exchange between disciplines, there is no excuse not to make such an attempt.

Conclusions

The future should be bright for this young field. The advances on both the physical sciences side of the equation through new instrumentation, better computation *etc.* are more than met by the advances made by the biologists as their knowledge of the genome grows. There is a wonderful opportunity for collaboration and cross-fertilisation, with a flow of knowledge in both directions being vital. Furthermore, with the area of the Life Sciences Interface already recognized by the EPSRC as being a challenging growth area, polymer scientists are exceedingly well placed to make their mark and create exciting new science.

References

1 F. Livolant, *J. de Phys.*, 1986, **47**, 1605.
2 D.G. Gray, *J. Appl. Poly. Sci. Symp.*, 1983, **37,** 179.
3 E. L. Wee, W. G. Miller, *J. Phys. Chem.*, 1971, **75**, 1446.
4 J. M. Davies, C. Viney, *Thermochima Acta*, 1998, **315**, 39-49.
5 P. J. Jenkins, R. E. Cameron, A. M. Donald, *Stärke*, 1993, **45**, 417-20.
6 M. Warner, *Mol. Cryst. Liq. Cryst.*, 1988, **155**, 433-442.
7 T. A. Waigh, P. J. Jenkins, A. M. Donald, *Far Disc*, 1996, **103**, 325-337.
8 S. E. C. Whitney, J. E. Brigham, A. H. Darke, J. S. G. Reid, M. J. Gidley, *Carbohydrate Research,* 1998, **307**, 299-309.
9 C. M. Durrani, D. A. Prystupa, A. M. Donald, A. H. Clark, *Macromols.,* 1993, **26**, 981-7.
10 H. R. Tromp, R. A. L. Jones, A. R. Rennie, *Macromols.,* 1995, **28**, 4129-38.
11 F. S. Bates, P. Wiltzius, *J. Chem. Phys.*, 1989, **91**, 3258-74.
12 A. J. Owen, R. A. L. Jones, *Macromlecules*, 1998, **31**, 7336-7339.
13 S. J. M. Yu, V. P. Conticello, G. H. Zhang, C. Kayser, M. J. Fournier, T. L. Mason, D. A. Tirrell, *Nature,* 1997, **389**, 167-70.
14 A. Panitch, K. Matsuki, E. J. Cantor, S. J. Cooper, E. D. T. Atkins, M. J. Fournier, T. L. Mason, D. A. Tirrell, *Macromolecules*, 1997, **30**, 42-49.
15 A. Aggeli, M. Bell, N. Boden, J.N. Keen, P.F. Knowles, T.C.B. McLeish, M. Pitkeathly, S.E. Radford, *Nature,* 1997, **386**.
16 J. Hermans, *J. Coll. Sci.*, 1962, **17**, 638.
17 T. T. Perkins, S. R. Quake, D. E. Smith, S. Chu, *Science,* 1994, **264**, 822-6.
18 S. R. Quake, H. Babcock, S. Chu, *Nature,* 1997, **388**, 151-154.
19 L. Tskhovrebova, J. Trinick, J.A. Sleep, R. M. Simmons, *Nature*, 1997, **387**, 308-312.
20 M. Rief, M. Gautel, F. Oesterhelt, J. M. Fernandez, H. E. Gaub, *Science,* 1997, **276**, 1109-1112.
21 M. S. Z. Kellermayer, S. B. Smith, H. L. Granzier, C. Bustamante, *Science,* 1997, **276**.
22 T. J. McMaster, M. Berry, A. P. Corfield, M. J. Miles, *Biophysical Journal,* 1999, **77**, 533-541.

23 A. Naber, H.-J. Maas, K. Razavi, U. C. Fischer, *Rev. Sci. Instrum.*, 1999, **70**, 3955-3961.
24 R. Micheletto, M. Denyer, M. Scholl, K. Nakajima, A. Offenhauser, M. Hara, W. Knoll, *Applied Optics,* 1999, **38**, 6648-52.
25 I. C. Bache, A. M. Donald, *Cer. Sci.*, 1998, **28**, 127-33.
26 K. Gribble, V. Sarafis, J. Nailon, P. Holford, P. Uwins, *Plant Cell Reports,* 1996, **15**, 771-776.
27 B. L. Thiel, A. M. Donald, *Annals of Botany,* 1998, **82**, 727-33.
28 S. J. Perkins, A. W. Ashton, M. K. Boehm, D. Chamberlain, *Int. J. Biol. Mac.*, 1998, **22**, 1-16.
29 A.P. Maierhofer, M. Lieb, D.G. Bucknall, T.M. Bayerl, *Biophys. J.* 1999, **76**, A211.
30 J. R. Lu, T. J. Su, R. K. Thomas, *Colloid and Interface Science,* 1999, **213**, 426-437.
31 R. J. C. Gilbert, R. K. Heenan, P. A. Timmins, N. A. Gingles, T. J. Mitchell, A. J. Rowe, J. Rossjohn, M. W. Parker, P. W. Andrew, O. Byron, *J. Mol. Biol.*, 1999, **293**, 1145-1160.
32 P. J. Jenkins, R. E. Cameron, A. M. Donald, W. Bras, G. E. Derbyshire, G. R. Mant, A. J. Ryan, *J. Poly. Sci. Phys. Ed.,* 1994, **32**, 1579-83.
33 C. Riekel, M. Müller, *Macromols.,* 1999, **32**, 4464-6.
34 T. A. Waigh, I. M. Hopkinson, A. M. Donald, M. F. Butler, F. Heidelbach, C. Riekel, *Macromols.,* 1997, **30**, 3813-20.
35 M. Müller, C. Czihak, G. Vogl, P. Fratzl, H. Schober, C. Riekel, *Macromols.,* 1998, **31**, 3953-7.
36 A. M. Donald, T. A. Waigh, P. J. Jenkins, M. J. Gidley, M. Debet, A. Smith, *Internal structure of starch granules revealed by scattering studies*; Ed. A. M. Donald, T. A. Waigh, P. J. Jenkins, M. J. Gidley, M. Debet, A. Smith, Royal Society of Chemistry, London, 1997.
37 P. J. Jenkins, A. M. Donald, *Carb. Res.*, 1997, **308**, 133-147.

13

Green Polymers for the 21st Century: Real Prospects and Virtual Realities

Simon B. Ross-Murphy[1] and Robert F. T. Stepto[2]

[1] BIOPOLYMERS GROUP, DIVISION OF LIFE SCIENCES, KING'S COLLEGE LONDON, FRANKLIN-WILKINS BUILDING, 150 STAMFORD STREET, WATERLOO, LONDON SE1 8WA, UK
[2] POLYMER SCIENCE AND TECHNOLOGY GROUP, MANCHESTER MATERIALS SCIENCE CENTRE, UMIST, GR

1 Introduction

One of the major emerging themes in polymer science for the 21st Century is the appreciation that our overwhelming reliance on so-called 'petropolymers' will have to change, and that renewable resource natural, or biopolymers will become of increasing importance. Coupled with this is the understanding that many of these polymers are implicitly 'green' and consequently are simply and efficiently biodegraded by naturally occurring agents. One obvious problem is therefore to control biodegradability so it occurs after, rather than during, the required lifetime of any natural polymer product.

There is much literature on biopolymer systems going back, for example to Ancient Egyptian times when both polysaccharide (starch) and proteinaceous polymers (as animal glues) were employed. Subsequent to that, much of Staudinger's early work on the 'polymer hypothesis' exploited naturally occurring polymers. However, it is probably true to assert that over the last 50 years the science of biopolymers and that of synthetic polymers have developed separately. Consequently, workers whose background is predominantly the latter, do not always appreciate the subtleties and specificities of the former. The result is that the apparent prospects for green polymers have sometimes run far ahead of the current climate, both of understanding and of economics. The current article, for the meeting 'Emerging Themes in Polymer Science' will try and temper imagination with a realistic assessment of prospects. It will also, we hope, serve both as a review of past and current work and introduce future lines of research.

2 Biopolymers – Theory and Practice

Biopolymers can be obtained from a variety of sources and, for example, starch polymers, extracted from corn or potatoes, can be extruded into loose fill packaging that literally dissolves and washes away with the next rain. Using a more sophisticated process, a thermoplastic glassy material can be obtained which has greater long-term stability. Some strains of bacteria produce biopolymers and there are several high molecular weight materials of interest. For example, the polysaccharide xanthan is of use in a considerable number of applications as a thickener and stabiliser. The material, known as Biopol, a copolymer polyhydroxyalkanoate, can also be obtained from bacteria. Cellulose derivatives are also important, for example, cellulose acetate derivatives are quite suitable for packaging materials, and have been exploited for almost 100 years. Other cellulose derivatives are also important, and protein-synthetic polymers such as casein-formaldehyde resins also date back to the beginning of the last Century. Some proteins can be formed into extremely strong fibres, and both gelling agents (gelatin) and primitive adhesives (animal glues) produced.

A major problem still facing the development of many biopolymers as industrial commodities is their price, ranging, typically from £2 per kg up to £30 per kg. In the short to medium term (say up to 2050) oil will continue to be relatively inexpensive, making competing materials like polyethylene available for a fraction of this price. Moreover, only ~4% of petroleum supplies are currently used as feed-stock for petrochemicals and synthetic resins. The majority is used for energy generation. However, it is important to realise that not all biopolymer sources are so expensive. For example, starch, of which there is an abundance world-wide, is cheaper than polyethylene.

In the US, where traditionally petro-polymers have held sway there is, nevertheless, now a major effort to develop certain biopolymers, such as polylactic acid and polyhydroxyalkanoates, into commodity materials. This is intriguing in itself, because some of the early starch-based packaging materials were a product of the US Department of Agriculture's Peoria, Illinois laboratory. It seems, however, that the US biotechnology industry, having apparently failed to persuade much of Europe to accept Genetically Modified Foods is now trying an alternative approach. Many of us will already have heard of inventive concepts such as the transfer of genes for polyhydroxyalkanoate formation into plants, and then the suggested 'harvesting' of the resultant polymers. This concept will be considered only briefly here, because it seems to the present authors that this is very much a case of the biotechnology tail wagging the polymer materials dog, whereas it is the converse which is required.

In succeeding sections we will consider the different types of biopolymer, and how their properties reflect their polymeric nature. The major groups of biopolymers are proteins, polysaccharides and nucleic acids. The latter are not to be considered as polymer materials in the present context, although they are the source of genetically engineered materials. Several other green polymers make up minority groups such as the bacterial polyesters or polyhydroxyalkanoates already mentioned, lactic acid polymers and glycol polyesters. Some of these are not, in our view, biopolymers *per se* but will be considered briefly under the more general heading of green polymers.

2.1 Hierarchical Structures and Distance Scales

The physical techniques that have been applied to macromolecular and supramolecular biomaterials may be divided into those that in turn examine the details at the level of individual chemical groups, whole macromolecules, and over much longer distances. Many biological systems occurring *in vivo* are structural composites, whether we consider plant cell wall structures, mammalian skeletal tissue or marine algae.[1] However, the supramolecular structure can be regarded as a simpler material composite, made up, for example of a biopolymeric matrix 'filled' with fibrous material. Although this approach might appear simplistic, it has had some success. In order to understand the structure–property relationships of such materials, an essential first step in controlling and improving them, a multi-technique approach is essential. Indeed it must be stressed that no one technique is capable of providing sufficient information to understand such hierarchical structures.

For example, measuring polymer flexibility by NMR, without employing other techniques is an ambitious strategy, but one some biochemists have adopted. This obviously reflects their concentration of interest at molecular length scales. Polymer scientists would appreciate intuitively that what is being measured is only a local parameter. Figure 1, an amended version of a figure originally published in 1987,[2] lists the techniques used to probe over different distances scales, together with an (albeit subjective) definition of what is meant by 'molecular', 'macromolecular' and 'supramolecular' distance scales. If there is one overwhelming message to be taken forward from this paper, it is that biomaterials and biopolymers are intrinsically far more complex that synthetic macromolecules, and a multi-distance approach is nearly always essential. One of the few counter examples is the case of thermoplastic starch polymers (TSPs) discussed in more detail below. Much of the best recent literature on starch polymers is concerned with relating the branched amylopectin macromolecule to the crystalline packing of the starch granule.[3] This has, however, very little to do with TSPs, because here the most useful processes are those which completely destroy the granular and sub-granular structure, and reduce the polymeric structure to its simplest polysaccharide chain form.

3 Proteins

Proteins are formally copolymers of amino acids, linked by peptide bonds. Indeed Carothers' classic syntheses of the nylons were an attempt to produce fibrous 'pseudo-proteins' by reacting a diamine with a diacid. There are only around 20 common amino acids in nature, but since a typical protein has a sequence of say 500 of these, then there are potentially around 20^{500} ($\sim 10^{650}$)

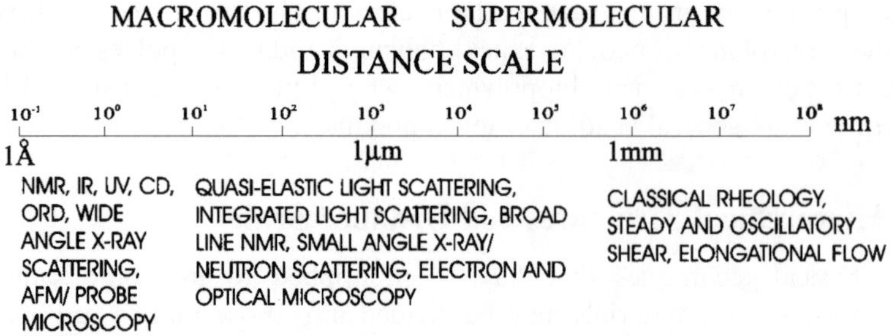

MACROMOLECULAR SUPERMOLECULAR

DISTANCE SCALE

10^{-1} 10^{0} 10^{1} 10^{2} 10^{3} 10^{4} 10^{5} 10^{6} 10^{7} 10^{8} nm

1Å 1μm 1mm

NMR, IR, UV, CD, ORD, WIDE ANGLE X-RAY SCATTERING, AFM/ PROBE MICROSCOPY

QUASI-ELASTIC LIGHT SCATTERING, INTEGRATED LIGHT SCATTERING, BROAD LINE NMR, SMALL ANGLE X-RAY/ NEUTRON SCATTERING, ELECTRON AND OPTICAL MICROSCOPY

CLASSICAL RHEOLOGY, STEADY AND OSCILLATORY SHEAR, ELONGATIONAL FLOW

Figure 1 *Macromolecular and supermolecular distance scales, and the techniques used to probe them*

possible such protein macromolecules. This is considerably larger than current estimates of the number of molecules in the universe, $\sim 10^{85}$ at the last count! Nature is not so profligate as to use more than a very small fraction of these. Further the function of any particular protein is governed by its secondary (α-helix, β-sheet, so-called 'random coil') building blocks and its tertiary and quaternary (coiled coil, multiple-helical, fibrous, globular, multi-sub unit) structures.

Within the context of proteins as polymer materials the number is still further limited, since only very few are available in sufficient bulk at low extraction cost to consider post-processing them into useful materials. More particularly, the fibrous proteins, such as collagen, certain plant proteins such as gluten, the component of wheat responsible for giving the elastic properties to bread doughs, and proteins produced from soy have been exploited to a limited degree, as we shall see below. In recent years there has also been renewed interest in fibrous silk proteins, from silk worms, spiders (as web-silk) and also from bioengineering routes.

3.1 Collagen

Collagen itself is the most abundant protein in mammals,[4] and occurs as the main constituent of connective tissue. One of the major products of collagen is gelatin, a polypeptide obtained by hydrolytic degradation. Commercially, both acid and alkaline hydrolysis routes are employed, and most commercial samples are extracted from the collagen in bovine or porcine bones. It can form insoluble fibres of high tensile strength made up of microfibrils of tropocollagen. The basic molecular unit of the latter is a triple helical rod, and gelatin production involves the hydrolysis of the initial collagen down to the individual chains making up this tropocollagen triple helix. This consists of three α-chains arranged in a left-handed axis, while the whole structure is coiled into a right-handed super-helix. Gelatin gelation then involves transformation of the hydrolysed 'collagen chains' by a mechanism which, at least partly, involves reformation of triple helical 'junction zones' (renaturation) separated along any particular chain contour by regions of flexible polypeptide chain. It is now accepted that cold set gelation proceeds *via* a coil-helix transition induced by cooling a warmed solution of these chains.[5–9]

The properties of resulting gelatins are influenced by the initial collagen, and by the precise treatment process. The precise amino acid content and sequence varies from one source to another, but collagen is unusual in that it always consists of large amounts of the peptides proline, hydroxyproline and glycine. Moreover, it is well known that the precise content effects the coil–helix transition temperature, which for mammals is (fortuitously!) a few degrees above their typical body temperature. The proline content plays a particular role in the stability of gelatin gels, since it promotes formation of the polyproline II helix, which in turn determines the form of the tropocollagen trimer. Above ~40 °C a solution of flexible, random, lightly cross-linked β and γ-chains is formed, together with non-covalently linked α-chains. On cooling to below the coil-helix transition temperature, a transparent, thermoreversible gel is obtained, provided the concentration is greater than some critical concentration, C_0, typically 0.4 to 1.0%. As a gelling agent gelatin is still widely employed, and in the photographic industry there is no interest in developing alternatives. Collagen based glues, produced by more extensive hydrolysis are, however, distinctly limited compared to ‘modern’ synthetic adhesives. There is also interest in exploiting non-mammalian collagens producing, for example, the low melting gelatins extracted from codfish collagen.[10]

3.2 Gluten Polymers

There have been several efforts to employ gluten and soy proteins as alternative feed-stocks for polymeric materials. The former consist of soluble (gliadins) and insoluble (glutenin) fractions, and are the cereal proteins remaining when starch is removed from wheat flour. Glutenin is the generic name for the complex mixture of glutenin polymers. It is possible using genetic engineering techniques to produce, albeit in small amounts, say 10 mg, the pure constituent glutenins, and carry out rheological characterisation of these at different moisture levels.[11]

It is generally accepted that the glutenin polymers have essentially a coiled-coil quaternary structure, and the current view is that its inherent elasticity is somewhat akin to stretching a helical spring and releasing it. Indeed, in connection with this, the interesting work by Urry and co-workers[12] has examined the properties of synthetically produced coiled-coil protein springs. However, early understanding of the elasticity of the cross-linked biopolymer *cis*-polyisoprene, produced by vulcanising natural rubber, also favoured the mechanical spring analogue. It seems that the properties of such ‘coiled-coil’ materials may hide a wealth of future interest.[12,13] Indeed the idea of biological ‘machines’ based upon the reversible release of coiled spring constraints has been suggested. More routinely, perhaps, at lower than ambient (say 20%) moisture levels the crude gluten passes through its glass transition, as do gelatin, starch and many of the other ‘amorphous’ polymers discussed here. In its glassy state, gluten has been suggested as a natural ‘barrier’ polymer.

3.3 Soy Proteins

Soy proteins are less interesting structurally, since they are essentially multi-subunit globules. The interest in exploiting both gluten and soy reflects their intrinsic renewability and comparative cheapness. However exploiting them sometimes requires chemical treatments, which are themselves not very 'green'. Although quite acceptable polymeric materials can result, all current proteinaceous materials are relatively unstable at high temperatures. This reflects their existence in a 'liquid water' world, and it requires quite drastic chemical treatment to extend the useful temperature of use to say >80 °C or to extremes of pH. It is not clear that those proposing the use of such materials have really appreciated this, since most are biochemists who appear to have been translated into polymer science by economic necessity.

It appears there are possibilities for future understanding however, since one of the precepts of the (still to be proven) prion hypothesis, which has decimated the UK beef industry, is that the prion protein, identified by the hypothesis as the causative agent of both bovine spongiform encephalopathy (BSE) and CJD, can survive high temperatures and autoclaving procedures. If, as also discussed below, we were to understand physico-chemically how this might come about, it might serve as a guide to developing high temperature stable proteinaceous materials.

3.4 Other Fibrous Proteins

Silk is a fibrous protein produced by several insect species. Commercially, silk is produced from the cocoon stage larvae of the moth caterpillar *Bombyx mori*, as it has been, in China, for some 4500 years. A single cocoon produces a continuous thread up to 1 km in length, and the protein fibroin contains large amounts of glycine, alanine, tyrosine, proline and serine The peptide chains are arranged in anti-parallel β-sheets which make up the hierarchical structure of the crystalline silk fibres. A number of spiders also produce silk webs, although the fibroin structure is rather different to that from silk worms.

Indeed, as the work of Gosline and co-workers has shown,[14] spiders produce up to seven types of web-silk. One of the best studied, called dragline is used as a safety line by most kinds of spiders. Spider dragline is known for its high strength and exceptional toughness, and the structural aspects which bring about these properties are of great interest. This is true both for macromolecular scientists, and also for the spider, since if the dragline were to fail, certain species of these are sufficiently heavy that impact with the ground could be fatal. Measurements of both mechanical properties and birefringence have been performed, both wet and dry. Dragline from some species exhibits higher birefringence and is stiffer and less extensible than that from others, and the properties, just as in collagen, can be correlated with the proline content. Lower proline increases the crystalline order structure, giving rise to greater stiffness and increased molecular order. Such studies may be valuable precursors to more systematic synthetic investigations.

Recent physico-chemical investigations of such materials are somewhat contradictory. Japanese workers, with a strong tradition in such work, suggest that certain silkworm fibroins are branched,[15] whereas different workers,[16,17] stress the liquid-crystalline nature of other materials, such as spider silks, and the parallels with linear synthetic liquid-crystalline polymers. Perhaps it is fair to say, as in other areas discussed in this article, that there is still a severe shortage of basic information.

3.5 Protein Fibrils in Disease States

Recent groundbreaking biomedical research has been concerned with essentially insoluble protein fibrils being formed and laid down in disease states. This includes not only the prion hypothesis already mentioned, but also Blobel's 1999 Nobel Prize winning work, which considers the deposition of so-called β-amyloid fibrils.[18] These exist, as one example, in the brains of patients with Alzheimer's disease, although similar structures are formed in other dementias such as Huntington's chorea. Amyloid is a term which describes an abnormal deposition of dense, and insoluble fibrillar protein, which damages organs or tissues. Small protein fragments assemble, by processes not yet fully understood, to produce the fibrils, although X-ray evidence[19,20] suggests that they are formed by several twisted β-sheets running parallel to the major fibre axis, and with the individual β-strands running perpendicular. Changing pH in the presence of certain metal ions, such as zinc and aluminium (hence the past scare concerning aluminium saucepans) and the presence of other proteins appears to encourages fibril formation.

Similar fibrils can be formed by lowering pH and then heating a number of globular protein systems to temperatures above the physiological, say 70 °C. Indeed structures like the particulate gels formed by boiling an egg, can be induced to form thin fibrillar networks by adjustment of parameters such as pH and ionic strength.[21–27] What is not clear, however, despite the biochemical and structural work which has characterised such systems, is the precise mechanism of self-assembly. It cannot be simply biochemical, or else the heat set non-physiological structures would not be formed. There is considerable scope for this generic science of protein fibril self-assembly. Indeed if it could be reduced or eliminated, there would be new treatments for a number of fatal diseases. On the other hand, if it could be controlled, then protein fibrils could be produced from bulk materials such as soy proteins, and without the disastrous heat and environmental sensitivity, which apparently destroyed the technology of spun protein textiles in the early 1960s.

4 Polysaccharides

Polysaccharides are the most abundant group of biopolymers on earth, since they make up a large proportion of all plant life, and they also form a major part of many marine and bacterial organisms.[28] The two major plant polysaccharides of interest are cellulose and starch, but chitin, which forms the

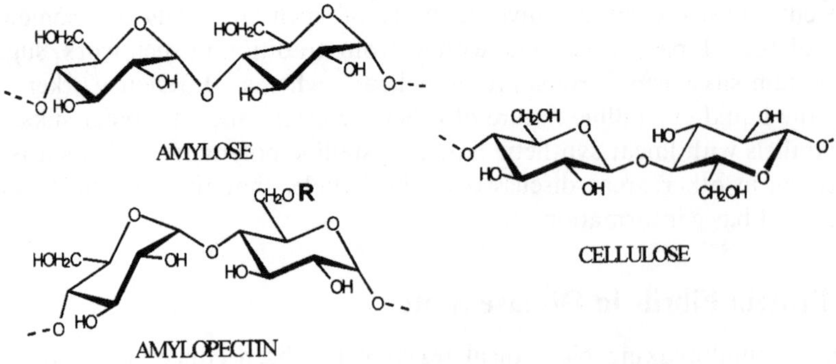

AMYLOSE

AMYLOPECTIN

CELLULOSE

Figure 2 *Glucose polymers – amylose (top) and its branched form amylopectin have α-1,4 glycosidic links, whereas cellulose is linked β-1,4*

exoskeleton of insects and crustacea, is also of interest. Both cellulose and starch are polymers of glucose, the only difference being that one, starch, is linked α (either α-1,4 or α-1,6), whereas cellulose is linked β-1,4, as illustrated in Figure 2. As we shall see this difference in linkage pattern, and the branched structure of amylopectin make crucial differences in all aspects of their respective applications. There are many other polysaccharides of commercial interest, including those from marine algae, such as the alginates, carrageenans and agaroses, from plants, pectins, and galactomannans, and from bacteria, of which the major commercial products are xanthan, gellan and dextran.

4.1 Cellulose

Cellulose is, of itself, the most abundant organic compound in the biosphere, and has, of course, as paper been used as a polymeric packaging material since the early days of Imperial China. The technology of paper making is, however, based upon the difficulty in obtaining cellulose as a simple high molecular weight polymeric species. Indeed, the very great propensity for cellulose polymer chains to remain mutually hydrogen bonded into supramolecular fibrous aggregates gives cellulose fibres their strength and flexibility. However, major industrial products require treatments to break the fibrillar structure and produce soluble products, cellulose ethers and esters, for a wide range of applications. Also, nature has its own way of producing a 'soluble cellulose' as xanthan gum, discussed below.

While extraction of cellulose from plant sources is still of major importance, the recent rediscovery of bacterial cellulose, by Japanese workers, has potential for controlling the synthesis and resultant properties. The technology is also well suited to developing countries. Bacterial cellulose was originally discussed by Brown more than 100 years ago, and involves the bacterium *Acetobacter xylinium* which also produces vinegar. The work by Iguchi and Indonesian co-workers using coconut milk as the growth medium, is a model study, since both microbiological and polymer physical methodologies were employed.[29]

4.2 Starch

In polymeric terms, a main distinction between starch and cellulose is that the former contains highly branched molecules whereas the latter contains linear molecules. The branching means that crystalline sequences are shorter in starch and fibres do not form. Accordingly, native starch is more readily destructured than native cellulose. Such destructuring is, of course, the basis of much food preparation and, hence, the processing of starch dates back several millennia into human history. In addition, starch adhesives were known by 3500 BC.

The conventional processing of starch, including food-processing and processing to give pastes thickeners and adhesives, is in the presence of heat and *excess* water. Initially, a process occurs that is termed gelatinisation[30,31] resulting in a breaking down of the structures in the starch granules to different extents, depending on the starch and the processing conditions. The structure of the starch granule is very complex and hierarchical, and it is partly crystalline. Various detailed models of starch granules are in the literature. Importantly, the crystallinity and the supramolecular structure are based on the amylopectin component, and not the amylose. The initial granules can be up to 100 μm in diameter. In excess water, the amylopectin crystallinity is lost, some hydrolytic degradation occurs, granules swell and eventually disappear, and the linear amylose molecules diffuse into solution. On ageing, the starch solution or suspension undergoes so-called retrogradation to a swollen network material with a structure now based principally on associations between sections of amylose molecules.

From a structural polymer materials point of view, the preceding conventional processing of starch uses too much plasticiser (water) and eventually lays emphasis on the wrong component, namely, the lower molar mass amylose, with native $M_n < 10^6$ g mol^{-1}. Native amylopectins, on the other hand, can have M_n and M_w in excess of 10^6 g mol^{-1} and 10^8 g mol^{-1}, respectively. Superior mechanical properties of amorphous materials will be obtained if the molecular, solid or network structure is formed at lower water contents and is based on the branched component of higher molar mass. In this respect break-throughs occurred from the early 1980s[32–34] culminating in the thermoplastics processing of starch at approximately its natural water content ($\approx$15 %), in a closed volume at temperatures above 100 °C. Using conventional injection moulding, glassy, amorphous, thermoplastic starch polymers (TSPs) were obtained ($T_g \approx 60$ °C), with moduli similar to those of poly(propylene) and high-density polyethylene and yields points at between 5 to 10 % extension. The difference between processing in excess water and at the natural water content is illustrated in Figure 3.[35,36] The endotherm below 100 °C at 42 % water is typical of gelatinisation or, at least, extrusion cooking, and that above 100 °C at 12 % water is characteristic of thermoplastic starch formation, being a thermal destructuring of the granules not accompanied by swelling.

Products based on injection-moulded TSPs are gradually appearing, for

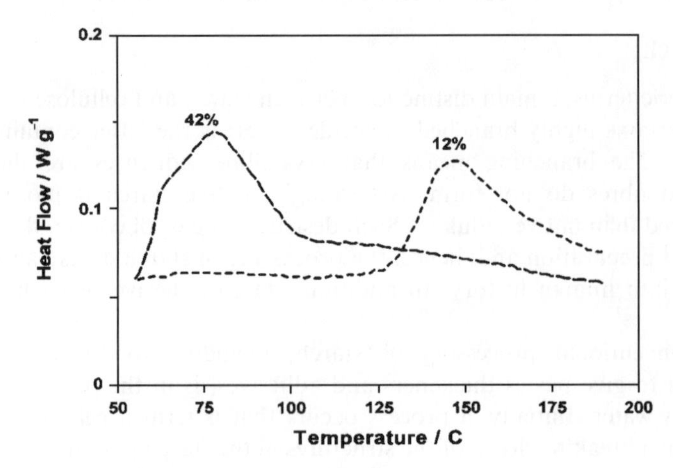

Figure 3 *DSC endotherms for a potato starch at 42 % and 12 % water content* ($=100\, W_{H_2O}/(W_{H_2O}+W_{starch})$)

example, medicinal capsules,[35] golf tees, cutlery, food containers. In addition, extrusion has been applied to produce rigid foams, suitable for loose-fill packaging. Generally, the polymers are stable under ambient, indoor conditions and are completely biodegradable. They break down in water. Hence, TSPs can be considered as a new class of inexpensive, green polymers that can be returned to the natural cycle, with no pollution, after use. Further developments beginning to appear in the patent literature are the thermoplastic processing of blends of starch with hydrophilic synthetic polymers[37] and the thermoplastic processing of starch in the presence of non-aqueous plasticisers, to give materials with improved mechanical properties, lower water sensitivity, but also lower biodegradability. Currently little detailed work appears in journals, although we expect more systematic and fundamental studies to be pursued in the future. In comparison with those of synthetic polymers, studies of processing-structure-property relationships for TSP-based materials are in their infancy.

Like cellulose, starch has for many years been used as an inexpensive filler for synthetic thermoset and thermoplastic materials. One recent example has been the development of starch-filled polyethylene films, which, due to their starch content, are partially biodegradable. In these materials, the starch granules are not destructured. Another class of starch-based materials is inorganic-filled starch composites. To achieve funding acceptability, such materials now have to be known as nanocomposites, but materials based on homogeneously dispersing silicates, extracted from commercial clays, in starch *via* polymer melt processing techniques have been discussed. (These may not be thermoplastic starch in the terms described above.) These are intended to produce biodegradable food packaging materials with low, but permanent atmospheric gas and water permeability. Of course the technology of melt processed particulate- and fibre-filled synthetic polymeric materials is well developed, but it is not so clear that the ideas have been fully exploited so far.

When polymer processors and modellers become directly involved with such materials and in TSPs, we expect a greatly increased interest. However, it is our personal experience that, so far, neither UK industry, nor the Research Councils, see non-food starch processing and the properties of the resultant materials as priority areas. By its nature, as we have stressed above, such work is highly interdisciplinary, and tends to cross the borders between physical and engineering sciences and the biological sciences.

4.3 Chitin and Chitosan

Chitin, generally extracted from South Atlantic krill, is poly-[1-4]-β-(*N*-acetyl-D-glucosamine) and closely resembles cellulose in form, the difference being at carbon C2, which can be aminated and acetylated.[28,31,38] Removal of the acetyl group yields chitosan, which is regarded as a copolymer of acetylated residues with sufficient deacetylation residues, to afford solubility in dilute acid solutions. Generally, commercial samples are 60% to 98% deacetylated. As might be anticipated, molar mass distribution and the pattern of both acetylated and deacetylated glucosamine residues are important, although there seem to be few methodical studies. Most industrial chitosan processes, employ heterogeneous deacetylation steps which yield block-like copolymers, with $MW_w > 3 \times 10^5$ approximately, although methodical studies appear limited by difficult chemistry. The same is true for other polysaccharide derivatives, as in the production of 'soluble celluloses', such as carboxymethyl-cellulose, CMC. Chitosan is currently used in a number of biomedical applications, including wound-healing, and in membrane filter production.

4.4 Other Polysaccharides of Commercial Interest

A number of polysaccharides of interest occur outside the cells of certain cultured microbes, either covalently attached or secreted into the growth media. These are the microbial exopolysaccharides, and over the past few years a great number of these have been described. At the moment, on a volume production basis, the two major members of this group are gellan, an anionic polysaccharide produced by *Auromonas elodea*, and xanthan, also anionic, from *Xanthomonas campestris*.[39] Gellan has a complex tetrasaccharide repeat unit, and gels in the presence of multivalent cations, *via* a double helical intermediate, in a way analogous to the gelling carrageenans. The bulk mechanical properties are sensitive to the degree of acylation of the chain. Xanthan gum, a high molecular weight stiff chain soluble polymer, has an acylated, and hence anionic, pentasaccharide sequence, based around a cellulose backbone.[31,40] It forms true gels only under specific conditions, for example addition of a high concentration of aluminium ions.[41] It is, however, the archetypal structured liquid, or so-called 'weak gel' structurant, and has been employed in a number of industries because of its rheological properties.

The most important of these gel forming polysaccharides are (ι- and κ-) carrageenan, agar(ose), and the alginates, all of which are extracted from marine algae (seaweeds). Much evidence suggests that the first two of these form thermoreversible gels by an extension of the gelatin mechanism. Although some details are still disputed, the general principles are as below. On heating above the helix–coil transition temperature they disorder. For the charged carrageenans the temperature required depends crucially on ionic strength and cation species but lies typically in the range 20–50 °C. On recooling they partly revert to a double helix, and for agarose there is then substantial side-by-side aggregation (confirmed by measurement of R_c by SAXS).[42] For the carrageenans gelation is known to depend crucially on the cations present, for Na^+ little change occurs, whereas high modulus gels are formed for example with K^+ and Ca^{2+}. This is consistent with the 'domain' model proposed by Morris and co-workers,[43] in which junction-zone formation involves ion mediated aggregation of double helical regions (Figure 4). The precise details of network formation in these systems is still somewhat controversial and would merit further studies.

Alginate gels are not thermoreversible, in fact they appear heat stable up to >100 °C, and their formation can only be induced by certain, specifically divalent, cations. Alginates are anionic block copolymers composed of two

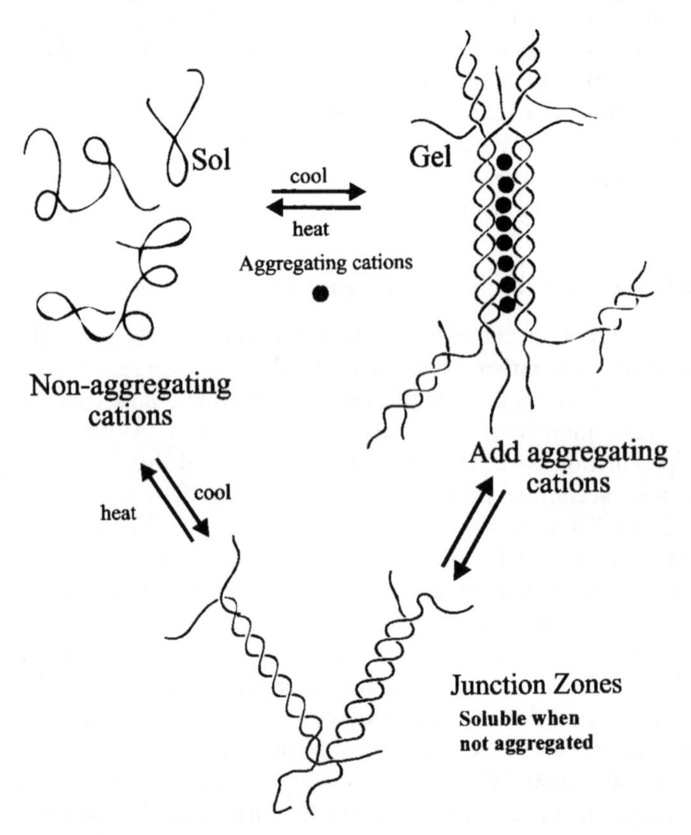

Figure 4 *Proposed mechanism for the gelation of carrageenans (after Robinson and co-workers)*[43]

very similar saccharide units, guluronate (G) and mannuronate (M). If Ca^{2+} ions are introduced into a solution of sodium alginate, gelation occurs extremely rapidly. Gelation is induced by specific ion binding accompanied by conformational change, and circular dichroism evidence implies that Ca^{2+} ions bind cooperatively to G blocks. In one model the junction zones involve two chains and chelated ions giving the so-called 'egg-box' structure. Only recently, have detailed SAXS techniques been applied to this system, and the precise details are still not completely clear.[44,45] Further scattering investigations of these and other polysaccharide systems will prove of great value in the future.

Pectin consists predominantly of sequences of galacturonic acid residues (which are quite similar to the G units in alginate), with occasional interruptions by rhamnose residues. At least some of the carboxyl groups are methyl esterified, the precise distribution depending upon the plant source and age, and an important aspect still not fully understood. Reasonably in view of their structural similarity, pectins of low degree of esterification behave like alginates, and gel with divalent ions. The more esterified materials gel under conditions of low pH and decreased water activity, *i.e.* where intermolecular electrostatic repulsions are reduced; in this case the junction zones are thermoreversible at, say, 40 °C.

What is of interest in the present context is that almost all the original complex egg-box and helical structures were deduced from (molecular distance scale) chiroptical techniques, such as circular dichroism, CD. More recently, probe microscopy has been employed,[46–48] and this has produced many interesting extra features. However light scattering, which would otherwise be of great value, is generally limited by the profound tendency of these systems to aggregate. Here, the pressure cell heating solubilisation method of Vorwerg and co-workers may be usefully employed.[49–51] SAXS is also of great interest, although, despite a promising start,[44,52] very much more needs to be done. Overall, this is very much a case of matching the experimental technique to the appropriate distance scale, in this case 10–100 nm.

5 Other Biopolymers of Potential Importance

5.1 Polylactides

Polylactides or lactic acid polymers are made from lactic acid, which is in turn made from lactose (or milk sugar) obtained from sugar beet, potatoes, wheat, maize, *etc.*[53–55] Polylactides are water resistant and can be formed by injection moulding, blowing and vacuum forming. Polylactides decompose harmlessly in the human body and have therefore been used for some time in biomedical applications, but their high price has until recently, made them uneconomic to use, for example, in packaging. Despite this, there is pressure in the US to employ them further, and studies of processing and material properties are beginning to appear, although these are still clearly in their infancy, and much more work can be expected in the future.

5.2 Polyhydroxyalkanoates (PHAs)

These should need little introduction to workers in the UK, since much effort in the pre-demerged ICI Corporation was based upon developing more efficient ways of producing what became known as Biopol.[56,57] The original work was performed on polyhydroxybutyrate, (PHB) but soon moved to polyhydroxyvalerate (PHV) and copolymers of the two, such as P3HV, P4HV. This is why the more general name polyhydroxyalkanoate is now more usually applied. The starting material for any of these polymers is either sucrose or starch, and again bacterial fermentation is employed. Varying the nutrient composition of the bacteria produces differences in the end product. This makes it possible to tune the properties of the material, *e.g.* its moisture resistance. The polymer can be processed by injection, extrusion, blowing and vacuum forming. For many years the cost was still far too high to make it of more than marginal use, for example, in high cost medical packaging.

For the last few years research both in the US and in the Netherlands has considered the production of PHAs in plants.[57,58] In the Netherlands, this has already resulted in the generation of PHA-producing *Arabidopsis thaliana* (a form of cress widely used in plant breeding studies) and other model plants. More recently starch-storing crop plants, such as potatoes, more suitable for the large-scale accumulation of PHAs, have been suggested.[58] Of course this approach is potentially very valuable, but it is not clear that 'green' pressure will allow the technology to be exploited, even if were to prove economically viable. Big players, such as Monsanto, are involved, and they may well be successful, but the present authors remain to be convinced. Indeed, it may be that such work will remain at the level of 'Tomorrow's World' science, rather than practicality! Nevertheless, it does suggest the intriguing possibility of plants producing starch-PHA blends, which may have attractive combinations of properties. The gene-based technology has, of course, the potential to produce proteins and polypeptides, and even alkenes, so perhaps a true silk-producing spider plant or polybutadiene producing rubber plant will be available in 100 years.

6 Conclusions

It is clear that 'green' polymers, as defined by their biodegradability, are almost exclusively biopolymers. The major classes of biopolymer of interest here are proteins and polysaccharides, naturally occurring biopolymers, and these are subdivided into various sub-classes, with different applications, as described above. Other polymers of interest are the bacterial polyesters and polylactides. All of these polymers have the potential to be processed into new materials, but clearly not all of these will have either attractive properties or be economically viable materials.

In native polymer materials, hierarchical structures are essential for their constitution, and hence functional properties. Hence, to develop and apply such materials, it is important to understand these structures and how to change them in a controlled fashion, as exemplified by the well-established technologies of paper and gelatin manufacture, by the recent development of TSP, and the exploitation of microbial polysaccharides such as xanthan gum. However, the complex organisation of native materials means that processing and chemical modification to give desirable molecular and supramolecular (morphological) structures and properties are generally more difficult and subtle than for synthetic petropolymers.

Even though some technologies for processing natural polymer materials have been established for centuries, it is clear that to date, only a very small fraction have been explored. As nature herself has shown, the potential for building different polymer-based materials is almost limitless, and properties and functions can far exceed those attainable with synthetic polymers. Hopefully over the next 50 years the need for green polymers will provide the impetus for the much greater research efforts that are needed to accelerate the man-made 'evolution' of materials based on natural, renewable resource polymers.

References

1 E. Baer, A. Hiltner and R. J. Morgan, *Physics Today*, 1992, **45 (10)**, 60.
2 A. H. Clark and S. B. Ross-Murphy, *Adv. Polym. Sci.*, 1987, **83**, 57.
3 R. E. Cameron and A. M. Donald, *J. Polym. Sci.: Part B: Polym. Phys.*, 1993, **31**, 1197.
4 L. Stryer. 'Biochemistry', W.H. Freeman and Co., San Francisco, USA, 1981
5 M. Djabourov, J. Maquet, H. Theveneau, J. Leblond and P. Papon, *Brit. Polym. J.*, 1985, **17**, 169.
6 M. Djabourov, J. Leblond and P. Papon, *J. Phys. (Paris)*, 1988, **49**, 333.
7 M. Djabourov, J. Leblond and P. Papon, *J. Phys. (Paris)*, 1988, **49**, 319.
8 K. te Nijenhuis, *Adv. Polym. Sci.*, 1997, **130**, 1.
9 S. B. Ross-Murphy, *Polymer*, 1992, **33**, 2622.
10 P. M. Gilsenan and S. B. Ross-Murphy, *J. Rheol.*, 2000, **44**, 871.
11 J. J. A. Greenfield, S. B. Ross-Murphy, L. Tamas, F. Bekes, N. G. Halford, A. S. Tatham and P. R. Shewry, *J. Cereal Sci.*, 1998, **27**, 233.
12 D. W. Urry, *Biopolymers*, 1998, **47**, 167.
13 D. W. Urry, *Scientific American*, 1995, **272**, 64.
14 J. M. Gosline, M. E. Demont and M. W. Denny, *Endeavour*, 1986, **10**, 37.
15 K. S. Hossain, N. Nemoto and J. Magoshi, *Langmuir*, 1999, **15**, 4114.
16 S. Carmichael, J. Y. J. Barghout and C. Viney, *Int. J. Biol. Macromol.*, 1999, **24**, 219.
17 J. Perez-Rigueiro, C. Viney, J. Llorca and M. Elices, *Journal of Applied Polymer Science*, 1998, **70**, 2439.
18 B. Austen and M. F. Manca, *Chemistry in Britain*, 2000, **36 (1)**, 28.
19 M. Sunde, L. C. Serpell, M. Bartlam, P. E. Fraser, M. B. Pepys and C. C. F. Blake, *J. Mol. Biol.*, 1997, **273**, 729.
20 M. Sunde and C. C. F. Blake, *Quart. Rev. Biophys.*, 1998, **31**, 1.
21 M. Stading, M. Langton and A. M. Hermansson, *Food Hydrocoll.*, 1992, **6**, 455.
22 P. Aymard, D. Durand and T. Nicolai, *Int. J. Biol. Macromol.*, 1996, **19**, 213.

23 P. Aymard, J. C. Gimel, T. Nicolai and D. Durand, *J. Chim. Phys. Physico-Chim. Biol.*, 1996, **93**, 987.
24 A. H. Clark. *(*Hill S.E., D. A. Ledward and J. R. Mitchell Eds.) 'Gelation of Globular Proteins', 'Functional Properties of Food Macromolecules', Aspen Publishers, Gaithersburg, USA, 1998, p. 77.
25 A. Tobitani and S. B. Ross-Murphy, *Macromolecules*, 1997, **30**, 4845.
26 W. S. Gosal and S. B. Ross-Murphy, *Current Opinion in Colloid & Interface Science*, 2000, **5**, in press.
27 G. M. Kavanagh, A. H. Clark and S. B. Ross-Murphy, *Int. J. Biol. Macromol.*, 2000, **in press**.
28 Q. Li, E. W. Grandmaison, M. F. A. Goosen and E. T. Dunn, *Journal of Bioactive and Compatible Polymers*, 1992, **7**, 370.
29 A. Budhiono, B. Rosidi, H. Taher and M. Iguchi, *Carbohydr. Polym.*, 1999, **40**, 137.
30 R. L. Whistler, J. N. BeMiller, E. F. Paschall, 'Starch Chemistry and Technology', (R. L. Whistler, J. N. BeMiller and E. F. Paschall Eds. Academic Press, New York, NY, USA, 1984
31 R. Lapasin and S. Pricl. 'Rheology of Industrial Polysaccharides: Theory and Applications', Blackie Academic and Professional, Glasgow, Scotland, 1995
32 L. Eith, R. F. T. Stepto, I. Tomka and F. Wittwer, *Drug Development and Industrial Pharmacy*, 1986, **12**, 2113.
33 R. F. T. Stepto and I. Tomka, *Chimia*, 1987, **41**, 76.
34 R. F. T. Stepto and.B. Dobler *UK Patent No.* 87 15941, 1988.
35 R. F. T. Stepto, *Polymer International*, 1997, **43**, 155.
36 R. F. T. Stepto, *Macromol. Symp.*, 2000, **152**, 73.
37 G. Lay, J. Rehm, R. F. T. Stepto, M. Thoma, J.-P. Sachetto, D. J. Lentz and J. Sibiger *USA Patent No.* 5095054, 1992.
38 K. I. Draget, *Polym. Gels Network.*, 1996, **4**, 143.
39 I. W. Sutherland, *Biotechnology Advances*, 1994, **12**, 393.
40 K. P. Shatwell, I. W. Sutherland, I. C. M. Dea and S. B. Ross-Murphy, *Carbohydr. Res.*, 1990, **206**, 87.
41 H. Nolte, S. John, O. Smidsrod and B. T. Stokke, *Carbohydr. Polym.*, 1992, **18**, 243.
42 M. Djabourov, A. H. Clark, D. W. Rowlands and S. B. Ross-Murphy, *Macromolecules*, 1989, **22**, 180.
43 E. R. Morris, D. A. Rees and G. Robinson, *J. Mol. Biol.*, 1980, **138**, 349.
44 B. T. Stokke, K. I. Draget, Y. Yuguchi, H. Urakawa and K. Kajiwara, *Macromol. Symp.*, 1997, **120**, 91.
45 B. T. Stokke, K. I. Draget, O. Smidsrod, Y. Yuguchi, H. Urakawa and K. Kajiwara, *Macromolecules*, 2000, **33**, 1853.
46 A. R. Kirby, A. P. Gunning and V. J. Morris, *Trends in Food Science & Technology*, 1995, **6**, 359.
47 M. Miles, *Science*, 1997, **277**, 1845.
48 M. J. Ridout, G. J. Brownsey, A. P. Gunning and V. J. Morris, *Int. J. Biol. Macromol.*, 1998, **23**, 287.
49 T. Aberle, W. Burchard, W. Vorwerg and S. Radosta, *Starch-Starke*, 1994, **46**, 329.
50 Q. Wang, P. R. Ellis, S. B. Ross-Murphy and W. Burchard, *Carbohydr. Polym.*, 1997, **33**, 115.
51 W. Vorwerg and S. Radosta, *Macromol. Symp.*, 1995, **99**, 71.
52 J. H. Y. Liu, D. A. Brant, S. Kitamura, K. Kajiwara and R. Mimura, *Macromolecules*, 1999, **32**, 8611.

53 S. Ferguson, D. Wahl and S. Gogolewski, *Journal of Biomedical Materials Research*, 1996, **30**, 543.
54 W. Weiler and S. Gogolewski, *Biomaterials*, 1996, **17**, 529.
55 M. Yin and G. L. Baker, *Macromolecules*, 1999, **32**, 7711.
56 P. A. Holmes, *Physics in Technology*, 1985, **16**, 32.
57 Y. Poirier, D. E. Dennis, K. Klomparens and C. Somerville, *Science*, 1992, **256**, 520.
58 H. E. Valentin, D. L. Broyles and others, *Int. J. Biol. Macromol.*, 1999, **25**, 303.

Rheology and Processing

14

Emerging Themes in Polymer Rheology

Ian Hamley[1], Tom McLeish[2] and Peter Olmsted[2]

[1]SCHOOL OF CHEMISTRY AND [2]DEPARTMENT OF PHYSICS,
UNIVERSITY OF LEEDS, LEEDS LS2 9JT, UK

1 Introduction

The gentle (or not-so-gentle) art of polymer rheology is presently undergoing a rapid revolution. Driven by beautifully-controlled new materials,[1] new techniques, *e.g.* Ref. 2, and significant developments in theory,[3] the focus is moving increasingly from macroscopic to microscopic and molecular thinking. It is no longer the case that the distribution of molecular architectures in industrial polymer melts will necessarily be an unknown – as the new metallocene polyolefins illustrate.[4] The combination of scattering and optical techniques alongside classical rheological measurements are opening a window into the molecular conformations under flow. Indeed, no modern rheological study, particularly of non-linear behaviour, can be considered adequate without accompanying structural information. Finally, major shortcomings are becoming apparent in the standard phenomenology of viscoelastic fluids, and quantifiable connections between molecular structure and fluid viscoelasticity are sharpening.

We divide our review by material: there are important current advances in all three subfields of homopolymers, multicomponent polymer fluids, and self-assembled fluids.

2 Current State of the Art

2.1 Homopolymers

2.1.1 Entangled Linear Polymers. Here the paradigm of the tube model introduced by de Gennes, Doi and Edwards in the 1970s[5] is proving increasingly fruitful. The notion that strongly-overlapped polymer chains "reptate" under mutual constraints equivalent to tubelike traps along their own contours (Figure 1a) was originally successful in describing the molecular weight

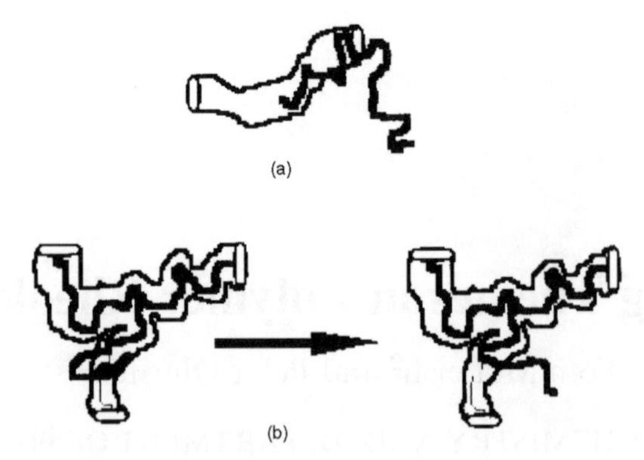

Figure 1 (a) *Reptation of a linear polymer molecule in a tube,* (b) *Arm retraction mechanism in the tube model for a star polymer*

dependence of melt viscosity qualitatively, and the step-strain non-linear dependence of stress quantitatively. However, a legacy of remaining discrepancies have gradually been investigated since.

Firstly, the essential correctness of the tube picture has only recently been established in a remarkable series of experiments. The complex monomer diffusive self-correlation predicted has now been seen in field-gradient NMR.[6] Reptative motion across an interface was the only successful explanation of time-resolved neutron reflectivity.[7] Neutron Spin Echo (NSE) can now be extended in time sufficiently to identify the tube diameter directly.[8] A series of massive many-chain numerical simulations have shown tube-like constraints with sizes identical to those obtained by rheology (*via* the plateau modulus G_0 and NSE).

Meanwhile, two long-standing apparent anomalies have been resolved and shown to be related. The first is the well-known difference between the observations of the molecular weight (M) dependence of melt viscosity, $\eta \sim M^{3.4}$ and the asymptotic tube model result for long chains $\eta \sim M^3$. For some time it has been suspected that the resolution lay in the need to consider fluctuations in the entangled path length of the chains.[10] But application of an analytic result for fluctuations derived in the case of star polymers (see below) demonstrated that the "3.4-law" and the form of the whole rheological spectrum emerged from a such a fuller treatment of the tube model, without the addition of extra parameters. Results from such a calculation are included in Figure 2.[11]

The second emerged as an apparent difference in the dependence of the self-diffusion constant D on M in melts and concentrated solutions. A series of studies covering limited degrees of entanglement in the 1970s and 1980s suggested a dependence of M^{-2} in melts and $M^{-2.5}$ in concentrated solutions. Very recently a comprehensive review of the data with the addition of a new set covering an unprecedentedly wide range of M[11] has shown that in *both* cases $D \sim M^{-2.4}$ for $M/M_e < 200$. The no-fluctuation tube model predicts M^{-2}, but current indications that fluctuations renormalise this exponent to ~2.5 need to be made precise.

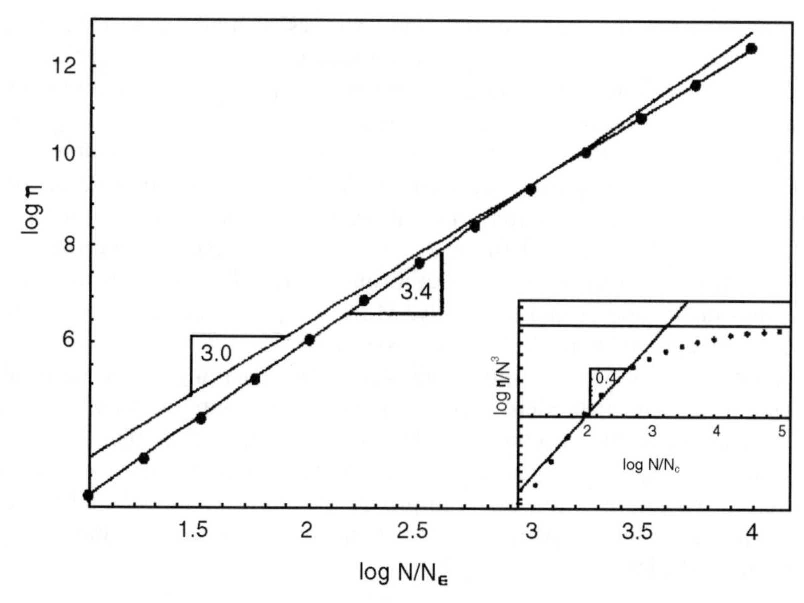

Figure 2 *Viscosity as a function of chain length, reduced with entanglement degree of chain length.*[11] *The points correspond to the theory allowing for fluctuations in the entangled path length of chains showing without adjustable parameters an effective exponent of 3.4. The line with a slope of 3.0 corresponds to classical reptation theory.*[5] *Inset: Plotted as* ηN^3 *versus* N/N_e, *crossover occurs at around* N/N_e *~200, as in experiment*[15]

Another important current research programme revolves around the careful compilation of data on many different chemistries of polymer chain and identifying correlations between values of the entanglement molecular weight M_e and critical molecular weight M_c.[12] A very strong correlation emerges of both quantities (but with different dependencies) on the "packing length" l_p. l_p Is a molecular-weight independent quantity defined as the ratio between the volume per molecule and the radius of gyration in the melt. The observations for M_e support an earlier conjecture[13] that the volume spanned by a single tube diameter contains a universal number of different chains.

2.1.2 Complex Architectures. Perhaps the most significant recent advances in molecular understanding of polymer melts have emerged from the study of branched polymer architectures. We have noted above how a tube theory for star-polymers provided the means to treat fluctuations in entangled path length in linear polymers (see Figure 1b). This is simply due to the complete suppression of reptation in star polymers: without fluctuation there is no stress-relaxation at all!

The programme on branched polymers is driven both by long-standing puzzles in the extension-hardening (yet shear thinning) of commercial branched polymers[2] and the radically different rheology seen in model mono-disperse branched melts such as stars.[15] Recent extensions of the tube model have now successfully treated more complex cases such as star star[16] and star linear[17] blends. In such systems where no molecule contains more than one branch point, nothing of the characteristic extension-hardening of commercial polymers emerges. However, in a recent study of H-polymer,[18] covering linear and non-linear flow, the predicted hardening was observed.[19] The study also employed small angle neutron scattering (SANS) to identify the role of small displacements of branch points under flow, responsible for both hardening, then subsequent free-flow, of the melt. In parallel, advances in the theory of SANS from polymer melts out of equilibrium[20] have allowed the technique to assess molecular models quantitatively. More complex controlled-architecture polymers are coming with reach of exact synthesis.[21]

A very recent move to apply the structures emerging from molecular rheology to non-linear models of polydisperse complex-architecture melts has met with considerable success.[21–24] The simple insight that the stress is a composite, not a structural, variable, with orientational and scalar components of different relaxation times, vastly improves the ability to model LDPE melts quantitatively. It also explains how such melts may be shear thinning yet extension-hardening.

2.1.3 Fast Flows and Constraint Release. A glaring shortcoming of the original tube model was the prediction that monodisperse polymer melts would exhibit a maximum in shear stress as a function of shear rate in steady flow.[5] Consequent instabilities are not seen in experiments (although the situation was unclear for a long time as other viscoelastic instabilities do arise). Early conjectures identified a possible resolution in the transient nature of the tube constraints.[25] Explicit calculation of the "constraint release" of retracting neighbouring chains is required for a quantitative account of linear rheology,[14] but additional contributions to chain-retraction arise in fast flows – termed "convective constraint release" (CCR). Since the overestimation of shear-thinning arises from a strong alignment of chains into the flow direction, it is expected that release of the constraints that align the chains would actually increase the shear stress of the model.

An approximate theory of the effect considered that CCR modified the configurational relaxation time of entire chains.[26,27] It also included a treatment of chain stretching and end-fluctuation, capturing some subtle non-monotonicities in reversing flows. However, the central problem of the shear-stress maximum was not removed. A series of simulations has suffered from the same drawback,[28] but does not lose information on chain conformations. Another analytic approach, treating the tube itself as a Rouse-like polymer with local reorganisations driven by CCR has the advantage of accounting for the structure factor of the chains, while finally showing that CCR can lead to a monotonic stress-shear rate curve.[29] Initial comparisons with limited SANS data on flowing melts support the conclusion that the spatially *local* nature of constraint release is an essential contribution to the melt rheology. Representative results from the CCR model are shown in Figure 3, which shows predicted steady-state shear stress as a function of shear rate as well as

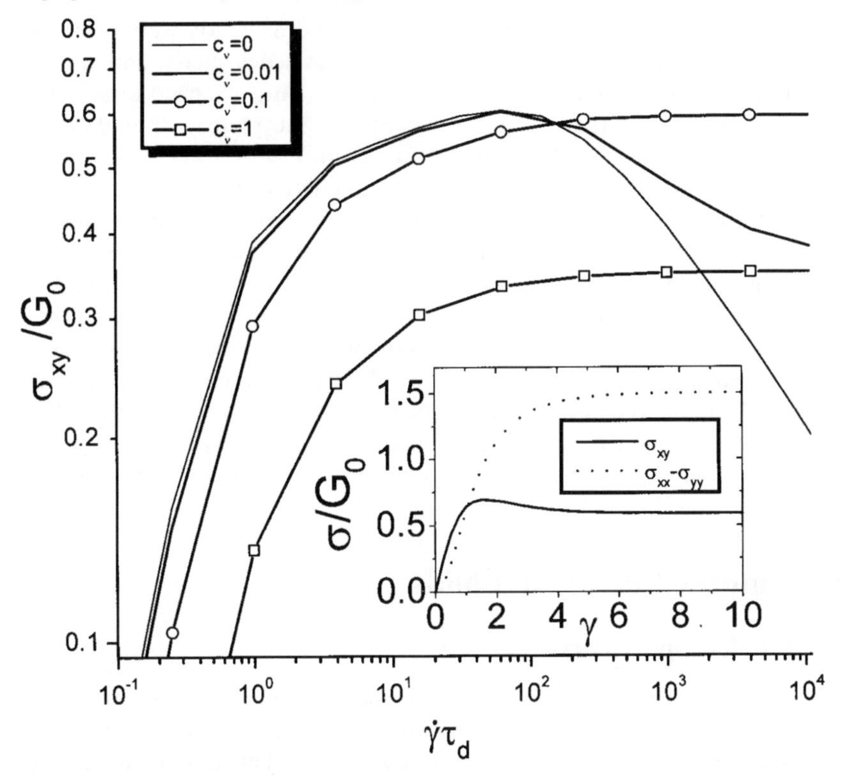

Figure 3 *Predicted steady-state shear stress as a function of shear rate for values of the CCR rate paramenter $c_v = 0, 0.01, 0.1, 1$. The inset displays the shear stress and normal stress transients in the limit of high shear rate (within the zero-stretch regime). The results displayed are extrapolations from numerical calculations on finite chain lengths n to the limit of large n*[29]

normal stress transients in the limit of high shear rate (within the zero-stretch regime).

2.1.4 Semiflexible Chains. Historically, polymer rheology has been motivated by synthetic polymers which are, in the main, flexible on a scale of tens of angstroms and amenable to Gaussian statistics. However, semiflexible polymers are no less important, as evidenced by the biopolymers DNA and *F*-actin. There has been a spate of recent interest in the rheology of semiflexible biopolymer solutions, resulting in new techniques for measuring the very small amounts of material typically available, including torsional oscillator[30] and light scattering[31,32] techniques. Such microrheological techniques obviate the need for large samples and eliminate the problem of wall slip, and can provide local measurements of viscoelasticity.

With the emergence of high quality data on model semiflexible polymer solutions came the realization that the successful tube model for flexible polymer chains should be suitably generalized to describe entangled solutions of stiff polymers. Early calculations of Odijk and Semenov were extended by MacKintosh and co-workers to describe the viscoelasticity and an elastic plateau modulus due to a tangential tension that unravels bend undulations,[33] while Maggs argued that on longer timescales the plateau should be determined by transverse relaxation (tube deformation).[34]

More recently Morse produced a complete microscopic tube theory for stiff polymers that successfully interpolates between the rigid-rod and flexible chain limits.[35] This theory explains many features of semiflexible polymer rheology, including the two mechanisms for plateau moduli described above (which depend on a comparison of timescales), with the tube diameter being the sole fitting parameter as in the Doi-Edwards theory. More recently, Morse successfully *computed* a tube diameter from two different approaches (self-consistent binary collision and continuum effective medium) that give similar results, *e.g.* modulus $G \sim \rho^{7/5}$ and $G \sim \rho^{4/3}$ respectively). An elastic network approximation to collect the important features of each approach compares surprisingly well with recent experimental results.[36]

3 Complex Polymeric Fluids

3.1 Block Copolymers

The rheological behaviour of block copolymers is immensely rich, largely due to their ability to form microphase-separated structures.[37,38] A given microphase separated structures has mechanical anisotropy that reflects its structural anisotropy. Thus a lamellar phase is liquid-like within the lamellar planes but solid in the direction perpendicular to it. Hexagonal-packed cylinder and cubic (micellar or bicontinuous) phases will behave as two and three-dimensional solids respectively. Recent work has probed the distinct frequency dependence of the dynamic shear moduli for aligned lamellar,[39,40] hexagonal[41] and cubic[42] phases oriented at different angles with respect to the shear direction. The resulting scalings have been compared with the predictions of theory, both strong-segregation polymer theory[43] and weak segregation Landau theory.[44]

Recent work has examined the effect of large amplitude shear on phase transitions. Large amplitude oscillatory shear has been shown to reduce the order–disorder transition for lamellar and hexagonal mesophases.[45,46] The results are in qualitative agreement with a mesoscopic mean field theory.[47,48]

Another important issue of current interest is the effect of mechanical contrast of the blocks on the rheological response.[49] This can be important in controlling the orientation of microphase separated structures following large amplitude oscillatory shear. Experiments and theory have largely focussed on the simplest mesophase, the lamellar phase.[38]

3.2 Wormlike Micelles and Shear-Banding

Living polymers, exemplified by surfactant wormlike micelles, continue to prove fruitful for experimental rheology of transitions between different behaviours, and a testing ground for ideas about entanglement common to both living and quenched length distributions. The non-linear rheology is spectacular, with various kinds of flow-induced phase separation, or "shear banding", observed for many systems.[50] Despite a decade of work, there are still no satisfactory explanations for the molecular mechanisms underlying most non-linear behaviours, including shear thinning, shear thickening, and shear banding of various forms. It is evident that the instability of semidilute entangled non-nematic micelles is roughly represented by the Cates model.[51] This predicts a stress maximum as a function of strain rate, but a molecular understanding of the stabilization of the high strain rate branch is lacking, and may require some of the emerging ideas of convected constraint release. In more concentrated micelles the transition is nematic in character (as observed by optical and neutron techniques),[52] and both polymeric (entanglement) and liquid crystalline effects compete in many systems. More recent work has revealed slow kinetics of transitions between flow-induced phases,[53–55] bearing tantalizing resemblance to the kinetics of phase separation in non-driven systems but quite obviously not describable by the same principles. NMR visualization studies of the velocity field have corroborated optical and rheologically-inferred conclusions of shear banding,[56] with bands for the most part stacked in the flow gradient direction, but newer results demonstrate the intriguing possibility of banding along both the gradient and *vorticity* directions,[57] as illustrated in Figure 4. This is consistent with recent theoretical suggestions of such behaviour in the Doi model for rigid rod suspensions,[58] but whether the micelle behaviour is due to flow geometry (the observations were in cone and plate) or bulk constititutive behaviour characteristic of nematics or entangled living polymers is unknown.

The shear-thickening micellar systems are proving even more enigmatic. Recent systematic studies show macroscopic coexistence of a gel-like phase[59] of undetermined structure[60] and, although molecular mechanisms are unknown and may relate to charge effects, a simple toy model[61] provides a rationalization based on the stability of an interface. A similar class of micellar system also displays a thickening instability, but with banding along the vorticity direction and oscillating in time.[62,63] In both cases the coexisting flow-induced phases appear to have different concentrations, implying that stress-diffusion couplings play a role. Inhomogeneous flows also lead to shear banding in cubic micellar phases formed by triblock[64] and diblock[65] copolymers in solution. A representative flow curve, obtained from steady state creep experiments is shown in Figure 5. The plateau corresponds to coexistence between BCC phases with different orientations.[65]

3.3 Blends and Polymer Solutions

Polymer blends and solutions are subjected to nonlinear flows when processed, and this can have important effects on the lengthscale of the ultimate morphology. Flow-concentration couplings in polymer solutions due to molecular deformation is an old problem, and much is known experimentally and

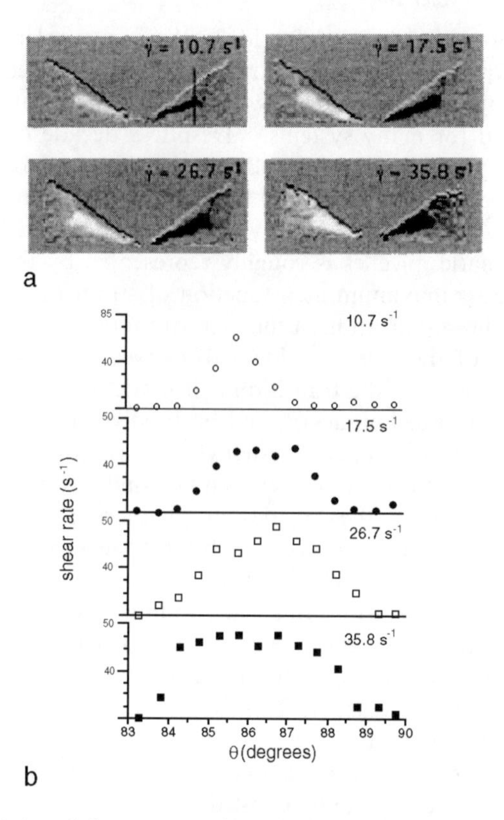

Figure 4 *Images (a) and shear rate profiles (b) from a cetylpyridinium chloride/ sodium salicylate solution of wormlike micelles at different imposed average shear rates in the cone-and-plate geometry. The dark and bright bands show regions of fast moving approaching and receding regions of fluid (from ref. 57)*

theoretically about flow-induced phase separation in solutions and blends near a liquid–liquid phase boundary. Newer work combines rheological and structural characterization, including light[66] and neutron.[67] Relevant related effects include stress-diffusion coupling,[68] viscoelastic contrast between components,[69,70] and molecular stretching, and there have been numerous phenomenological and molecular bead-spring-type studies.[71] A related phenomena of shear-thickening or apparent gelation can occur at higher strain rates: while there have been many experimental studies of this behaviour,[72–74] there has been little conclusive theoretical work. This phenomenon may appear in disguise the shear thickening of living polymer solutions (wormlike micelles), and is an open field. Stress-diffusion coupling due to the macromolecular nature of polymers[75–77] can induce concentration gradients due to polydispersity in curved flows, quite far removed from any demixing line, and such effects undoubtedly combine with shear thickening and phase separation in real systems.

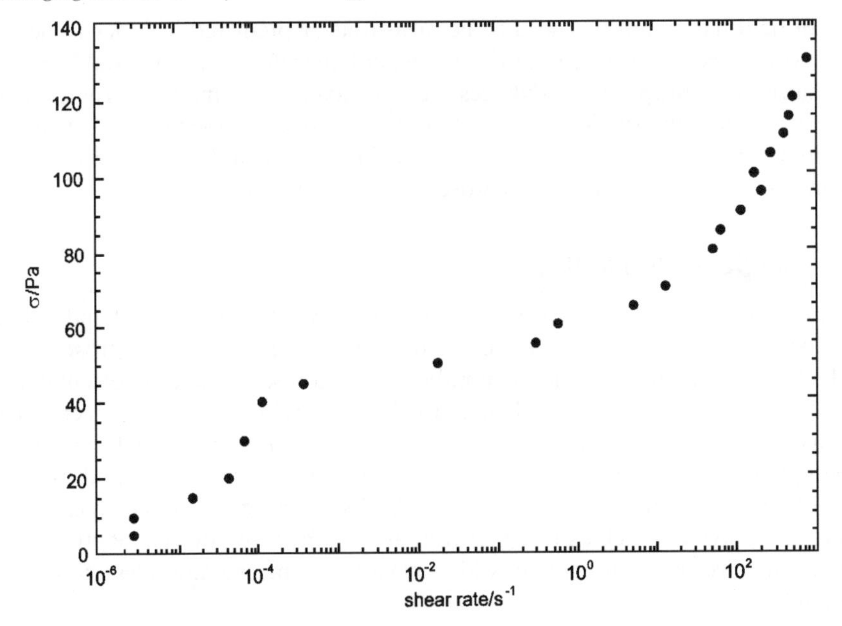

Figure 5 *Flow curve for a 6 wt% solution of diblock $E_{315}B_{17}$ (E = oxyethylene, B = oxybutylene, subscripts = number of repeats) in water at 25 °C. Every point was measured from an individual creep curve on fresh samples, to eliminate orientation effects. The plateau at around $\sigma = 50$ Pa indicates inhomogeneous flow, probably shear banding*[65]

4 Future Potential

4.1 Modelling Dynamics of Entangled Chains

Even the best-defined and controlled industrial polymerisation processes result in melts with significant variety of molecular weight and, if branched, topological architecture. Yet processes such as single-site catalysis do extend the promise the possible calculation of the variety of forms from knowledge of the reaction conditions. In this case, might it be possible to extend the molecular rheology structure-property relations developed for exact structures to the polydisperse case? Although not at all straightforward, there are signs that this might indeed be possible by careful parallel programmes of theory and experimental rheology, working up through cases of increasing complexity. On the hopeful side, it is likely that only a subset of all possible statistical descriptions of the variety of molecular structures is responsible for the rheology.[78] On the other hand, reflection on the accuracy of any pre-averaging of stresses carried by segments in melts under very strong flows seem at present to require heavy simulation at the very least.

More challenging still is the interaction of flow with temperature-dependent transitions such as crystallisation. It is not even known how what an adequate set of field-like quantities need to be measured or predicted in order to define the post-process state of a semi-crystalline polymer in a way that would predict its mechanical properties. Advances are most likely to be made in programmes that work initially with small quantities of structurally well-controlled chemical architecture and polydispersity, processed in very small quantities but with a high degree of molecular and macroscopic interrogation.

4.1 Single Molecule Rheology

Increasingly, techniques will be able to focus on the response of individual molecules in flow. This will either be done in advanced video microscopy in which image analysis of large numbers of fluorescing macromolecules are analysed for the statistics of their individual behaviour,[79] or true mechanical measurements on single molecules. Two routes for individual molecular manipulation are currently being explored: optical tweezers[80] and treated AFM tips.[81] At present, the resolution limits of these techniques mean that giant biopolymers such as DNA or titin are the favoured objects of study, but in future we may anticipate wider classes of macromolecules becoming candidates.

4.2 Flow Visualization and Prediction

An important aspect of flow behaviour of polymers is neglected in current rheological measurements, that is wall effects. This is important in processing (for example it is the origin of the sharkskin morphology of extruded thermoplastics) as well as being of academic interest. New methods for probing flow profiles in polymer melts and solutions such as laser doppler velocimetry[82–85] are in the early stages of development. Another promising technique is field gradient NMR on samples under shear, for instance in a Couette cell.[86–87] NMR imaging permits visualization of the flow velocity profile. Microfocus small-angle x-ray scattering will also allow profiling of flows by rastering across a flowing material and collecting SAXS patterns every 1 μm.[88–90]

The implications for modelling and theory of these advances are challenging, yet it is not impossible that entire complex flow fields of molecular variables may be within the range of numerical prediction in the near future (see chapter on Theory and Modelling in this volume).

4.3 Complex Fluids

The correlation between rheology and thermodynamics is likely to prove a fruitful area for investigation in the future. Very little is as yet known about the detailed mechanisms of non-linear viscoelastic flows, such as those involved in large-amplitude oscillatory shear. Mesoscopic modelling will no doubt throw light on the role of defects in such flows. This is likely to involve both analytical models, and mesoscopic simulation techniques such as Lattice

Boltzmann and cell dynamics simulations. Non-linear flows are experienced when processing block copolymer thermoplastic elastomers, and this subject is thus likely to attract industrial and academic interest. Molecular models will provide insight into the viscoelasticity at higher frequencies.

4.4 Technique developments

Fourier transform rheology is a new technique, which has as yet not been widely employed.[91] In the future, it will provide quantitative information on non-linear viscoelasticity, in particular on the onset of nonlinearity. Combined with structural probes such as small angle scattering, this will enhance enormously our understanding of nonlinear viscoelasticity in soft polymeric materials.

Rheometers are currently under development that will enable the anisotropic stress tensor of anisotropic complex fluids such as block copolymer melts and solutions to be probed, even during large amplitude shear. Here, a small amplitude probe waveform is applied "orthogonal" to the primary large amplitude shear flow.[92] This could provide the linear dynamic modulus of an anisotropic system under nonlinear flow.

References

1 G. Quack, N. Hadjichristidis, L.J. Fetters, R.N. Young, *Ind. Eng. Chem. Prod. Res. Dev.*, 1980, **1**, 587.
2 H.M. Laun and H. Schuch, *J. Rheol.*, 1989, **33**,119.
3 T.C.B. McLeish and S.T. Milner, *Adv. Polym. Sci.*, 1999, **143**, 195.
4 J. B. P. Soares and A.E. Hamilec, *Macromol. Theory Simul.*, 1996, **5**, 547.
5 M. Doi and S.F. Edwards, *The Theory of Polymer Dynamics*, Oxford, 1986.
6 M.E. Komlosh and P.T. Callaghan, *J. Chem. Phys.*, 1998, **109**, 10053.
7 K.A. Welp, R.P Wool, G. Agrawal, S.K. Satija, S. Pispas, J. Mays, *Macromolecules*, 1999, **32**, 5127.
8 P. Schleger, B. Farago, C. Lartigue, A. Kollmar and D. Richter, *Phys. Rev. Lett.*, 1998, **81**, 124.
9 M. Pütz, K. Kremer and G.S. Grest, cond-mat/9906204 v2 15 July (1999); M. Rubinstein, *Phys. Rev. Lett.*, 1987, **59**, 1946.
10 S. T. Milner and T. C.B. McLeish, *Phys. Rev. Lett.* 1998, **81**, 725.
11 T. P. Lodge, *Phys. Rev. Lett.*, 1999, **83**, 3218.
12 L.J. Fetters, D.J. Lohse, S.T. Milner and W.W. Graessley, *Macromolecules*, 1999, **32**, 6847.
13 T.A. Kavassalis and J. Noolandi, *Phys. Rev. Lett.*, 1987, **59**, 2674.
14 S.T. Milner, T.C.B. McLeish, *Macromolecules*, 1997, **30**, 2159.
15 L.J. Fetters, A.D. Kiss, D.S. Pearson, G.F. Quack, F.J. Vitus, *Macromolecules*, 1993, **26**, 647.
16 B. Blottière, T. C. B. McLeish, A. Hakiki, R. N. Young and S. T. Milner, *Macromolecules*, 1998, **31**, 9295-9309.
17 S. T. Milner, T. C. B. McLeish, R. N. Young, A. Hakiki and J. M. Johnson, *Macromolecules*, 1998, **31**, 9345-9353.
18 J. Roovers, *Macromolecules*, 1991, **24**, 5895.

19 T.C.B. McLeish, J. Allgaier, D.K.Bick, G. Bishko, P. Biswas, R. Blackwell, B. Blottière, N. Clarke, B. Gibbs, D. J. Groves, A. Hakiki, R. Heenan, J. M. Johnson, R. Kant, D.J. Read, R. N. Young, *Macromolecules*, 1999, **32**, 6734-6758.
20 D. Read, *Eur. Phys. J. B*, 1999, **12**, 431.
21 L.A. Archer and S.K. Varshney, *Macromolecules,* 1998, **31**, 6348.
22 T.C.B. McLeish and R.G. Larson, *J. Rheol.*, 1998, **42**, 81.
23 N.I. Inkson, O.G. Harlen, D.J. Groves, D.J. and T.C.B. McLeish, 1999.
24 R. J. Blackwell, T.C.B. McLeish and O.G. Harlen, *J. Rheol.,* 2000,121.
25 J.L. Viovy, M. Rubinstein and R.H. Colby, *Macromolecules*, 1991, **24**, 3587.
26 G. Marrucci, *J. Non-Newt, Fluid Mech*., 1996, **62**, 279.
27 D.W. Mead, R.G. Larson and M. Doi, *Macromolecules*, 1998, **31**, 7895.
28 C.C. Hua, J.D. Schieber and D.C. Venerus, *J. Chem. Phys*., 1998, **109**, 10018 and 10028.
29 S.T. Milner, T.C.B. McLeish and A.L. Likhtman, *J.Rheol*., in press,.
30 J. Kas *et al., Biophys. J.*, 1996, **70**, 609.
31 F. Gittes *et al., Phys. Rev. Lett.*, 1997, **79**, 3286.
32 T. G. Mason *et al., Phys. Rev. Lett.*, 1997, **79**, 3282.
33 F. C. MacKintosh, J. Kas, and P. A. Janmey, *Phys. Rev. Lett.*, 1995, **75**, 4425.
34 A. C. Maggs, *Phys. Rev. E,* 1997, **55**, 7396.
35 D. C. Morse, *Macromolecules,* 1998, **31**, 7044; 1998, **31**, 7030; 1999, **32**, 5934.
36 D. C. Morse, *Phys. Rev. E*, in press.
37 I.W. Hamley, *The Physics of Block Copolymers,* Oxford University Press, Oxford, 1998.
38 See U.Wiesner, *Macromol. Chem. Phys.*, 1997, **198**, 3319 for a recent review.
39 K.A. Koppi *et al., J. Phys. II (France)*, 1992, **2**, 1941.
40 T. Tepe *et al., J. Rheol.* 1997, **41**, 1147.
41 C.Y. Ryu, M.S. Lee, D.A. Hajduk and T.P. Lodge, *J. Polym. Sci. B,* 1997, **35**, 2811.
42 M.B. Kossuth *et al., J. Rheol.,* 1999, **43**, 167.
43 M. Rubinstein and S.P. Obukhov, *Macromolecules,* 1993, **26**, 1740.
44 K. Kawasaki and A. Onuki, *Phys. Rev. A,* 1990, **42**, 3664.
45 K.A. Koppi, M. Tirrell and F.S. Bates, *Phys. Rev. Lett*., 1993, **70**, 1449.
46 K.Almdal *et al., J.Phys. II (France)*, 1996, **6**, 617.
47 C. Marques and M.E. Cates, *J.Phys. (France)*, 1990, **51**, 1733.
48 M.E. Cates and S.T. Milner, *Phys. Rev. Lett.*,1989, **62**, 1856.
49 G.H. Fredrickson, *J. Rheol.*, 1994, **38**, 1045.
50 H. Rehage and H. Hoffmann, *Mol. Phys*. 1991, **74**, 933.
51 M. E. Cates, *J. Phys. Chem*., 1990, **94**, 371.
52 J. F. Berret, G. Porte, and J. P. Decruppe, *Phys. Rev. E,* 1997, **55**, 1668.
53 C. Grand, J. Arrault, and M. E. Cates, *J. Phys. II (France)*, 1997, **7**, 1071.
54 J. F. Berret, *Langmuir,* 1997, **13**, 2227.
55 J. F. Berret and G. Porte, *Phys. Rev. E,* 1999, **60**, 4268.
56 P. T. Callaghan, *Rep. Prog. Phys.*, 1999, **62**, 599.
57 M. M. Britton and P. T. Callaghan, *Eur. Phys. J. B,* 1999, **7**, 237.
58 P. D. Olmsted and C.-Y. D. Lu, *Phys. Rev. E,* 1999, **60**, 4397.
59 Y. T. Hu, P. Boltenhagen, and D. J. Pine, *J. Rheol*., 1998, **42**, 1185.
60 S. L. Keller, P. Boltenhagen, D. J. Pine, and J. A. Zazadzinski, *Phys. Rev. Lett.* 1998, **80**, 2725.
61 J. L. Goveas and D. J. Pine, *Europhys. Lett*., 1999, **48**, 706.
62 P. Fischer and H. Rehage, *Rheol. Acta*, 1997, **36**, 13.
63 E. K. Wheeler, P. Fischer, and G. G. Fuller, *J. Non-Newt. Fl. Mech*., 1998, **75**, 193.

64 E. Eiser, F. Molino, G. Porte and O. Diat, *Phys. Rev. E,* 2000, **61**, 6759.
65 P. Holmqvist, C. Daniel, I.W. Hamley, W.Mingvanish and C. Booth, in preparation.
66 S. Saito, K. Katsuzaka and T. Hashimoto, *Macromolecules,* 1999, **32**, 4879.
67 H. Gerard and J. S. Higgins, *Phys. Chem. Chem. Phys.*, 1999, **1**, 3059.
68. S. T. Milner, *Phys. Rev. E,* 1993, **48**, 3674 and chapter in *Theoretical Challenges in the Dynamics of Complex Fluids*, T.C.B. Mcleish ed., Kluwer, Dordrecht,1997.
69 H. Tanaka and T. Araki, *Phys. Rev. Lett.*, 1997, **78**, 4966.
70 N. Clarke and T. C. B. McLeish, *Phys. Rev. E,* 1998, **57**, R3731.
71 R. G. Larson, *Rheol. Acta*, 1992, **31**, 497.
72 J. W. Vanegmond, *Curr. Op. Coll. Int. Sci.*, 1998, **3**, 385.
73 E. P. Vrahopoulou and A. J. McHugh, *J. Non-Newt. Fl. Mech.*, 1987, **25**, 157.
74 P. Moldenaers *et al., Rheol. Acta*, 1993, **32**, 1.
75 A. N. Beris and V. G. Mavrantzas, *J. Rheol.*, 1994, **38**, 1235.
76 M. J. MacDonald and S. J. Muller, *J. Rheol.*, 1996, **40**, 259.
77 J. L. Goveas and G. H. Fredrickson, *J. Rheol.*, 1999, **43**, 1261.
78 D.K. Bick and T.C.B. McLeish *Phys. Rev. Lett.,* 1996, **76**, 2587.
79 T.T. Perkins, D.E. Smith, R.G. Larson and S. Chu, *Science*, 1995, **268**, 83.
80 T.T. Perkins, D.E. Smith and S.Chu, *Science*, 1997, **276**, 2016.
81 H. Li, M. Rief, F. Oesterhelt and H.E. Gaub, *Appl. Phys. A Mat. Sci Proc.*, 1999, **68**, 407.
82 E. Wassner, M. Schmidt, H. Munstedt, *J. Rheol.*, 1999, **43**, 1339.
83 T. Yamamoto, M. Morimoto and K. Nakamura, *Rheol. Acta,* 1999, **38**, 384.
84 M.K. Lyon and L.G. Leal, *J. Fluid Mech.,* 1998, **363**, 57.
85 P. Pakdel and G.H. McKinley, *Phys. Fluids,* 1998, **10**, 1058.
86 P.T. Callaghan, M.L. Kilfoil and E.T. Samulski, *Phys. Rev. Lett.*, 1998, **81**, 4524.
87 P.T. Callaghan, *Rep. Prog. Phys.*, 1999, **62**, 599.
88 C. Riekel, M. Muller and F. Vollrath, *Macromolecules,* 1999, **32**, 4464.
89 C. Riekel, P. Engstrom and C. Martin, *J. Macromol. Sci., Phys. B,* 1998, **37**, 587.
90 C. Riekel, A. Cedola, F. Heidelbach and K. Wegner, *Macromolecules,* 1997, **30**, 1033.
91 M. Wilhelm, D. Maring and H.-W. Spiess, *Rheol. Acta,* 1998, **27**, 399.
92 J. Vermant, P. Moldenaers, J. Mewis, M. Ellis and R. Garritano, *Rev. Sci. Instrum.*, 1997, **68**, 4090.

15

Emerging Themes in Polymer Processing

Arthur N. Wilkinson

DEPARTMENT OF CHEMISTRY & MATERIALS,
MANCHESTER METROPOLITAN UNIVERSITY,
CHESTER ST., MANCHESTER M1 5GD, UK

1 Introduction

The previous century saw the developments of the major moulding and extrusion processes for polymers, used to produce the vast array of polymer products which surround us in modern-day life. The main reason for the ubiquitous presence of polymers is their inherent versatility, conferred by their unique combination of physical and chemical properties and, perhaps most of all, their ease of processing *via* automated mass production techniques. Thus, compared to traditional engineering materials, polymers exhibit a superior balance of price, performance and processability, and are therefore well suited to the rapid production of complex, light-weight components. This advantage has been further enhanced by the development of variants of the basic moulding and extrusion processes, designed to enable the production of ever more complex components, and several of the most noteworthy are described in the following sections. Polymers tend to exhibit strong processing–structure–property relationships,[1,2] with their final properties often being determined by the deformation-temperature history experienced during processing. Thus, polymer processing can be viewed as a delicate balance of achieving appropriate conditions of flow (and reactivity in reactive processing) to produce a good quality product at an economic production rate, without either adversely affecting microstructural development during processing or inducing degradation, excessive orientation or premature solidification of the polymer with subsequent reductions in material properties. For consistent processing it is therefore vital to impose a reproducible deformation-temperature history onto a polymer during processing, and process control technology is constantly evolving towards this end. The current level of process control available to the processor has been achieved by a combination of empirical experimentation and mathematical modelling. In recent years, it is the latter which has contributed most significantly to our understanding of the complexities of polymer processing,[3–8] and which offers the possibility for process control to transcend merely attempting to maintain given levels of process parameters.

2 Polymer Processing – Current State of the Art

2.1 Extrusion Processes

Most extrusion processes[9] involve plasticating extrusion in which a solid polymer particulate is fed into an extruder, which plasticates the solid into a melt and pumps the melt through a shaping die to create a continuous profile. The extrudate may then undergo a number of post-extrusion processes, such as 'sizing' (a further shaping process) and solidification (cooling for thermoplastics, crosslinking for rubbers) before passing to a haul-off device and product handling system. Control systems for extrusion lines usually measure some dimension of the extrudate with feedback to alter either the dimensions of the die exit, extruder output (by varying the screw speed) or the haul-off velocity (to alter the drawing-down of the extrudate). The most common extruded products are fibres (polyamides, polyesters, polypropylene), film (polyolefins, PVC, polyamides, polyesters), pipes/tubes (PVC, polyolefins) and u-PVC profiles used mainly in the construction industry. Composite extrudates may be produced by coextrusion, in which two or more extruders feed a single die where the polymer melt streams are layered together. This technique has been applied to most, if not all, extrusion processes to incorporate layers of recycled, cellular or barrier polymers into extrudates or produce profiles with flexible coextruded seals.

Controlled molecular orientation within an extrudate results in a significant increase in stiffness and strength. Techniques have therefore been developed for thin-gauge products to impart controlled uniaxial orientation to fibres, monofilaments and tapes, and biaxial orientation to film.[9,10] For example, the most common method of polyethylene film production is the film-blowing (or lay-flat) process[11] shown in Figure 1a. The extruder forces melt through an annular die to create a thin-walled tube which is drawn away from the die at a controlled rate by a pair of nip rolls, thinning down the film and inducing axial molecular orientation. In addition, compressed air is fed into the tube *via* the centre of the die and it inflates to a controlled diameter, thinning down the film and inducing circumferential orientation. The major advantage of film blowing is the ease with which this biaxial orientation can be induced, producing a more 'balanced film' with reduced anisotropy. The level of orientation is supposedly controlled *via* manipulation of the ratio of the final tube diameter to that of the die (the blow-up ratio, BR), and the ratio of the velocity of the film at the nip rolls to the average melt velocity at the die exit (the draw-down ratio, DR), but the stress ratio in the melt upon solidification can be vastly different from the macroscopic DR/BR ratio. Crystallisation of the oriented melt generates 'shish-kebab' structures consisting of twisted

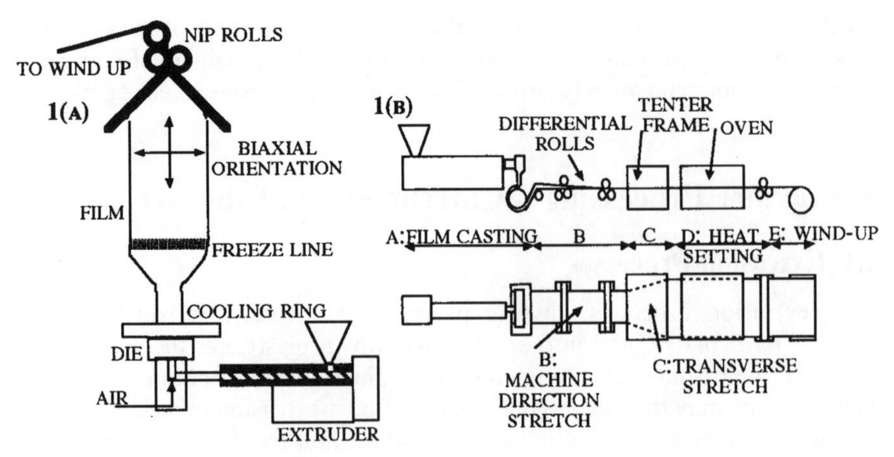

Figure 1 (a) *Blown film extrusion.* (b) *Linear, two-stage production of biaxially-oritented PP*

lamellae, whose chain direction is determined by the stress ratio, overgrown with row-nucleated chain-folded crystals. In contrast, biaxially-oriented films of PP, polyamides and polyesters are typically produced by linear processes (Figure 1b), in which film is chill-roll cast then fed to a system of differential draw rolls where the film is axially stretched at a draw ratio in the range 4–10:1. After leaving the draw rolls the film is fed into a stenter frame, which consists of two divergent endless belts fitted with clips. These clips grip the film so that as it travels forward through a heated zone it is drawn to the required draw ratio in the transverse direction.

In thicker-gauge products, the slow cooling of the melt results in relaxation of the majority of molecular orientation, although a number of techniques to produce oriented products have been developed. In extruded tubes/pipes molecular orientation occurs predominantly along the product axis, and specialised techniques have been developed to impart circumferential orientation in order to improve hoop stiffness and strength. There have been a number of studies of pipe dies with rotating sections to impart hoop orientation to unfilled and fibre-filled melts and more recently[12] LCP. The Scorex process,[13] developed at Brunel University in the UK, is designed to work with both unfilled and fibre-filled melts and involves the use of two or more hydraulic pistons placed radially around a water-cooled mandrel (Figure 2). Sequential operation of the pistons pushes the melt circumferentially around the annular die gap, the melt then flows between the water cooled mandrel and an external cooling sleeve which together form the die exit, and the imparted orientation is retained as the melt solidifies. Biaxial orientation can also be imparted to pipes using die-drawing techniques.[14] In this process, pipe is extruded and sized in the normal way and allowed to cool to just below its melting point. The pipe is then drawn over a large-diameter heated mandrel, which significantly enlarges the pipe diameter and induces biaxial orientation. Pipes made in this manner show improved stiffness, burst strength, chemical

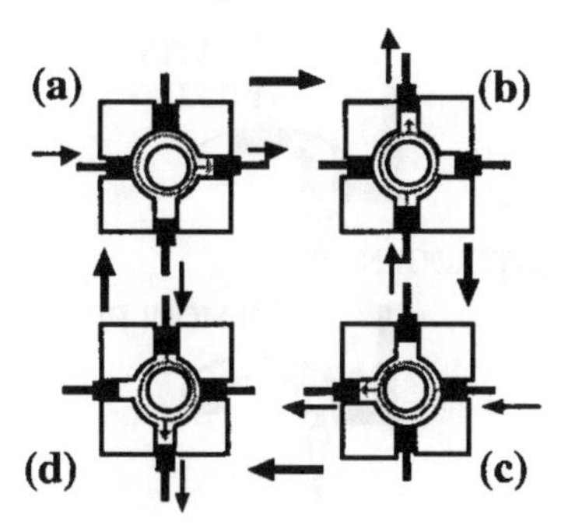

Figure 2 *Shear controlled orientation extrusion (SCOREX)*

resistance and barrier properties. The Netlon® process is a technique for the continuous extrusion of oriented meshes, in which the polymer (usually PE or PP) is extruded through slots cut into two counter-rotating die rings. As the rings rotate, wherever the slots coincide the extruded strands fuse together to form a junction point for the mesh, which is extruded in a tubular form. Biaxial orientation is imparted to the mesh by means of the haul-off unit, which stretches the tubular mesh over a large diameter mandrel to provide elongation in both the axial and circumferential directions. Highly-oriented profiles are usually produced *via* solid-state extrusion,[15] in which a polymer billet is forced through a converging die while it is below its melting point, inducing a high degree of uniaxial molecular orientation. HDPE profiles produced in this way can exhibit levels of tensile modulus and tensile strength similar to those of aluminium and carbon steel, respectively.

Reactive extrusion (REX) processes[16] are the subject of intensive research and development, particularly in the areas of polymerisation, polymer modification and the reactive compatibilisation of polymer blends. The significant growth in REX processes for the production and modification of polymers reflects the high efficiency of extruders in handling high viscosity liquids, which allows chemical reactions to be conducted in the bulk phase. Extruders also make highly versatile chemical reactors, capable of solids conveying, melting, pumping, mixing and devolatilisation, and can be easily optimised to suit the requirements of a particular reaction.

2.2 Moulding Processes

2.2.1 Injection Moulding. Injection moulding (IM) involves the rapid injection of a metered portion of polymer melt into a closed mould, where it is held under pressure until solidification occurs. Process control for injection moulding is more highly developed than any other polymer process. Typically, a cycle is optimised by experienced personnel and set values of key process parameters are fed into a microprocessor-based control system. Using input data from a range of transducers, this system attempts to maintain melt temperatures, melt velocity during filling and cavity packing pressures. Thus,

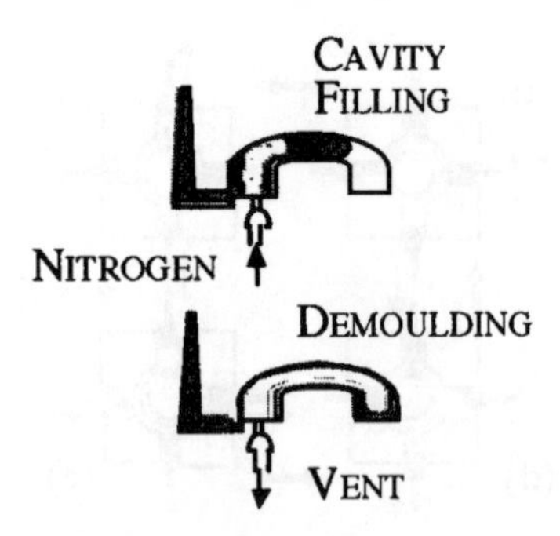

Figure 3 *Gas-assisted injection moulding (GAIM)*

IM is an efficient and highly versatile process, capable of rapid production of complex components within relatively tightly-controlled dimensional tolerances. A number of variants of the conventional IM process have been developed either to address various processing problems or to mould products of increased complexity.

A number of gas-assisted injection (GAIM) techniques (Figure 3) have been developed which involve the controlled injection of an inert gas (almost exclusively nitrogen) into the melt as it enters the mould.[17] The point of gas entry can be the melt injection nozzle, the feed system or thick sections within the cavity, depending on the technique used and the product geometry. The gas does not mix but rapidly expands within the melt, creating a hollow core and aiding the filling and packing of the mould cavity. Thus, both the pressures required to fill and pack the mould are reduced, allowing the use of lower capacity clamping systems, and pressure distribution in the cavity is more uniform, reducing distortion. Additional benefits include weight reduction, the elimination of shrinkage 'sink' marks in ribs or thick-sections, and shorter solidification times resulting from both the reduced mass of the moulding and improved heat-transfer within the cavity as the moulding maintains contact with the cavity wall during shrinkage. The first applications for gas injection techniques were mainly for thick-section mouldings, such as grab-handles, and hollow components, such as shower-heads and spray-gun bodies, which were originally produced by joining two conventional mouldings. Increasingly, however, GAIM is being used to produce complex, large-area thermoplastic mouldings such as TV cabinets and car bumpers. These products have long flow-paths, and gas-injection allows the use of thick-section channels within the mould (flow leaders) to rapidly transport melt to the extremities of the cavity, reducing both the time and pressure required to fill the cavity and significantly improving the surface finish of the moulding. The flow-leader channels are subsequently hollowed-out by the gas, creating a box-beam section.

Fusible-core moulding[18] is another IM method which produces hollow-section components. In this process (Figure 4) alloy cores are used to form the complex internal geometry of the moulding; these are die-cast from a low

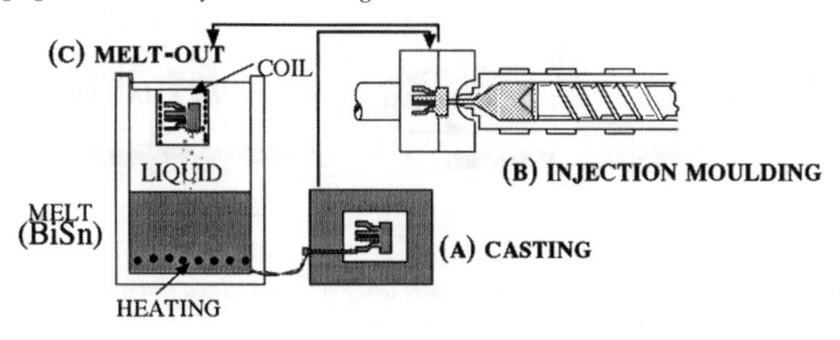

Figure 4 *The main steps of fusible-core injection moulding*

melting point tin-bismuth alloy. The cores are then placed in the cavity of an injection mould and encapsulated with polymer melt. Once the melt has solidified, the alloy-filled moulding is removed from the mould and heated to melt out the alloy core. For thermoplastics, such as PA 6,6, the alloy typically has a melting point of $\approx 140\,^{\circ}C$, considerably lower than that of the polymer (T_m 264 °C). This is achievable as during mould filling a thin layer of the incoming polymer melt immediately solidifies on the surface of the core, insulating it from the heat of the bulk of the melt. Fusible-core techniques have been used for some time to mould articles with a complex shape and undercut internal contours, such as tennis rackets and water taps. A more recent development is the moulding of complex, integrated pump components and large automobile engine components, especially inlet manifolds. The latter were traditionally made by casting of metal alloys, however, polymeric materials offer the advantages of reduced weight (up to 60% in some cases), no post-mould machining and improved engine performance (due to superior air flow resulting from the smoother inner surface, and lower incoming-air temperature).

There a number of multicomponent IM techniques, in which more than one polymer melt is injected into a mould cavity. Sandwich moulding (SM) utilises the 'fountain flow' behaviour of polymer melts as they fill a cavity to produce laminate mouldings in which a 'skin' material totally encapsulates a 'core' material. Typically, SM utilises machines equipped with two injection units, the first unit injects the skin material into the cavity, followed immediately by injection of the core material from the second unit which forces the skin material against the mould wall. SM has allowed the properties of the skin and core to be optimised for a wide range of applications, including: the use of solid polymers for the skin and foamed material for the core; expensive, high-performance polymers for the skin and cheaper commodity, or recycled material for the core; flexible material for the skin, to impart a 'soft-feel' to the surface, and a rigid core material. A recent development of SM is 'granular injected paint technology', designed for automotive components, in which a thermosetting solid paint formulation is used as the skin material eliminating the need for post-mould painting. In contrast to SM, in the vast majority of multicomponent mouldings all the different materials are visible in precisely

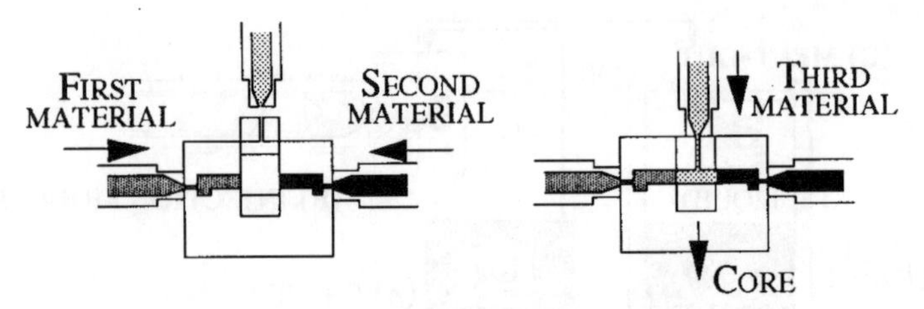

Figure 5 *Three component injection moulding using a mould with a sliding core*

defined areas of the moulding – a process sometimes referred to as 'over-moulding'. This may be achieved by transferring moulded preforms from their set of cavities to another, where they are over-moulded with a second component, or the use of moulds with moving cores, in which the injection operations are performed in series. With three components (Figure 5) a moving core separates two sections of the cavity, once these are filled with the first two components the core retracts to reveal a connecting section to be filled with the third component. The applications for over-moulding are very wide, the most common being automobile rear-light lenses and, probably the most sophisticated, whole keypad assemblies with up to six colours. In most over-moulding processes the polymers are carefully selected to give a good bond. However, the use of highly immiscible polymers, which do not bond, allows articulated mouldings to be produced. For example, articulated toy figures have been produced with ball and socket joints moulded from polyoxymethylene and polyamide, respectively. The ball and socket do not bond, therefore the limbs and head of a figure are free to move. Directable interior air vents for cars may also be produced using over-moulding of incompatible polymers, eliminating the need for post-moulding assembly.

A number of the moulding faults which can occur during IM can be addressed using controlled melt-orientation techniques. Such faults include weld-lines, points of weakness formed where two melt fronts meet within the cavity and do not bond together fully, voids in thick-sections of a component due to shrinkage, and orientation of fibrous reinforcements leading to severe anisotropy in the final moulding. By manipulating the flow of melt within the mould cavity, using techniques such as 'shear controlled orientation IM' (SCORIM)[19] and 'push-pull' IM, these faults may be significantly reduced or eliminated. SCORIM was developed at Brunel University in the UK, and in the double-live feed configuration (Figure 6a) involves placing a hydraulically-operated packing block between the injection unit of a conventional machine and the mould. This block splits the injected melt into two streams, both of which can be independently pressurised for packing by a hydraulic piston. During filling the pistons in the packing block are stationary, with one of the two pistons blocking the feed channel from the injection unit. Therefore the melt enters the cavity *via* only one of the two gates, proceeding through the cavity and out the other gate until it contacts the other piston. Once the mould is full the packing cycle can begin, in which the pistons are pumped back and

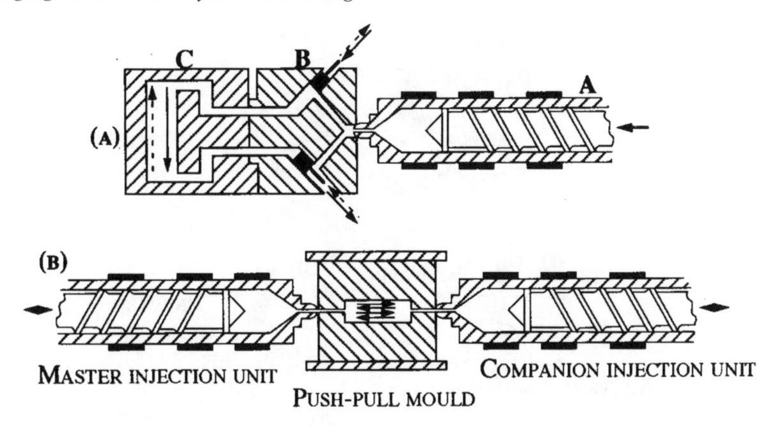

Figure 6 (a) *Shear controlled orientation injection moulding (SCORIM).* (b) *Klöckner 'puch-pull' injection moulding*

forth at the same frequency but 180° out of phase. This oscillates the melt in the cavity back and forth as it is solidifying, eliminating weld lines, inhibiting the formation of voids in thick sections remote from the gate and aligning reinforcing fibres with the direction of the oscillating flow. SCORIM may be extended to quadruple-live feed, in which moulds are provided with four gates, consisting of two pairs with each pair being fed from a packing block, allowing orientation of fibres in two directions. Thus, with fibre-reinforced materials and liquid crystal polymers SCORIM provides a high degree of 'orientation management' within the cavity, allowing the optimisation of the physical and mechanical properties of the moulding. Push-pull moulding, uses a twin injection unit machine with the units 180° out of phase to feed a twin-gated mould (Figure 6b). As the master injection unit injects melt through one gate, the overfill flows out through the other gate and into the companion injection unit, where the screw retracts to accommodate the material. The injection sequence then reverses creating oscillating flow of the melt in the cavity, and any weld lines are eliminated.

2.2.2 Blowing Moulding. Blow moulding (BM) is a generic term for a range of processes[20] for the production of hollow thermoplastic products, in which a tubular preform (or parison) is produced *via* either extrusion or IM and inflated to take up the internal contours of a mould. The basic BM technique was developed from the glass blowing method for the production of glass bottles and packaging is still the largest application area for blow mouldings, mainly as bottles, jars and drums. The IM-based techniques are restricted to relatively small components, and find limited application outside of packaging. However, the extrusion-based technology has now expanded into mouldings for industrial applications, such as spoilers, air ducts and fuel tanks for the automotive industry. Multi-layer mouldings containing recycled and/or barrier polymers may also be produced using either coextrusion or sandwich moulding to produce the preforms.

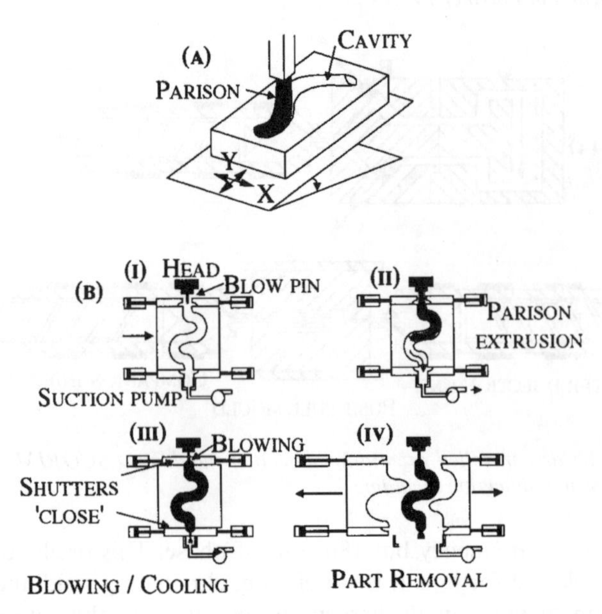

Figure 7 *3-D blow moulding technologies:* (a) *mould manipulation,* (b) *'suction blowing'*

Extrusion BM techniques can be used to produce multi-dimensionally bent components (or '3-dimensional blow mouldings') such as complex ducting and shaped piping assemblies.[21] Some 3-D systems use robot manipulators to seal, cut and transfer an extruded parison to the cavity of an open mould. Alternative methods include controlled manipulation of the mould (Figure 7a) or guiding the parison through the mould cavity on a cushion of air, which is the basis of the 'suction blowing' technique (Figure 7b). As the parison is extruded, a pneumatic suction pump draws air from the bottom of the partially opened mould. This creates an air flow through the mould cavity which guides the parison around the contours without touching the walls. Once the parison has reached the end of the cavity, the shutters close to seal the parison and form the ends of the cavity. A relatively recent development for blow moulding is sequential coextrusion, in which two extruders sequentially feed different polymers to a die to form discrete axial sections of the parison. The initial target product for this development is the automotive air duct, which traditionally has consisted of a number of rigid sections connected with flexible gaiters. The idea of sequential coextrusion is to extrude a parison with alternating sections of rigid and flexible polymers, which can then be 3-D blow moulded to form a complete ducting assembly.

2.2.3 Reactive Moulding Techniques. Reactive processing utilising low-viscosity monomers/oligomers offers a number of advantages over traditional melt processing techniques, particularly for the production of large-area and/or thick-section components. Reactive techniques operate at significantly lower pressure levels allowing the use of cheaper, less robust equipment, and there is little need for rapid control of heat transfer for either feed plastication or product solidification. The two most common reactive moulding techniques are reaction injection moulding[22] (RIM) and resin transfer moulding (RTM).[23] Both are closed-mould, net-shape processes in which a reactive mixture of monomers/oligomers is injected into a mould cavity where they polymerise and solidify. However, these processes are used to produce very different products. In RTM the reaction mixture encapsulates mats of fibrous reinforcements (usually glass) which are pre-placed in the mould cavity, and cures to form the matrix of a structural composite component. The polymers used in RTM tend to be relatively slow reacting, heat-activated systems such as unsaturated polyester resins, phenolics and polyepoxides. In RIM, highly reactive mixing-activated systems are used. These are rapidly mixed using high-pressure impingement (≈ 10 MPa) but then flow under relatively low pressures (≤ 1 MPa) to fill the mould cavity. The main polymers used in RIM are copolyurethanes and related copolymers, which have found widespread application for the moulding of shoe soles and for the production of large-area automotive mouldings. However, the use of RIM for the latter application is now in decline, due to both competition from thermoplastic IM and the reduction in 'legislative push' in the USA. In contrast, RTM appears poised to move beyond the production of panels for low-volume vehicles and break into the main stream car market.

2.2.4 Modelling of Polymer Processes. Computer modelling of polymer processes developed significantly over the final two decades of last century. In research, analyses of viscous non-isothermal flow for 3-D geometries are well established.[3–8] However, due to problems of computation time and memory requirements most commercial programs are somewhat simpler. For example, commercial CAE injection moulding packages such as Moldflow[24] and C-mold[25] use the 2.5D approach,[26] in which the pressure equation is solved in 2-D, using the lubrication approximation, whereas the temperature and velocity fields are solved in 3-D. A proposed mould design is simulated by the computer and different material grades, cavity geometries, runner and gate sizes and positions may be tried to predict optimum filling and packing conditions, whilst minimising material costs and cycle times. Ideally, the process analysis provided by these programs should be incorporated into a concurrent engineering approach to the complete process of producing a component (*i.e.* component design-mould design-mould manufacture-injection moulding).[27] Thus, in addition to filling simulations sophisticated CAE packages also provide an analysis of the cooling stage of the process[28] which will aid in the design of the cooling system *during* the development of the mould design rather than afterwards, which all too often was the case in the past. Similarly, a range of commercial CAE packages are available to provide accurate simulations of viscous non-isothermal flow in complex extrusion die geometries,[7] which allow the performance of a prospective die design to be evaluated with a range of materials and processing conditions.

3 Future Developments

3.1 New Processing Technologies

In the first century of the new millennium we can expect the development and establishment of processing technologies which expand the market penetration of polymers by increasing the complexity of components which can be produced (allowing increased integration of component assemblies), or which generate new applications.

The first step to creating 'new' technologies is often the combination of existing techniques. For example, sandwich moulding (SM) in combination with gas-assisted injection moulding (GAIM) is beginning to be used to produce mouldings with a hollow core and a two-layer skin. SCORIM is currently being studied in combination with SM for the production of mouldings with alternating polymer layers with controlled orientation, and in combination with GAIM to form polymer/starch/hydroxyapatite compounds into an oriented 3-D 'scaffold' for tissue growth which mimics the mechanical properties of bone.[29]

Rapid-prototyping (RP) techniques are already widely used to speed the component design process, both to produce prototype polymer components from CAD data and to produce prototype moulds (rapid tooling techniques). Common RP techniques include stereolithography, in which a laser sequentially cures thin layers of photocurable resin to build up a 3-D component, selective laser sintering of thin layers of powered polymer and 3-D 'printing' which uses modified ink-jet printing heads through which molten thermoplastic is sprayed. Development of polymer systems for RP with improved properties will see some of these techniques evolve into niche-production technologies providing components for low-volume applications – such as head lamps for speciality cars which have often had to compromise on their design in order to accommodate readily available lamp components.

Stereolithography and laser sintering are also currently undergoing development for 'micromoulding' applications, to produce plastic components with structural dimensions in the range 50–100 μm.[30] Micro injection-moulding techniques for thermoplastics,[31] and for compounds with metal or ceramic powders – which have their thermoplastic binder removed before sintering, are also under development. Such micro-components are predicted to find extensive applications in electronics, minimally-invasive surgery, watch gearing and integrated analytical chemistry.

A number of processing techniques are currently under development which enhance material properties by generating thin polymer layers within components, either *via* the formation of microcellular foams or microlayered structures. Microcellular foamed polymers contain high concentrations ($\geq 10^8$ bubbles cm^{-3}) of gas bubbles of the order of 1–10 μm in diameter.[32,33] Such materials can provide a reduction in density of 20 to 40% in combination with increased fatigue and impact resistance without significantly compromising modulus and yield strength[1,32,33] and will be processed by either extrusion[34] or injection moulding.[35] Microlayered polymer–polymer composites consisting of thin (1–0.1 μm) lamellae of two or more polymers can provide synergistic property enhancements, combining improvements in modulus, strength, impact resistance, barrier and electrical properties, depending on the combination of polymers used.[36] The 'lamellar' injection[37] and extrusion[38] technologies developed to produce microlayered components use multicomponent IM and coextrusion techniques in combination with a series of layer-multiplying elements, which split a multicomponent flow then recombine the two streams to double the number of layers (Figure 8).

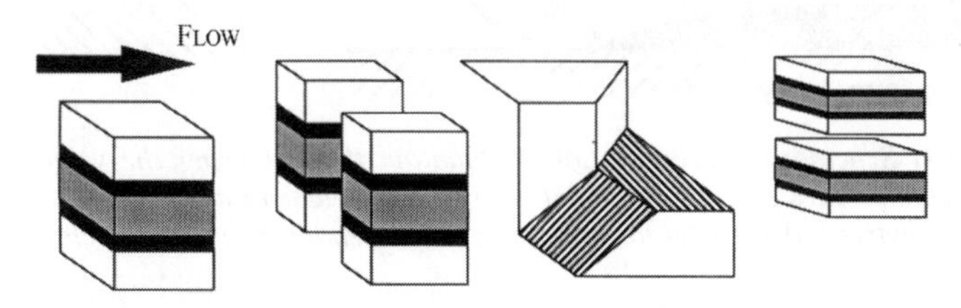

Figure 8 *Schematic showing layer mulitplication in a three component system. An ABCBA flow is sliced vertically, spread horizontally, then recombined to double the original number of layers*

3.2 Process Modelling and Control

The most significant advances in polymer processing are probably to be expected in the areas of process modelling and control. A significant trend in current modelling is the prediction of polymer *structure,* given the grade of polymer, processing conditions and the product design. The final aim is to predict and control product *properties*, *via* manipulation of processing-structure-property relationships. However, analyses of structure development during polymer require the complete mapping of the deformation-temperature history of every element within a component model, and therefore present a number of significant challenges in combining complex macroscopic flow behaviour, involving both shear and elongational deformations, with structure development on a local scale. Meijer[1] outlined these challenges as both the development of suitable constitutive equations and the determination of experimental input data for strong, complex viscoelastic flows, 3-D temperature and density profiles and polymer crystallisation under conditions of complex flow and rapid, non-isothermal cooling. Consequently, significant experimental efforts are being applied to the non-invasive measurements of complex temperature[39] and flow fields[40] and the collection of PVT data under typical melt processing conditions.[41,42]

A good example of the challenges to be overcome in process modelling is injection moulding (IM), which exhibits very distinct processing-structure relationships resulting from strong, complex flows (fountain flow – Figure 9a) under highly non-isothermal conditions. In particular, the rheology of crystallising melts plays a major role in structure development during mould filling.

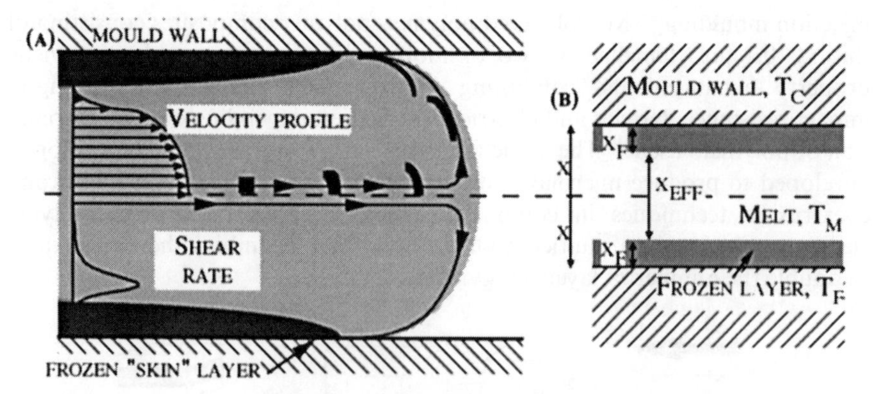

Figure 9 (a) *Schematic representation of fountain flow, showing the velocity and shear rate profiles and the deformation of a cubic element of melt as it approaches the flow front.* (b) *Model for growth of the frozen layer in a mould cavity*

Elongational flow at the flow front and shear flow between the frozen skin layers generate molecular orientation-induced nuclei which accelerate crystallisation and create anisotropy by dictating the crystal growth direction.[43] Flow-induced crystallisation in polymers is promoted by high molar mass fractions,[44] which are easily oriented in an applied flow field. Crystallisation also induces network formation in an entangled melt, which in iPP[45] generates a 'gel' point at very low levels of crystallinity ($\leq 2\%$). However, rheological studies of crystallising melts under more typical IM processing conditions are required to provide data on both structure development and to improve models of the growth of the thickness (X_f) of the frozen layers which form at the mould walls during mould filling, reducing the effective cavity thickness (X_{eff}) (Figure 9b). Currently, the rate of thickening is calculated using the concept of the melt cooling to a 'no-flow' temperature (T_f)[46] at which the viscosity of the melt has increased to such a degree that it cannot flow, but has not yet fully solidified as a result of vitrification or crystallisation. For amorphous polymers, T_f is usually between 20 to 70 °C above T_g (*i.e.* in the thermoelastic region) whereas, due to supercooling, T_f values for semi-crystalline plastics are usually between 10 to 80 °C *below* the crystalline melting point, depending on crystallisation behaviour. Although not a fundamental physical property of a material, the concept of a no-flow temperature reflects actual behaviour during mould filling and is incorporated into a number of common injection moulding flow-analysis software packages.[24] Similarly, models to simulate the development of shrinkage and residual stresses for a wide range of materials and processing conditions during injection moulding are still at a relatively early stage of development.[47,48] Thus for simulation of packing and cooling, commercial CAE packages typically utilise rudimentary expert systems. These correlate processing conditions within each mesh element of the cavity, such as the temperature and pressure distributions calculated from the filling analysis, with experimental measurements of PVT data.[24] A similar approach may be applied to calculation of warpage, in which calculated temperature and pressure distributions are correlated with experimental shrinkage data, measured both parallel and perpendicular to the flow direction. The resulting differential shrinkage values for each mesh element are then input to a finite-element stress analysis programme which calculates the deflection of each node, and hence the overall warpage of the moulding. The capability of predicting fibre orientation in an injection moulding has been available in CAE software packages for several years. However, all these programmes calculate the velocity field in the cavity using the 2.5D approximation[26] combined with a planar orientation analysis, such as that developed by Advani and Tucker[49] to determine the fibre orientation. However, this approach ignores fountain flow and breaks down in complex features such as ribs, corners and bosses, and full 3-D analyses are currently an area of basic research.[50]

As an alternative to prediction and optimisation of structure development *via* process modelling, melt manipulation techniques such as SCORIM offer the opportunity to *impose* molecular and fibre orientation distributions, as the influence of flow during mould filling is overwhelmed by the flow imposed during solidification. Thus, modelling of strain-induced crystallisation of shish-kebab structures during SCORIM[51] will point the way to the molecular design of more suitable polymers and ways in which to optimise of the imposed deformation-temperature history. Eventually, such studies may help to unlock some of the molecular potential of polymers, in terms of modulus and strength, which are currently only realised in gel-spun fibres and solid-state extrudates. In polymer–fibre composites controlled fibre orientation distributions may be imposed. For example, with plate mouldings the fibres can be oriented at a set angle in alternating layers, to optimise the flexural and impact properties of the component or, with more complex mouldings, may be applied only to selected areas where the process models struggle to provide a guide to optimisation.

A recent interesting development in process control was the linking of Moldflow™ CAE process simulation software to the control system of an injection moulding machine to provide 'intelligent' process control or IPC.[52] Sensors on the machine provided measurements of injection ram velocity, melt/mould temperatures and cavity pressures during processing. From this data, the simulation software provided an insight into what was happening in the cavity during the moulding cycle. Initially, the IPC system would propose intial conditions to set up the machine based on an optimised simulation, using an expert system which takes into account machine and polymer characteristics to eliminate defects. During production the system controlled the injection ram, and maintained both the required travel of the melt flow front in the cavity during filling, and the cavity pressure profile during packing. The emphasis was therefore shifted from controlling the *machine* to controlling the *melt* in the cavity. This pilot project was probably ahead of its time and a scaled-down version of IPC, more suited to the current needs of industry, has now been released.[53] However, the computational capacity available for IM process control will continue to increase and the concept of IPC points the way to control of complex polymer processes in the future. Thus, it may be envisaged that once suitable models of polymer structure development are developed for incorporation into CAE simulation software, systems of this type will provide the basis for optimising and controlling the development of polymer structure and properties during injection moulding.

4 Conclusions

The key opportunity to emerge in polymer processing in the current century will be the development of process control to achieve a desired *property profile*. Thus, improvements in process measurements and modelling will drive the evolution of process equipment design towards the provision of control of deformation-temperature history during processing, allowing the manipulation of polymer processing–structure–property relationships.

References

1 H.E.H. Meijer, Processing for Properties, Chap. 1 in 'Processing of Polymers', H.E.H. Meijer (Ed.), Vol. 18 of 'Materials Science & Technology', R.W. Cahn, P. Haasen and E.J. Kramer (Eds.), Wiley-VCH, Weinheim, 1999.
2 A.N. Wilkinson and A.J. Ryan, 'Polymer Processing & Structure Development', Kluwer, Dordrecht, 1999.
3 J.R.A. Pearson and S.M. Richardson (Eds.), 'Computational Analysis of Polymer Processing', Applied Science, London, 1983.
4 C.L. Tucker III (Ed.), 'Fundamentals of Computer Modelling for Polymer Processing', Hanser, Munich, 1989.
5 A.I. Isayev (Ed.), 'Progress in Polymer Processing', Hanser, Munich, 1990.
6 J.F. Agassant, P. Avenas, J.-Ph. Sergent, and P.J. Carreau, 'Polymer Processing: Principles and Modelling', Hanser, Munich, 1991.
7 K.T. O'Brien (Ed.), 'Applications of Computer Modeling for Extrusion and Other Continuous Polymer Processes', Hanser, Munich, 1992.
8 T. Kanai, and G.A. Campbell (Eds.), 'Film Processing', Hanser, Munich, 1999.
9 F. Henson (Ed.), 'Plastics Extrusion Technology', Hanser, Munich, 2nd. Ed., 1997.
10 P.J. Mills, Chap. 9 in 'Structure and Properties of Oriented Polymers', Chapman & Hall, London, 2nd. Edn., 1997.
11 T. Kanai, Chap. 3.1 in 'Film Processing', T. Kanai, and G.A. Campbell (Eds.), Hanser, Munich, 1999.
12 R.W. Lusignea, *Polym. Eng. Sci.*, 1999, **39**, 2326.
13 P.S. Allan, M.J. Bevis, J.R. Gibson, C.J. May and I.E. Pinwill, *J. Mat. Process. Technol.*, 1996, **56**, 272.
14 A.K. Taraiya and I.M. Ward, *J. App. Polym. Sci.*, 1996, **59**, 627.
15 G. Capaccio, A.G. Gibson and I.M. Ward, in 'Ultra-High Modulus Polymers', A. Ciferri and I.M. Ward (Eds.), Applied Science, London, 1979.
16 M. Xanthos (Ed.), 'Reactive Extrusion', Hanser, Munich, 1992.
17 H. Eckardt, Chap. 1 in 'Innovations in Polymer Processing: Molding', J.F. Stevenson (Ed.), Hanser, Munich, 1996.
18 C. Hauck and A. Schneiders, Chap. 4 in 'Innovations in Polymer Processing: Molding', J.F. Stevenson (Ed.), Hanser, Munich, 1996.

19 P.S. Allan and M.J. Bevis, SCORTEC pat. GB 2170-140-B, 1992.
20 D.V. Rosato and D.V. Rosato (Eds.) 'Blow Molding Handbook', Hanser, Munich, 1989.
21 S. Sugiura, Chap. 6 in 'Innovations in Polymer Processing: Molding', J.F. Stevenson (Ed.), Hanser, Munich, 1996.
22 C.W. Macosko, 'RIM Fundamentals', Hanser, Munich, 1989.
23 C.D. Rudd, A.C. Long and C.G.E. Mangin, 'Liquid Moulding Technologies', Woodhead, Cambridge, 1997.
24 P. Kennedy, 'Flow Analysis of Injection Molds', Hanser, Munich, 1995.
25 C.A. Hieber, L.S. Socha, S.F. Shen, K.K. Wang and A.I. Isayev, *Polym. Eng. Sci.*, 1983, **23**, 20.
26 C.A. Hieber and S.F. Shen, *J. Non-Newtonian Fluid Mech.*, 1980, **7**, 1.
27 C. Austin, Chap. 12 in 'Innovations in Polymer Processing: Molding', J.F. Stevenson (Ed.), Hanser, Munich, 1996.
28 G. Menges and P. Mohren, 'Injection Molds', Hanser, Munich, 1992.
29 R.L. Reis, A.M. Cunha, P.S. Allan and M.J. Bevis, *Advances in Polymer Technology*, 1997, **16**, 263.
30 R. Wechsung, N. Ünal, J-C.Ely and H. Wicht, 'Proc. Micro System Technol. 98', H. Reichl and E. Obermeier (Eds.), VDE-Verlag, Berlin, 1998.
31 Anon, British Plastics & Rubber, Oct. 1999, 19.
32 J.E. Martini, F.A. Waldman and N.P. Suh, SPE Technical Papers, 1982, **28**, 674.
33 V. Kumar and J.E. Weller, Chap. 7 in 'Polymeric Foams: Science & Technology', ACS Symp. Series 669, K.C. Khemani (Ed.), ACS, Washington DC, 1997.
34 C.B. Park and N.P. Suh, in 'Cellular Polymers', V. Kumar and S.G. Advani (Eds.) Vol. 38, ASME, New York, 1992.
35 N.P. Suh, Chap. 3 in 'Innovations in Polymer Processing: Molding', J.F. Stevenson (Ed.), Hanser, Munich, 1996.
36 C. Mueller, J. Kerns, T. Ebeling, S. Nazarenko, A. Hiltner and E. Baer, 'Polymer Process Engineering '97', P.D. Coates (Ed.), 137, University Press, 1997.
37 M.A. Barger and W.J. Schrenk, Chap. 8 in 'Innovations in Polymer Processing: Molding', J.F. Stevenson (Ed.), Hanser, Munich, 1996.
38 C. Mueller, S. Nazarenko,T. Ebeling, A. Hiltner and E. Baer, *Polym. Eng. Sci.*, 1997, **37**, 355.
39 E.C. Brown,T.L.D. Collins, A.J. Dawson, P. Olley and P.D. Coates, *J. Reinf. Plast. Comp.*, 1999, **18**, 331.
40 A.N. Hrymak and P.E. Wood, *Proc. XV Ann. Mtg. of the Polymer. Proc. Society*, 'sHertogenbosch, The Netherlands, 1999, 378.
41 G. Menges and P. Heinal, *Polym. Eng. Sci.*, 1997, **11**, 758.
42 C.S. Brown and C. Hobbs, *Proc. XV Ann. Mtg. of the Polym. Proc. Society*, 'sHertogenbosch, The Netherlands, 1999, 71.
43 P. Jerschow and H. Janeschitz-Kriegl, *Rheologica Acta*, 1996, **35**, 127.
44 P. Jerschow and H. Janeschitz-Kriegl, *Intl. Polym. Proc.*, 1997, **12**, 72.
45 N.V. Pogodina and H.H. Winter, *Macromolecules*, 1998, **31**, 8164.
46 I.T. Barrie, in 'Polymer Rheology', R.S. Lenk (Ed.), Applied Science, London, 1978.
47 W.F. Zoetelief, F.A. Douven and A.J. Ingen Housz, *Polym. Eng. Sci.*, 1996, **36**, 1886.
48 X. Guo and A.I. Isayev, Intl. Polym. Proc., 1999, **14**, 377–386, 387–398.
49 S.G. Advani, C.L. Tucker, III, *J. Rheol.*, **1987**, 31, 751.

50 B.E. VerWeyst, C.L. Tucker III, P.H. Foss and J.F. O'Gara, *Intl. Polym. Proc.*, 1999, **14**, 409.
51 H. Zuidema, G.W.M. Peters and H.E.H. Meijer, *Proc. XV Ann. Mtg. of the Polym. Proc. Society*, 's Hertogenbosch, The Netherlands, 1999, 140.
52 C. Maier, *British Plastics & Rubber*, Nov. 1995, 4.
53 P.D. Coates, *British Plastics & Rubber*, July 1999, 16.

16

Shaping the Future in Polymer Processing

Malcolm Mackley

DEPARTMENT OF CHEMICAL ENGINEERING,
UNIVERSITY OF CAMBRIDGE, PEMBROKE STREET,
CAMBRIDGE CB2 3RA, UK

1 Introduction

This paper reviews the author's thoughts on how certain aspects of polymer processing may evolve over the next decade. The review is anecdotal and is limited to processing areas where the author has some experience. The paper is intended to provoke discussion rather than be considered a definitive declaration as to what will occur in the future.

2 Background

Polymer Processing is a mature, but still evolving business and science. Manufacturers have been processing polymers such as polyethylene for the last sixty years and scientists equally have spent that length of time trying to establish how these materials can best be processed. Advances in both manufacture and science have in general been incremental although there are a few notable exceptions. The fact that polymers are eminently processible has been a significant factor in the steady rise in usage of bulk polymers and the ability to form polymers into such diverse structural shapes such as tubes, mouldings and bottles has helped to drive the annual tonnage of bulk polymers ever upwards. Ease of processing has not been the only reason for this major growth over the last sixty years. End user properties are equally important and it is the delicate balance between ease of processing and product properties that has presented polymer scientists with one of the greatest challenges of the past, in terms of the manufacture and usage of bulk polymers.

Chemical Engineers are taught that scale matters and this has been true in relation to polymer manufacture and processing. There are now only a few very important world tonnage commodity polymers. Polyethylene, polypropylene, PVC, Nylon and PET dominate the market place. These polymers are made at levels of 50×10^6 tonnes/annum and as such the way in which they are manufactured and processed is highly developed and the processing behaviour is and continues to be extensively studied in commercial and academic laboratories. In this area of polymer processing even small advances in polymer processing can result in major economic improvement.

In addition to large-scale manufacture, polymer processing is playing an increasing role in terms of smaller volume 'high value' polymer product. Here there is perhaps even greater potential for step change advances in polymer processing and/or opportunities for polymer processing to be able to produce a shape or form that would otherwise be impossible or difficult to achieve.

Whether the product is large-scale, at say 50×10^6 tonnes per year, or 'high value' at 1 kg/day, the underlining process science is often the same. This paper reflects these factors and the following section examines possible future developments in the science of polymer processing, followed by two sections addressing bulk processing and 'high value', processing.

3 The Science of Polymer Processing

Polymers are generally, but not always processed as a bulk viscous liquid. The molten polymer will have a viscosity ranging from $1-10^5$ Pas and this factor alone dictates many of the aspects relating to its processibility. The high viscosity means that high Reynolds numbers are unlikely to be reached and consequently inertial effects do not in general play a complicating role. It is the rheology of the polymer melt that makes polymers both potentially difficult to process and fascinating to the applied mathematician and physicist to model. Not all of the polymer rheology necessarily works against the polymer processor. The fact that most polymers shear thin during simple shear processing is a bonus in that extrusion pressures are reduced below that of simple Newtonian behaviour. It is the viscoelastic nature of many polymers that often adds to the richness and difficulties of processing polymers.[1]

In the film blowing process where a continuous 'stable parison' is blown from an annular die, it is crucial that the molten polymer exhibits certain elastic extensional properties and it is here that the viscoelastic nature of the polymer is beneficial.[2,3] If, however, the manufacturer is concerned with profile and surface finish of an extrudate, viscoelastic effects of the polymer may well present difficulties. Both die swell[3] and most polymer extrusion instabilities are linked to viscoelastic effects and as such different levels of viscoelasticity give rise to different extrusion characteristics.

In the last twenty years, major advances in the characterisation of polymer melt viscoelasticity has taken place[4] and in addition applied mathematicians have produced numerical codes that enable viscoelastic fluids to be modelled for processing conditions.[5] Within the last few years it has become possible to reasonably accurately predict the way in which a viscoelastic polymer will flow into, within and out of an extrusion die.[6,7] The accurate prediction of die swell is nearly possible and advances are being made to predict the onset of extrusion instabilities.

An important aspect relating to the ability to predict extrusion behaviour is the recognition that all commercial polymers must be characterised by a spectrum of relaxation times.[8] A single relaxation time constitutive equation is inadequate to capture all necessary features. The detailed form of the non-linear component within the constitutive equation is also important and decisive in terms of accurate prediction and again much work has been done refining these parameters.[9]

In comparison to the rheological constitutive equation that has been used to model viscoelastic flows the appropriate wall boundary conditions for the flow is a much-neglected subject. In general a no slip boundary condition has been applied; however, there is a gradual recognition that this may not always be the case and important differences for example is the processibility of certain polyethylenes is now being associated with partial slip factors rather than bulk constitutive responses.[10–12] Recently the author and Dr Liang have discovered a method of introducing gas at the wall and thereby essentially providing a full slip boundary condition.

Establishing the link between molecular architecture and processibility remains a major objective for polymer physicists and technologists. Significant advances have been made in linking molecular chain behaviour to rheology and then to processibility.[13]

It is clear that developments described in this section will continue for many years to come. These developments can be summarised in terms of further improvements in the following areas.

1. *Experimental materials characterisation.* Linear and non linear viscoelastic response, simple shear, extensional flow and mixed shear behaviour.
2. *Rheological description of material.* Constitutive equations will become progressively more refined.
3. *Numerical prediction of processing flow.* Numerical techniques will be refined. Three dimensional, non-isothermal, time dependent and compressible flows will be solved for complex flow geometries.
4. *Boundary conditions.* Increasing attention will be paid to the importance of boundary conditions.
5. *Linking processing to molecular architecture.* Molecular mass distribution and chain architecture will be linked to a constitutive response. New chain architecture may be discovered which give processing advantage.

The integration of all these five factors will be important in relation to a deeper understanding of processing behaviour. All that has been said so far applies to a single-phase polymer, although it must be remembered that commercial polymers can have broad molecular mass distributions and there is scope for phase separation effects to be present even within a single species material. Polymer blends suspensions and foams all add to the additional complexity of the problem.

4 Bulk Processing Behaviour

At present the starting point for most bulk processing operations is a solid polymer particle and this is fed either into an extruder or injection-moulding machine to yield a particular shape. This pattern of manufacture is likely to continue, however, the recent development of Metallocene polymers,[14] offers interesting new possibilities. Metallocene polymers can be produced in solution and this opens up the possibility of solution processing direct from the reactor. In general, solution processing is expensive, in that solvent recovery is necessary; but if the reactor feedstock is already a solution there may be advantages in processing at the lower solution viscosities, particularly when working with very high molecular weight material.

The screw extruder is the workhorse of most bulk polymer processing and its position as the premier method for melting, mixing and pressure generation is likely to continue. Screw and die design will continue to evolve following generally engineering principles and input from the five scientific areas summarised in the previous section.

Novel extrusion technologies will continue to be developed. For example gas assisted injection moulding has become, in the last ten years, an important way of producing clever hollow section structures.[15] Novel ways of aligning composite fibres in flowing melts are being pioneered and there are continued developments in relation to improved speed and quality of production. The processing of composite structures is perhaps an area where there is greatest scope for the advances in bulk processing; automated continuous lay-up technology for anisotropic composite structures is in its infancy and future developments in this area are urgently needed.

The move to couple reaction and processing will continue. Reactive injection moulding is a good way to get round the 'high viscosity' processing problem in that the low viscosity material is fed into the mould and the reaction completed within the material. This of course can lead to other complicating issues.

Multiphase processing is an important area in every shape or form that is manufactured from polymers. Micron thickness extruded layer structures in polymer films and sub-micron droplet suspensions in mouldings all play a crucial role in product enhancement. These microstructures generally originate or are influenced by their process history.

5 High Value Processing

As our understanding of polymer processibility increases there is greater scope to produce increasingly complex and subtle structures. The trend has been towards a progressively finer scale of structure which in some cases, has moved well below the sub micron barrier.

Inkjet printing is an example where polymers are involved in high value processing.[16,17] There are now several different ways of producing micron-sized drops for printing. Many involve using a piezo-electric crystal to generate a forced frequency perturbation that in turn controls the break-up of a liquid stream as it emerges from a fine nozzle. Small volume fractions of polymer are added to the ink to improve the uniformity of the produced drops and the polymer is acting in this case as a vital processing aid. In other cases the polymer itself forms the final product and here thin film manufacture is a good example. Recently, light emitting polymers (LEPs)[18] were discovered and a necessary requirement for their successful performance is to manufacture LEPs with a film thickness of the order of 10^3 Å. At present this cannot be done by melt processing but if the viscosity is reduced by dissolving the LEP in a volatile solvent the essentially Newtonian polymer solution can be 'spin coated' to produce the necessary film thickness.[19]

Solution processing also plays a vital role in the manufacture of high value, lyotropic high modulus fibres[20] and also high modulus polyethylene fibres.[21] In all these cases the additional cost and complication of solvent recovery is offset by the high value of the produced product.

6 Conclusion

In order to fully exploit the maximum product potential of polymers it may be necessary to sacrifice some of their useful melt processibility aspects. At present most polymers are manufactured as a compromise between processing and product qualities and this will continue. If however, maximum product performance is required novel processing methods that do not use melt processing will need to be found. Nature makes extensive use of polymers for a very wide range of functions, but does not appear to use melt processing as a route to fabrication. We still have a lot to learn.

References

1 J. M. Dealy, K. F. Wissbrun, *Melt rheology and its role in plastic processing: theory and applications*, Chapman & Hall, London, 1995.
2 S. Eggen, and A. Sommerfeldt, *Polym. Eng. Sci.*, 1996 **36(3)**, 336–346.
3 R. J. Koopmans, *Polym. Eng. Sci.*, 1992 **32(3)**, 1750–1754.
4 M. Baumgaertal and H. H. Winter *Rheol. Acta*, 1989 **28**, 511–519.
5 M. J. Crochet and K. Walters, *Computional Rheology: A New Science*, Endeavour. New Series, 1993, **17(2)**, 64–77.
6 J. P. W. Baaijens, G. W. M. Peters, F. P. T. Baaijens and H. E. H. Meyer, *J. Rheol.*, 1995, **39(6)**, 1243–1277.
7 A. D. Goublomme, B. Draily and M. J. Crochet, *J. Non-Newt. Fluid Mech.*, 1992, **44**, 171–195.
8 R. Ahmed, R. Liang and M.R. Mackley, *J. Non-Newt. Fluid Mech.*, 1995, **59**, 2–3, 129–153.
9 M. H. Wagner, H. Bastian, P. Ehrecke, M. Kraft, P. Hachmann and J. Meissner, *J. Non-Newt. Fluid Mech.*, 1998, **79**, 2983–296.
10 S. G. Hatzikiriakos, J. M. Dealy, *J. Rheol.*, 1992, **36 (4)**, 703–741.
11 J. M. Piau, N. Kissi, *J. Non-Newt. Fluid Mech.*, 1994, **54**, 121, 142.
12 T. Person, M. M. Denn, *J Rheol.*, 1997, **41(2)**, 249–265.
13 T. C. B. McLeish and R. G. Larson, *J. Rheol.*, 1998, **42(1)**, 81–110.

14 G. J. P. Britovsek, V. C. Gibson and D. F. Wass, *Angew Chem. Int. Ed. Engl.*, 1999, **38**, 428–447.
15 L. S. Tung, *Advances in Polymer Technology*, 1995, **14(1)**, 1–13.
16 S. A. Curry and H. Portig, *J. Res. Develop.*, 1997, **21(1)**, 10.
17 R. G. Sweet, *Rev. Sci. Inst.*, 1965, **36**, 131.
18 W. C. Holton, *Solid State Technology*, 1997, 163–169.
19 W. W. Flack, D. S. Soong, A. T. Bell, and D. W. Hess, *Journal of Applied Physics*, 1984, **56(4)**, 1199–1206.
20 S. J. Picken, *Orientational order in aramid solutions*, PhD thesis, Scheikunde van de Rijksuniverstiteit Utrecht (1990).
21 R. Kirschbaum, J. L. J. van Dingenen, 'Advances in gel spinning technology and Dyneema fibre applications', *Integration of polymer science and technology*, 3rd Rolduc Conf. Elsevier Amsterdam (1988).

Theory and Modelling

17
Emerging Themes in Polymer Theory

Tom McLeish,[1] Peter Olmsted[1] and Ian Hamley[2]

[1]DEPARTMENT OF PHYSICS AND ASTRONOMY AND
[2]SCHOOL OF CHEMISTRY, UNIVERSITY OF LEEDS,
LEEDS LS2 9JT, UK

1 Introduction

Theoretical Polymer Physics has roots deep in the rich conceptual revolution in modern physics. The three central pillars set in place during the closing years of the 19th and early years of the 20th Century, geometrical gravitation, quantum mechanics and statistical mechanics, have all supplied our sub-field of Soft Matter Physics with powerful concepts and tools. Differential geometry supplies mathematical tools to describe the curvature and continuity of polymeric chains and membranes. A raft of techniques comes from quantum mechanics of rarified and condensed matter: renormalisation of fields, diagrammatic perturbation expansions, path integrals, Landau-Ginzburg theory for example. Finally the framework set up by Boltzmann and Gibbs provides the world in which these tools play out the game of building models of emergent phenomena from our notions of molecular and mesoscopic structures. Statistical field theory, stochastic dynamics, coarse-graining and increasingly powerful numerical simulation have brought us to our present provisional understanding of polymer equilibria, phase transitions and dynamics. Our choice of topic in the following in personal and does not claim to be comprehensive, but we hope that it is representative.

2 Current State of the Art

2.1 Single Chain Conformations

In the case of flexible polymer chains, the dominant problem for theory has been the understanding of the statistical properties of self-avoiding random walks in 2, 3 and 4 dimensions.[1,2] The case of $D>3$ arises of course because the self-avoiding interaction only becomes perturbative in 4 dimensions and greater. This has spawned a programme of renormalisation-group (RG) calculations and simulations, treating simple quantities such as the Flory exponent ν as well as more sophisticated statistics. Scaling approaches continue to be of value, such as in the swollen statistics of random branched molecules[3] or more regular star polymers[4] in solution.

Surprisingly the dynamics of non-interacting dilute chains in shear and extensional flows has been a tougher puzzle than in entangled, concentrated systems. Ideas of folded configurations in shear flow[5] are finding confirmation in simulations,[6] but while experiments on large biopolymers have shown quantitative agreement with theory, severe problems remain in the case of flexible chains, which seem to show anomalously small extensions.

2.2 Equilibrium Phase Behaviour in Blends and Block Co-polymers

The theory for the phase behaviour of block copolymers[7] is one of the most successful, in terms of quantitative predictive capability, of any for soft materials. In 1980, Leibler presented his mean field random phase approximation theory for weakly segregated block copolymers[8] and in 1985 Semenov[9] presented an analytical theory for strongly segregated block copolymers, building on earlier work by Helfand.[10] The current state of the art, however, is the numerical self-consistent (mean) field theory (SCFT) pioneered by Matsen.[11] This enables prediction of the thermodynamic state ranging from weak to strong segregations. In addition, detailed microscopic information such as composition profiles, interfacial widths and domain spacings can be computed. The theory, however, does not apply close to the order-disorder transition, where composition fluctuations have been shown to modify the mean field phase behaviour. Fredrickson's development of Brazovskii's theory is the state of the art here.[12] This has been developed to allow for chain stretching, an effect known to be important for block copolymers close to the ODT.[13,14] Systems that are more complex than the two component di- and triblocks that have been the focus to date, are now attracting attention. In particular, ABC triblocks can form a rich panoply of structures due to the competition between the microphase separation of the three components. Only a small number of the possible structures have recently been identified experimentally. The state of the art in theory for such structures is the SCFT theory of Matsen,[15,16] although simpler (brush-like) scaling theory works quite well in the strong-segregation limit.[17,18]

Thin films of block copolymers are likely to find many applications as nanostructured materials, due to the ability to tailor nanoscale 'dot' and 'stripe' patterns. Theory for microphase separation in thin films, especially the effect of confinement on structure orientation is now quite advanced.[18–22] Models for the effect of confinement on thermodynamics have also been developed,[23] although this aspect has attracted less attention.

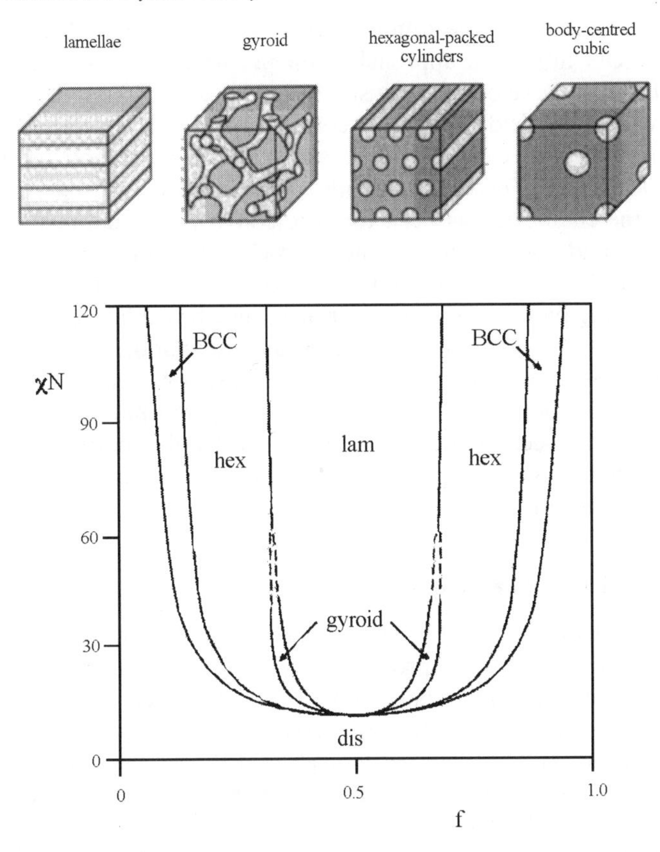

Figure 1 *Phase diagram for diblock co-polymers computed in SCFT*[11]

2.3 Kinetics of Phase Separation

We turn now to dynamic aspects of micromorphologies. The vital role of controlled morphologies in phase-separated structures, and the wonderful zoo of possible structures that arise has motivated a recent interest in the kinetics of phase separation. This is because the final morphology can be determined not only as a result of the equilibrium phases discussed above but also by the kinetic history of the fluid. This is especially true if a bulk flow is operating. Classical Flory-Huggins theory for monodisperse blends is still frequently used as the framework for discussions, but can be misleading – for example in the case of the high polydispersity generated by approaching gelation when the nucleation/growth regions can become very large.[24] More generally, phase separation controlled by polymerisation rather than by a temperature quench contains much greater structure.[25]

The kinetics of phase separation between polymers of widely differing mobilities generated a long debate over 'fast-mode' and 'slow-mode' views.[26,27] It is now recognised that, depending on degree of overlap and solvent quality, both modes combine in the mutual diffusion behaviour with a division of amplitude.[28] One molecular approach to the problem has introduced the useful concept of the 'tube velocity' against which individual components diffuse.[29] This approach has yielded initial theories for the role of viscoelasticity in phase-separation.[30] In early stages delays in phase separation result; in late stages the possibility of elastically-stabilised phase-inversion.[31]

In flow the challenge has been to write convincing equations that couple concentration and composition gradients to elastic stresses and the bulk flow field. When done within a two-fluid model for polymer solutions[32] transitions in light-scattering patterns seen in experiment may be explained. Extensions to polymer blends are potential candidates as explanations of shear-induced shifts of the spinodal and biphasic 'islands' seen experimentally.[33]

Kinetic transitions between well-defined micromorphologies are usually dominated not by second order but by first order thermodynamics. Recent ideas have shown how propagating fronts of the new nucleating phase may be responsible for the limiting rate.[34]

2.4 Entangled Linear Polymers

Here the paradigm of the tube model introduced by de Gennes, Doi and Edwards in the 1970s[35,36] is proving increasingly fruitful. The notion that strongly-overlapped polymer chains 'reptate' under mutual constraints equivalent to tubelike traps along their own contours was originally successful in describing the molecular weight dependence of melt viscosity qualitatively, and the step-strain non-linear dependence of stress quantitatively. However, a legacy of remaining discrepancies has gradually been investigated since. Key experiments are reviewed in the chapter on rheology in this volume. Theoretically, two long-standing apparent anomalies have been resolved and are clearly related. The first is the well-known difference between the observations of the molecular weight (*M*) dependence of melt viscosity, $\eta \sim M^{3.4}$ and the asymptotic tube model result for long chains $\eta \sim M^3$. For some time it has been suspected that the resolution lay in the need to consider fluctuations in the entangled path length of the chains.[36] But application of an analytic result for fluctuations derived in the case of star polymers (see below) demonstrated that the '3.4-law' and the form of the whole rheological spectrum emerged from a such a fuller treatment of the tube model, without the addition of extra parameters.[37]

The second emerged as an apparent difference in the dependence of the self-diffusion constant D on M in melts and concentrated solutions. A series of studies covering limited degrees of entanglement in the 1970s and 1980s suggested a dependence of M^{-2} in melts and $M^{-2.5}$ in concentrated solutions. Very recently a comprehensive review[38] of the data with the addition of a new set covering an unprecedentedly wide range of M has shown that in *both* cases $D \sim M^{-2.4}$ for $M/M_e < 200$. The no-fluctuation tube model predicts M^{-2}, but current indications that fluctuations renormalise this exponent to ~ 2.5 need to be made precise.[37]

In strong flows, the next main theoretical challenge seems to be that of a fully self-consistent tube theory in which constraints on each chain are released by the retraction of neighbouring chains. This effect in now termed 'Convective Constraint Release' (CCR); the chief goal of such a theory being to remove the unphysical stress-maximum predicted by standard theory for polymer melts as a function of shear rate. An approximate theory of the effect considered that CCR modified the configurational relaxation time of entire chains.[39,40] It also included a treatment of chain stretching and end-fluctuation, capturing some subtle non-monotonicities in reversing flows. However, the central problem of the shear-stress maximum was not removed. A series of simulations has suffered from the same drawback,[41] but does not lose information on chain conformations. Another analytic approach treats the tube itself as a Rouse-like polymer with local reorganisations driven by CCR. The main difference between Rouse-tube and classical Rouse motion is that chain-retraction within the tube maintains a constant path length. The reptation dynamics may also be represented (non-diagonally) in the Rouse-tube coordinates, so representing all dynamical processes together in the same formalism. This approach has the advantage of accounting for the structure factor of the chains, while finally showing that CCR can lead to a monotonic stress-shear rate curve.[43] The important physics seems to be that CCR operates *locally* on the chain, rather than globally. Initial comparisons with limited SANS data on flowing melts support the conclusion that the spatially *local* nature of constraint release is an essential contribution to the melt rheology.

2.5 Complex Architectures

Very rapid progress has recently been made on the theory of entangled polymers of complex topology in both linear and non-linear response. For a recent full review see ref. 44 and other chapters in that volume devoted to branched polymers. The most successful theoretical approaches have been based on the tube model (see section on linear entangled chains above), recognising that the role of long branches is to suppress reptation in favour of entanglement loss *via* retraction modes of entangled arms. Within the 2-parameter description, a quantitative theory of linear stress relaxation has now been developed to a series of increasingly complex systems, verified in turn by experiment. Monodisperse star polymers,[45] blends of stars and linears[46] and of stars and stars[47] have been treated successively. Very recently a full theory for a melt of monodisperse polymers each with two branch points has covered both linear and non-linear response[48] (see Figure 2). It seems that even subtle shifts of the branch points within the entanglement network can cause large effects in the local relaxation rate that are important for processing.

A full constitutive equation derived from this programme has recently been a focus of interest.[49] A rather simple but generic insight emerges: for long chain branching (LCB) melts it is important not to assume that the stress tensor is a primary dynamical variable, but at the very least be decomposed into a scalar (representing average segmental stretch) and a tensor (repre-

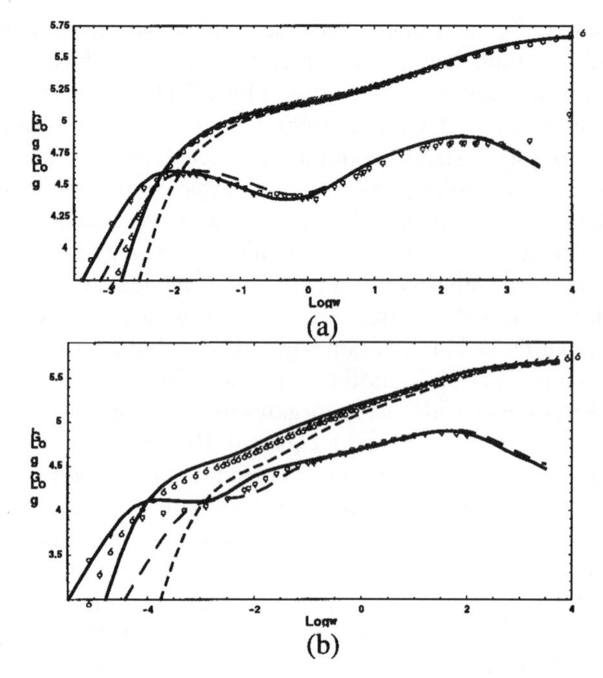

Figure 2 *Experimental and theoretical complex moduli for two H-polymers of different molecular weights. Fitting parameters of G_0 and τ_e were consistent with literature values, the number of entanglements along arm and crossbar, s_a and s_b together with their polydispersities ε_a and ε_b determined by GPC and SALS. Theoretical curves accounting for polydispersity are dashed; those without are solid. Feature arising from arm-retraction (mid-range) and cross-bar reptation (low frequency peak) are clear*[48]

senting average segment orientation). These quantities obey coupled differential equations with different relaxation times. A natural consequence is strain hardening in all extensional flows while retaining strong shear thinning. Polydisperse modes with the pom-pom structure can be employed to fit extremely large data sets for LDPE.[50,51]

2.6 Stiff Polymers

With the emergence of high quality data on model semiflexible polymer solutions came the realization that the successful tube model for flexible polymer chains should be suitably generalized to describe entangled solutions of stiff polymers. Early calculations of Odijk and Semenov were extended by MacKintosh and co-workers to describe the viscoelasticity and an elastic plateau modulus due to a tangential tension that unravels bend undulations,[52] while Maggs argued that on longer timescales the plateau should be determined by transverse relaxation (tube deformation).[53]

More recently Morse produced a complete microscopic tube theory for stiff polymers that successfully interpolates between the rigid-rod and flexible chain limits.[54] This theory explains many features of semiflexible polymer rheology, including the two mechanisms for plateau moduli described above (which depend on a comparison of timescales), with the tube diameter being the sole fitting parameter as in the Doi-Edwards theory. More recently, Morse successfully *computed* a tube diameter from two different approaches (self-consistent binary collision and continuum effective medium) that give similar results, *e.g.* modulus $G \sim \rho^{7/5}$ and $G \sim \rho^{4/3}$ respectively). An elastic network approximation to collect the important features of each approach compares surprisingly well with recent experimental results.[55]

2.7 Liquid Crystalline Polymers in Shear

Taking the early models of Doi as a starting point, one of the more remarkable recent success stories of theoretical molecular rheology must be found in the non-linear shear response of liquid crystalline polymers (LCPs). In these materials, the usual nematic-isotropic transition is seen, but the modifications under shear flow are particularly rich, as the orientational degrees of freedom are now *slow* variables, coupled to the viscoelasticity of the entanglement network. Even without the very careful treatment of Morse reviewed above, two cases clearly arise in the nematic phase. In the first the director field 'wags' about the flow direction periodically; in the second it 'tumbles' (see Figure 3). Approaches to a molecular understanding of the resulting oscillations of stress, and the remarkable change of sign of the first normal stress difference were made by Marrucci in two dimensions and Larson in three. The physics is reviewed in ref. 56.

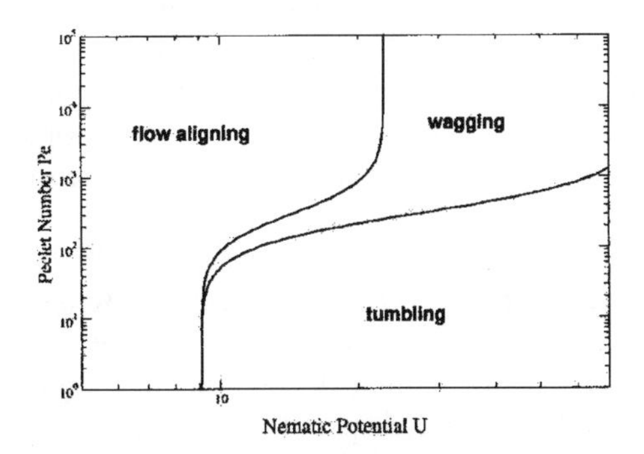

Figure 3 *Schematic 'Phase diagram' in flow-rate and nematic potential for the three molecular flow behaviours of sheared liquid crystals*

2.8 Structure Development During Quiescent Crystallisation

Polymer crystallisation is an old and tumultuous field that is experiencing a revival of interest,[57] particularly with the advent of time-resolved X-ray scattering studies of structure development in the early stages of crystallisation.[58] Older discussions revolved around establishing chain folding as the basic chain configuration in the crystalline state. More recently, several studies of the earliest stages of nucleation and growth have demonstrated that there is a clear induction time between the emergence of structure on the 100A scale and the identification of crystalline structure.[58–64] Polymers include polyethylene,[65,66] PET,[60–62] i-PP,[64,58,67] and polystyrene.[59] The nature of this structure has not been definitively established.

Possible explanations include: (i) nucleating crystallites of some form (either small and uncorrelated, or growing lamellar stacks) too small and few in number to scatter effectively at wide angles, due in part to the intrinsic lower sensitivity of wide angle detectors; (ii) some form of phase separation. Evidence in favour of the latter include a SAXS structure factor which grows exponentially in time with a characteristic wavevector q*, characteristic of growth of a conserved order parameter following a quench as predicted by Cahn-Hilliard theory. In standard spinodal decomposition during phase separation q* decreases as the system evolves, indicating a coarsening of the characteristic domain size $L \approx 1/q^*$. Such a decrease has been seen, along with an induction time, in PEKK,[68] PET[60] and *i*-PP.[58] Complementary explanations for this behaviour posit that the polymer density is coupled to conformation and orientation, as with liquid crystalline polymers.[69,70] The appearance of polymer crystals is a direct result of chain stiffening with decreasing temperature, which can equivalently be cast in terms of enhanced occupancy of preferred conformational (*e.g.* rotaional isomeric) states. Simple phenomenological models that couple either orientational order to density, as occurs in liquid crystalline systems, or conformational order to the density *via* the ability to pack, lead to the concept of a buried spinodal to liquid–liquid phase separation. The two liquids possess different densities and degrees of conformational (and possibly orientational) order.[70] Buried liquid–liquid spinodals have also been postulated to explain slow dynamics and density fluctuations in other molecular liquids in which orientational order influences packing, notably water.[71] This implies the existence, in the induction period, of orientational and/or conformational order. Such order has been observed in PET[60] and PS[59] by light scattering, and inferred in PE by FT-IR measurements.[65] Recent experiments on *i*-PP,[67] PE[66] and PEEK have shown a similar induction time, but in this case the characteristic wavevector *grows* with time, which obviously differs from standard spinodal decomposition. Moreover, Akpalu have suggested that polydispersity can influence whether or not a significant spinodal signal is seen.[66]

An important consideration in the existence of a spinodal is the prescribed experimental conditions. In a monodisperse melt, liquid–liquid coexistence can only occur along a line in the pressure–temperature $p-T$ plane. Hence, liquid–liquid phase separation under isobaric conditions can only be transient, before the entire phase reverts to the dense liquid. On the other hand, an isochoric quench would be expected to yield true spinodal-like behaviour. The true system is probably something between the two extremes, with volume leaving the system on some timescale. Based on estimates of thermal diffusivity in melts, the time to shrink is of order 10 s (based on a 1 m sample thickness). If

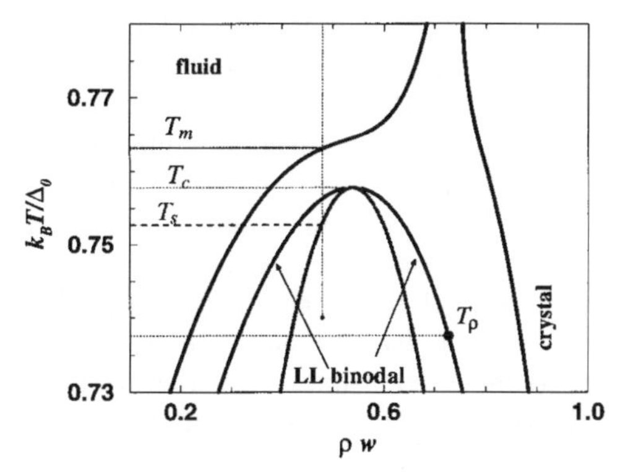

Figure 4 *Cartoon for hidden liquid–liquid spinodal in a polymer melt, calculated for a Flory Rotational Isomeric State chain with a simple coupling between density and conformational order: Δ is the trans-gauche energy gap and ρ_w a dimensionless density. Shown is the path of an isochoric quence into the unstable regime. T_s is the spinodal temperature, T_m the melting temperature, T_c the liquid–liquid critical point, and T_ρ the temperature at which the barrier between dense liquid and crystal is of order k_BT*[70]

the sample (typically a flat disc) is constrained in the thin direction, then volume changes occur on order 1000 s, close to typical induction times and ignoring (unknown) complications due to coarsening kinetics.

On the other hand, in a polydisperse melt a spinodal instability would be expected eventually to reach biphasic equilibrum, in which case standard Cahn-Hilliard kinetics would be expected. Phase separation could be due to several effects, including chemical polydispersity, coupling of mass density to the density of polymer ends, and conformation–density–orientation coupling, which has a natural prediliction for inducing phase separation (longer chains are, per unit monomer, more 'nematic' than their short chain cousins[72]). There are evidently several effects to sort out, and close collaboration between theory and experiment will help to attack the various mechanisms.

The current state of theory is quite crude: state-of-the-art density functional theory calculations (*e.g.* PRISM[73]) have yielded scant clues about the renormalisation of isomeric state potentials due to density, and phenomenological theories have introduced an admittedly *ad hoc* coupling.[70] Hence there is a clear need for theory which captures the minimal necessary molecular detail of orientation, conformation and packing. A possible route is simulation: computational speed can almost handle the necessary melt simulations, while simulations of solution crystallisation have demonstrated how orientational correlations grow as crystallisation in theta solvent conditions.[74,75]

2.9 Mesoscopic Modelling

Computer simulations in the mesoscopic regime are now possible using methods such as Lattice Bolztmann, Dissipative Particle Dynamics, MesoDyn and Cell Dynamics Simulations. MesoDyn is a commercial package (from

MSI) that uses the same time-dependent Ginzburg Landau kinetic equation as CDS, but starts from (arbitrary) bead models for polymer chains.[76,77] The methods have been summarized elsewhere.[78] Examples of recent applications include LB simulations of viscoelastic effects in complex fluids under oscillatory shear,[79] DPD simulations of microphase separation in block copolymers[80,81] and mesophase formation in amphiphiles,[82] and cell dynamics simulations applied to block copolymers under shear.[83,84] DPD is able to reproduce many features of analytical mean field theory[80] but in addition it is possible to study effects such as hydrodynamic interactions.[81] The use of cell dynamics simulations to model non-linear rheology (especially the effect of large amplitude oscillatory shear) in block copolymer miscrostructures is currently being investigated.[85,86]

2.10 Polyelectrolytes

Polyelectrolytes are of intense current interest, not only due to their applications in synthetic polymers but also because many biopolymers such as DNA are polyelectrolytes in solution. Understanding the effect of electrostatic interactions on conformation of such biopolymers is the focus of recent theoretical work from several groups.[87–89] Adsorption of polyelectrolytes is also attracting much theoretical interest,[90,91,88] again motivated by the role of protein adsorption in nature. The formation of complexes in bulk solutions of polyelectrolytes continues to fascinate, both from an experimental and theoretical perspective. More complex charged polymers such as polyampholytes also show intringuing aggregation phenomena, that are the subject of current theoretical work.[92] Gelation in polyelectrolyte solutions is another aspect of theory[93] with current commercial relevance. Finally, the rich complexation effects in ionomers are the focus of ongoing research.

3 Future Potential

3.1 Melt to Morphology and Multiscale Approaches

The last half-century of polymer science has seen huge strides of analysis of single phenomena. Much of the programme ahead must be to reintegrate analytical theory into more holistic pictures. The single polymer chain furnishes us with an excellent example: at very local scales libration of side groups, structure of non-bonding potentials, RIS theories and so on are essential to understand the NMR, DSC, NSE experiments that probe these length and timescales. On the other hand all this misses completely the nature of chain behaviour at the largest scales such as Rouse and reptation dynamics. Models of real systems, especially if applied to *ab initio* design of new materials must be able to integrate these length scales. A few current attempts are being made to link simulations at different length scales, passing data between coarse-grain/fine-grain jumps as necessary.[94]

A similar pattern of length–scale emergence of new phenomena appears in the case of multicomponent polymers. Local dynamics of block co-polymers provide the local basis for microphase separation. The morphology of the final phase separated state depends also on the initial noise, the processing and the dynamics of composition fields at longer length scales. Finally the bulk-scale mechanical properties of the material depends on subtle averaging of these local frozen morphologies. A large effort in Japan is currently focussed on this project under the direction of Doi.

A final example might be offered in the case of melt rheology. The local constitutive behaviour is complex enough as a function of molecular architecture and its distribution, but the behaviour of such a melt in a complex processing flow has as much effect on the properties *via* frozen-in orientation and predisposition of the semi-crystalline morphology. A multi-scale approach to this problem is the focus of a large UK-based collaboration.[95]

3.2 Nanostructure

Nanostructured polymers, in particular block copolymers, have immense potential in areas ranging from templating inorganic materials to permeation/filtration media, nano-assaying, nanolithography and high density data storage media.

Theories for block copolymers all suffer from the necessity for *a priori* assumptions about the microphase separated structure. To date, theories rely on comparison of free energies for assumed structures. Historically, this led to the failure to account for the bicontinuous gyroid phase, because it was not included as a possible trial structure in the free energy calculations (not being identified until 1997). This problem will be exacerbated for ABC triblocks where there is a far wider range of possible structures. Fortunately, the timely development of an *ab initio* method in SCFT looks extremely promising.[96] This combinatorial method has been shown to be able to predict morphologies for ABCA tetrablock copolymers, with no *a priori* restriction on space group symmetry.[96]

3.3 Glass Transitions

The ubiquitous set of phenomena collectively known as the 'glass transition' has been of long and widespread research interest spanning polymers,[97] organic liquids,[98] and magnetic systems.[99] Even when experiments are restricted to the temperature-dependence of transport phenomena, qualitative variety in the emergence of 'slow dynamics' appears as the class of 'fragile' and 'non-fragile' glasses.[98] In spite of the success of 'mode-coupling' theories,[100] the underlying co-operative molecular reasons for exponential slowing down are not well-understood. However, the evidence from simulations,[101] new experimental probes (*e.g.* NMR/rheology on 2-component blends[102]) and concepts (memory, ageing, dynamical scaling, fragility)[103] and a number of recent 'toy' theoretical models[104] suggest that the time is ripe for a renewed look at the glass transition. The future is far more likely to stress *variety* rather

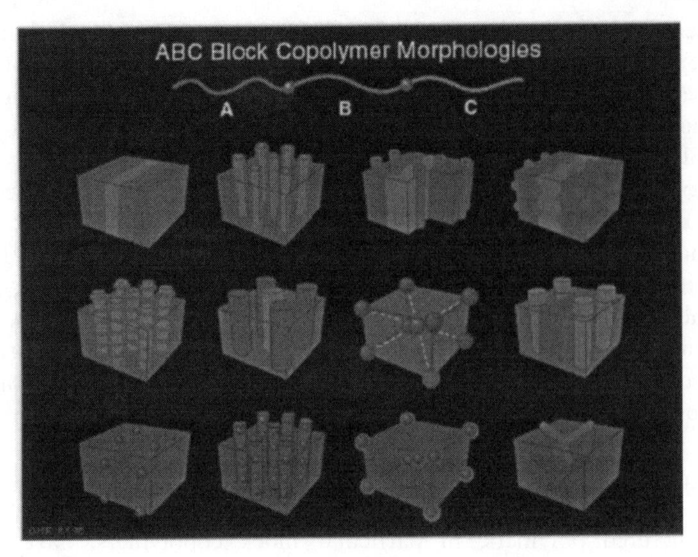

Figure 5 *Predicted morphologies from ABCA triblock co-polymers*[102]

than *uniformity* in approach. One sign of this that emerged in a very recent workshop[103] was that, as in Theme II, there is a great need for discussion and information exchange between communities working on polymers, granular matter, organic glasses and theory.

3.4 Flow-induced Crystallisation

Another old problem for which there is surprisingly little quantitative theory is flow-induced crystallisation.[105–107] Flow has a dramatic effect on crystalline order: put simply, polymer chains align, hence reducing the barrier to crystallization, and initiate crystallization depending on draw ratio, temperature, and other material and control parameters.[108] It has been established that in rapidly drawn melts some form of precursor threadlike 'mesophase', often liquid-crystalline in character, initiates the crystallization event. Phenomenological ideas have focussed on generalizing Avrami kinetics to flow conditions, with various assumptions about the dependence of nucleation rates on flow.[105] More recently, McHugh and co-workers have developed a more formal phenomenological model coupling crystallization kinetics to theories of polymer microstructure,[106,107] incorporating polymer deformation, orientation, and crystallinity as dynamical variables. Such a theory is a useful first step, but it is desirable to have a description that incorporates the minimal necessary molecular detail: including conformation, density, and orientation couplings to describe the appearance of a mesophase and subsequent crystallization, as well as faithful theoretical molecular (tube-model) rheology.

3.5 Protein folding and function (IWH)

Although much progress has been made in modelling protein folding (see refs 109–111 for reviews), there is no consensus on the best method. Most methods consider a protein folding energy landscape. The problem is that this is a rough surface, with many local minima, and it can often be hard to model the guiding forces that stabilize the native structure, and cause the free energy to adopt a 'funnel' landscape. Many minimalist models are based on computer simulations of particles on a lattice, these always being coarse-grained models.[110,112] Fully atomistic models seem some way off. Analytical methods have modelled the free energy landscape based on random energy models, the most current of which analyse the conformational transition in a random heteropolymer using spin-glass methods.[113–116] Mean field methods based on replica techniques will also be developed further. Some structural insights on protein conformational dynamics have emerged from steered molecular dyanmic simulations in which Monte Carlo moves are used as well as molecular dynamics.[117]

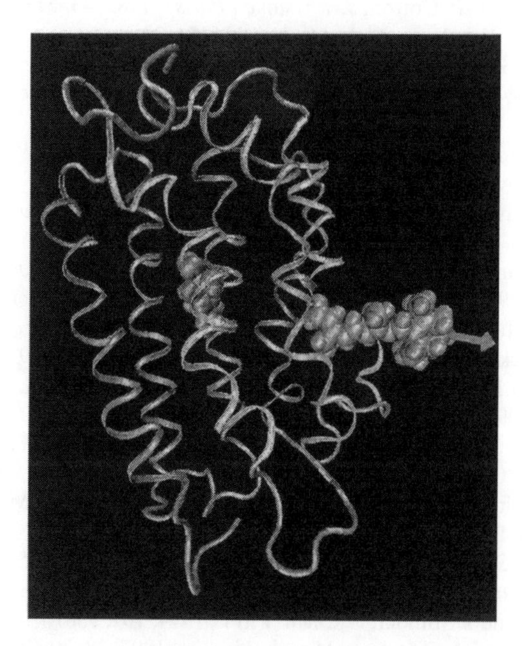

Figure 6 *Visualisation of a steered molecular dynamics simulation of the Avidin–Biotin complex*[117]

There are immense challenges also to model protein function, which will rely on better theoretical models for secondary structure formation (α helices, β sheets). Models presently used are molecular force field approaches, which are rather phenomenological. Realistic atomistic modelling is a long-term goal. In the meantime, energy landscape approaches should help us elucidate the detailed folding mechanisms that lead to protein function.

This vast challenge is only one of the many that lie ahead for theoretical polymer physics in the life sciences. It is certain that the quiverful of concepts and tools only touched on in the above will have some part to play in making progress *via* a closer relationship with biology. However, we are unused to the high degree of specificity found in biological systems, where both evolutionary adaption and the ubiquitous metabolism of living things takes us very far from familiar notions of equilibrium. Perhaps in the search for some physical understanding of examples such as protein folding, cellular mechanics, molecular motors, muscle elasticity *etc.*, a coherent research programme will emerge.

References

1 P.G. de Gennes, *Scaling Concepts in Polymer Physics*, Cornell (1978).
2 J. des Cloiseaux and G. Jannink, *Polymers in Solution, their Modelling and Structure*, Oxford, (1990).
3 M. Daoud and L. Leibler, *Macromolecules,* 1988, **21**, 1497.
4 M. Daoud and J.-P. Cotton, *J. Physique (Paris)*, 1982, **43**, 513.
5 E.J. Hinch, *J. Non-Newt. Fluid Mech.,* 1994, **54**, 209.
6 L. Li and R.G. Larson, *Macromolecules*, 2000, **33**, 1411.
7 I.W. Hamley, *The Physics of Block Copolymers*, Oxford University Press, Oxford, (1998).
8 L. Leibler, *Macromolecules,* 1980, **13**, 1602.
9 A.N. Semenov, *Sov. Phys. JETP,* 1985, **61**, 733.
10 E. Helfand and Z.R.Wasserman, in *Developments in Block Copolymers*, vol.1 (ed. I. Goodman), Applied Science, London, (1982).
11 M.W. Matsen, *J. Chem. Phys.* 1998, **108**, 785; M.W. Matsen and M. Schick, *Phys. Rev. Lett.*, 1994, **72**, 2660; M.W. Matsen and F.S. Bates, *Macromolecules,* 1996, **29**, 764; M.W. Matsen and M. Schick, *Curr. Opin. Colloid Interface Sci.*, 1996, **1**, 329.
12 G.H. Fredrickson and E. Helfand, *J. Chem. Phys.*, 1987, **87**, 697.
13 K. Almdal, J.H. Rosedale, F.S. Bates, G.D. Wignall and G.H. Fredrickson, *Phys. Rev. Lett.*, 1990, **65**, 1112.
14 D.M.A. Buzza, I.W. Hamley, A.H. Fzea, M. Moniruzzaman, J.B. Allgaier, R.N. Young, P.D. Olmsted and T.C.B. McLeish, *Macromolecules,* 1999, **32**, 7483.
15 T.A. Shefelbine, M.E. Vigild, M.W. Matsen, D.A. Hajduk, M.A. Hillmyer, E.L. Cussler, F.S. Bates, *J. Am. Chem. Soc.*, 1999, **121**, 8457.
16 V. Abetz, R. Stadler and L. Leibler, *Polym. Bull.,* 1996, **37**, 135.
17 R. Stadler *et al.*, *Macromolecules,* 1995, **28**, 3080.
18 M. Kikuchi and K. Binder, *J. Chem. Phys.*, 1994, **101**, 3367.
19 G.T. Pickett and A.C. Balazs, *Macromolecules,* 1997, **30,** 3097.
20 M.W. Matsen, *J. Chem. Phys.*, 1997, **106**, 7781.
21 G.G. Pereira and D.R.M. Williams, *Macromolecules,* 1999, **32,** 758; ibid 8115, 1999.
22 M.W. Matsen, *Curr. Opin. Colloid Interface Sci.,* 1998, **3**, 40.
23 G.H. Fredrickson, *Macromolecules,* 1987, **20**, 2535.
24 N. Clarke, T. C. B. McLeish and S. D. Jenkins, *Macromolecules,* 1995, **28**, 13, 4650.
25 D.J. Yontz, S.L. Hsu, D. Gier, *Abstr. Pap. Am. Chem.*, 1997, **S 214**: 322-PMSE.
26 E.J. Kramer, P.F. Green, C.J. Palmstrom, *Polymer,* 1984, **25**, 473.

27 F. Broachard and P.G. de Gennes, *Europhys. Lett.*, 1986, **1**, 221.
28 A.N. Semenov in *Theoretical Challenges in the Dynamics of Complex Fluids*, ed. T.C.B. McLeish, Kluwer, Dordrecht (1997).
29 M. Doi and A. Onuki, *J. Phys.* (France), 1992, **2**, 1631.
30 N. Clarke, T. C. B. McLeish, S. Pavawongsak, J. S. Higgins, *Macromolecules,* 1997, **30**, 15, 4459–4463.
31 H. Tanaka and T. Araki, *Phys. Rev. Lett.,* 1997, **78**, 4966.
32 S. T. Milner, *Phys. Rev. E,* 1993, **48**, 3674 and S.T. Milner in *Theoretical Challenges in the Dynamics of Complex Fluids*, ed. T.C.B. McLeish, Kluwer, Dordrecht (1997).
33 N. Clarke, T.C.B. McLeish, *Macromolecules,* 1999, **32**, 4447–4449.
34 J.L. Goveas and S.T. Milner, *Macromolecules,* 1997, **30**, 2605.
35 M. Doi and S.F. Edwards, *The Theory of Polymer Dynamics*, Oxford (1986).
36 M. Rubinstein in *Theoretical Challenges in the Dynamics of Complex Fluids*, ed. T.C.B. McLeish, Kluwer, Dordrecht (1997).
37 S.T. Milner and T.C.B. McLeish, *Phys. Rev. Lett.,* 1998, **81**, 725.
38 T. P. Lodge, *Phys. Rev. Lett.,* 1999, **83**, 3218.
39 G. Marrucci, *J. Non-Newt. Fluid. Mech.*, 1996, **62**, 279.
40 D.W. Mead, R.G. Larson and M. Doi, *Macromolecules,* 1998, **31**, 7895.
41 C.C. Hua, J.D. Schieber and D.C. Venerus, *J. Chem. Phys.,* 1998, **109**, 10018 and 10028.
43 S.T. Milner, T.C.B. McLeish and A.L. Likhtman, submitted (2000).
44 T.C.B. McLeish and S.T. Milner, *Adv. Polym. Sci.,* 1999, **143**, 195.
45 S.T. Milner, T.C.B. McLeish, *Macromolecules,* 1997, **30**, 2159.
46 S. T. Milner, T. C. B. McLeish, R. N. Young, A. Hakiki and J. M. Johnson, *Macromolecules,* 1998, **31**, 9345.
47 B. Blottière, T. C. B. McLeish, A. Hakiki, R. N. Young and S. T. Milner, *Macromolecules,* 1998, **31**, 9295–9309.
48 T.C.B. McLeish, J. Allgaier, D.K. Bick, G. Bishko, P. Biswas, R. Blackwell, B. Blottière, N. Clarke, D. J. Groves, A. Hakiki, R. Heenan, J. M. Johnson, R. Kant, D. Read and R. N. Young, *Macromolecules,* 1999, **32**, 6734–6758.
49 T.C.B. McLeish and R.G. Larson, *J. Rheol.,* 1998, **42**, 81.
50 N.I. Inkson, O.G. Harlen, D.J. Groves, D.J. and T.C.B. McLeish, *J. Rheol.,* 1999, **43**.
51 R. J. Blackwell, T.C.B. McLeish and O.G. Harlen, *J. Rheol.,* 2000, 121.
52 F. C. MacKintosh, J. Kas, and P. A. Janmey, *Phys. Rev. Lett.,* 1995, **75**, 4425.
53 A. C. Maggs, *Phys. Rev. E,* 1997, **55**, 7396.
54 D. C. Morse, *Macromolecules*, 1998, **31**, 7044; 1998, **31**, 7030; 1999, **32**, 5934.
55 D. C. Morse, *Phys. Rev. E* , 2000, in press.
56 G. Marrucci, in *Theoretical Challenges in the Dynamics of Complex Fluids*, ed. T.C.B. McLeish, Kluwer, Dordrecht (1997).
57 A. Keller, G. Goldbeck-Wood, and M. Hikosaka, *Far. Disc.*, 1993, **95**, 109.
58 N. J. Terrill, P. A. Fairclough, E. Towns-Andrews, B. U. Komanschek, R. J. Young, and A. J. Ryan, *Polymer,* 1998, **39**, 2381.
59 G. Matsuba, K. Kaji, K. Nishida, T. Kanaya and M. Imai, *Macromolecules,* 1999, **32**, 8932.
60 M. Imai, K. Kaji, T. Kanaya, and Y. Sakai, *Phys. Rev. B,* 1995, **52**, 12696.
61 M. Imai, K. Kaji, and T. Kanaya, *Macromolecules,* 1994, **27,** 7103.
62 B. S. Hsiao, Z. G. Wang, F. J. Yeh, Y. Gao, and K. C. Sheth, *Polymer,* 1999, **40**, 3515.

63 T. A. Ezquerra, F. Liu, R. H. Boyd, and B. S. Hsiao, *Polymer,* 1997, **38**, 5793.
64 A. J. Ryan, J. P. A. Fairclough, N. J. Terrill, P. D. Olmsted, and W. C. K. Poon, *Far. Disc.,* 1999, **112**, 13.
65 S. Sasaki, K. Tashiro, M. Kobayashi, Y. Izumi, and K. Kobayashi, *Polymer,* 1999, **40**, 7125.
66 Y. A. Akpalu and E. J. Amis, *J. Chem. Phys.,* 1999, **111**, 8686.
67 Z.-G. Wang, B. S. Hsiao, E. B. Sirota, P. Agarwal, and S. Srinivas, *Macromolecules,* 2000, **33**,978.
68 T. A. Ezquerra, E. Lopezcabarcos, B. S. Hsiao and F. J. Baltacalleja, *Phys. Rev. E,* 1996, **54**, 989.
69 T. Shimada, M. Doi, and K. Okano, *J. Chem. Phys.,* 1988, **88**, 7181.
70 P. D. Olmsted, W. C. K. Poon, T. C. B. McLeish, N. J. Terrill, and A. J. Ryan, *Phys. Rev. Lett.,* 1998, **81**, 373.
71 S. Harrington, R. Zhang, P. H. Poole, F. Sciortino, and H. E. Stanley, *Phys. Rev. Lett.,* 1997, **78**, 2409.
72 A. N. Semenov, *Europhys. Lett.,* 1993, **21**, 37.
73 J. D. Mccoy, K. G. Honnell, K. S. Schweizer, and J. G. Curro, *J. Chem. Phys.,* 1991, **95**, 9348.
74 C. Lui and M. Muthukumar, *J. Chem. Phys.,* 1998, **109**, 2536.
75 S. Fujiwara and T. Sato, *J. Chem. Phys.,* 1999, **110**, 9757.
76 N.M. Maurits *et al.*, *Macromolecules,* 1999, **32**, 7674.
77 B.A.C. van Vlimmeren *et al.*, *Macromolecules,* 1999, **32**, 646.
78 I.W.Hamley, *Introduction to Soft Matter,* John Wiley, Chichester (2000).
79 A. Malevanets and J.M.Yeomans, *Faraday Discuss.,* 1999, **112**, 237.
80 R.D. Groot and T.J. Madden, *J. Chem. Phys.,* 1998, **108**, 8713.
81 R.D. Groot, T.J. Madden and D.J. Tildesley, *J. Chem. Phys.,* 1999, **110**, 9739.
82 S. Jury, P. Bladon, M. Cates, S. Krishna, M. Hagen, N. Ruddock and P. Warren, *Phys. Chem. Chem. Phys.,* 1999, **1**, 2051.
83 H. Kodama and M. Doi, *Macromolecules,* 1996, **29**, 2652.
84 T. Ohta, Y. Enomoto, J.L. Harden and M. Doi, *Macromolecules,* 1993, **26**, 4928.
85 S.R. Ren, I.W. Hamley and P.D. Olmsted, submitted to *Phys. Rev. Lett.*
86 I.W. Hamley, submitted to *Macromol. Theor. Simul.*
87 F.J. Solis and M.O. de la Cruz, *Phys. Rev. E,* 1999, **60**, 4496.
88 J.F. Joanny, *Euro. Phys. J. B,* 1999, **9**, 117.
89 A.R. Khokhlov *et al.*, *Macromolecules,* 1992, **25**, 1493.
90 X. Chatellier and J.F. Joanny, *J. Phys. II (France)*, 1996, **6**, 1669.
91 R.R. Netz and J.F. Joanny, *Macromolecules,* 1999, **32**, 9013.
92 A.V. Dobrynin, M. Rubinstein and J.F. Joanny, *Macromolecules,* 1997, **30**, 4332.
93 M. Rubinstein, R.H. Colby, A.V. Dobrynin and J.F. Joanny, *Macromolecules,* 1996, **29**, 398.
94 K. Kremer in *Soft and Fragile Matter*, M.E. Cates, ed., SUSSP/NATO (2000).
95 EPSRC MaPEA project *Micro-scale Polymer Processing.*
96 F. Drolet and G.H. Fredrickson, *Phys. Rev. Lett.,* 1999, **83**, 4317.
97 D.J. Plazek *et al., Coll. Polym. Sci.,* 1994, **272**, 1430.
98 C.A. Angell, *Science*, 1995, **267**, 1924.
99 L.F. Cugliandolo and J. Kurchan, *Phys. Rev. Lett.,* 1993, **71**, 173.
100 W. Goetze and L. Sjogren, *Rep. Prog. Phys.*, 1992, **55**, 241.
101 C. Bennemann *et al.*, *Nature*, 1999, **399**, 246.
102 D. Vlassopoulos *et al.*, *J. Rheol,* 1997, **41**, 739.
103 M.E. Cates (ed.), *Soft and Fragile Matter*, NATO ASI, July 1999.

104 C. Monthus and J.P. Bouchaud, *J. Phys. A*, 1996, **29**, 3847.
105 G. Eder and H. Janeschitz-Kriegl, in *Materials Science and Technology*, edited by H. E. H. Meier (VCH Verlagsgesellschaft, Weinheim, 1997), Vol. 18, pp. 270–342.
106 A. K. Doufas, I. S. Dairanieh, and A. J. McHugh, *J. Rheol.* 1999, **43**, 85.
107 A. C. Bushman and A. J. McHugh, *J. Polymer Science Part B-Polymer Physics* 1997, **35**, 1649; erratum 1996, **34**, 2393.
108 A. Mahendrasingam, C. Martin, W. Fuller, D. J. Blundell, R. J. Oldman, D. H. MacKerron, J. L. Harvie, and C. Riekel, *Polymer,* 2000, **41**, 1217.
109 J.N. Onuchic, Z. Luthey-Schulten and P.G. Wolynes, *Annu. Rev. Phys. Chem.,* 1997, **48**, 545, and references therein.
110 C.M. Dobson, A. Salj and M. Karplus, *Angew. Chem. Intl. Ed.*, 1998, **37**, 868.
111 K.A. Dill *et al.*, *Protein Sci.*, 1995, **4,** 561.
112 M. Karplus and D.L. Weaver, *Protein Sci.,* 1994, **3**, 650.
113 S. Plotkin, J. Wang and P.G. Wolynes, *J. Chem. Phys.,* 1997, **106**, 2932.
114 V.S. Pande, A.Yu Grosberg, T. Tanaka and D.S. Rokhsar, *Curr. Opin. Struct. Biol.,* 1998, **8**, 68.
115 V.S. Pande, A.Yu Grosberg and T. Tanaka, *Biophys. J.,* 1997, **73**, 3192.
116 R. Du, A.Yu Grosberg and T. Tanaka, *Folding and Design,* 1998, **3**, 203.
117 S. Izrailev, S. Stepaniants, M. Balsera, Y. Oono, and K. Schulten, *Biophysical Journal,* April, 1997.

Measuring Structure and Dynamics

18

Polymer Theory and Modelling – Physical and Knowledge-based Approaches

Gerhard Goldbeck-Wood

DEPARTMENT OF MATERIALS SCIENCE AND METALLURGY, UNIVERSITY OF CAMBRIDGE, PEMBROKE STREET, CAMBRIDGE CB2 3QZ, UK*

1 Introduction

The remit of this review is to summarize the state of the art in theory and simulation of polymers and expand upon future directions. Since both the size of the review, and the knowledge of the author are limited, it is necessarily going to be a sketchy, biased picture. Polymer theory and simulation are going to be treated as one subject here which does not really do them justice, and the author being a 'modeller' more than a theoretician, the subject is going to be viewed mostly from the simulation angle.

As the subtitle suggests, the field can be viewed from two angles: that of physical modelling and that of knowledge-based modelling. While the former is concerned with the physical interactions of matter from the electronic all the way to the macroscopic level, the latter deals with statistical relationships between models of structure and property data. The emerging themes in each of these areas seem at first sight disparate.

In physical modelling one seems to be concerned with replacing experimentation. In some ways this means breaking the traditional link and conception that modelling follows on from and in fact depends upon experiment. As the precision of modelling keeps on increasing as well as the cost decreasing, it becomes attractive to replace more and more of experiment through modelling. The virtual polymer material, designed and tested by simulation alone, seems to come within reach. While it is neither likely nor desirable that all experimentation will be replaced, it nevertheless represents an inspiring goal. It requires a modelling hierarchy starting from *ab initio* quantum mechanical

* Present address: Molecular Simulations Ltd, 230/250 The Quorum, Barnwell Road, Cambridge CB5 8RE, UK

calculations *via* atomistic and mesoscale simulations all the way to the macroscopic behaviour. At each level of coarse graining the relevant parameters need to be identified and calculated from the finer scale, shedding all superfluous information. The reduction of the problem at each level is not only necessary to keep computation times under control, but also in fact a check that one understands the physics of the problem.

In contrast, the emerging theme in knowledge based modelling is an ever closer intertwining of modelling and experimentation. A powerful combination of three factors contribute to this theme: (i) structure-property relationships that are based on better models and more relevant structural descriptors; (ii) robotics techniques in experimentation, which lead to a real explosion in the amount of data available; (iii) informatics tools, which can handle the large amounts of data generated by both experiment and simulation.

2 Physical Modelling

Our understanding of nature in general and polymers in particular is largely based on a combination of experience and models, *i.e.* somehow simplified representations of reality. From experimentally derived experience we deduce models and from models we derive predictions about the outcomes of particular experiments, and so these two strands complement each other. Computer modelling is a relatively new third component which makes models less simplified and hence allows us to 'derive' prediction for more complex situations, but also merges model and experience: computer simulations are experienced as reality.

As the structure, dynamics and properties are determined by phenomena on many length and time scales physical modelling is subdivided into the quantum mechanical, atomistic, mesoscale, microscale and continuum levels, while research into the way in which these levels are linked is known as hierarchical or multiscale modelling. The typical structural levels arising in the polymer field are shown Figure 1.

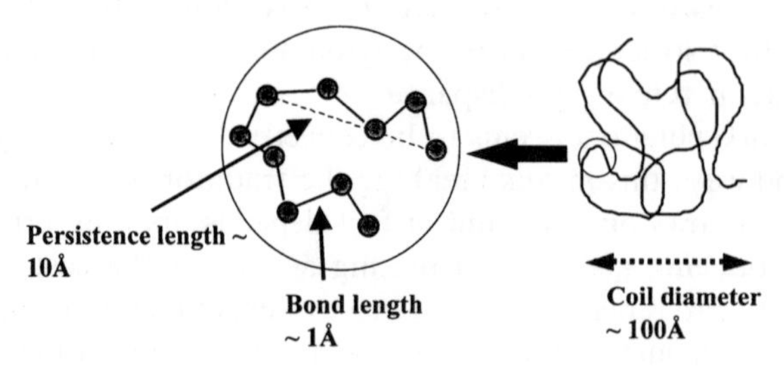

Figure 1 *Scales of polymer structure*

In the development of atomistic polymer models some of the milestones were Flory's rotational isomeric state (RIS) model[1] and the development of force field simulations of polymers which led to, and have been significantly developed in the context of commercial molecular modelling software.

The mesoscale on the other hand means coarse grained representations of polymers, *e.g.* a chain of beads. Arguably Kuhn[2] was the first to bridge the gap between the atomistic and the mesoscale, by deriving a way in which a real polymer chain may be approximated by an equivalent freely jointed chain. There are some lessons to be learnt from this. One is that the description at the mesoscale level is left with a certain degree of arbitrariness. The length of a segment can be chosen within a range of values which fulfil the condition that it is large enough that subsequent segments are uncorrelated in their direction, and small enough that the whole chain is still made up of enough segments for a Gaussian distribution approximation to hold (which is about 10). While exponents and scaling laws are not affected, the front factors are. Hence specificity tends to get lost. In the case in hand the segment length is fixed by the additional requirement that the contour length is preserved. In any case, there are no hard and fast rules. Rather the recipe depends on the specific method and application.

In microscale models the explicit chain nature has generally been integrated out completely. Polymers are often described by variants of models, which were primarily developed for small molecular weight materials. Examples include the Avrami model of crystallization,[3] and the director model for liquid crystal polymer texture.[4] Polymeric characteristics appear *via* the values of certain constants, *i.e.* different Frank elastic constant for liquid crystal polymers rather than *via* explicit chain simulations. While models such as the liquid crystal director model are based on continuum theory, they typically capture spatiotemporal interactions, which demand modelling on a very fine scale to capture the essential effects. It is not always clearly defined over which range of scales this approach can be applied.

Another point to note from a historical perspective is that there has been a divergence of the research communities. Flory and others during his time made major contribution to all of the above areas: the RIS model on the atomistic scale, the lattice models for solutions and blends on the mesoscale, and constitutive equations for networks on the continuum level, to name just a few examples. Since then the atomistic, mesoscale and continuum areas developed to a considerable extent independently from each other. The atomistic level was significantly driven by contributions from the biopolymer community (*e.g.* the modelling of polymer crystal structures and the helical structures of biopolymer). The mesoscale branch, on the other hand, was the realm of the physical sciences, see, *e.g.*, Doi and Edwards.[5] The continuum level, at the border between science and engineering developed ever more complex constitutive equations, *e.g.* to describe the rheology of LCPs. While the atomistic level, driven by the pharmaceuticals industry, aimed for quantitative modelling of specific substances, the mesoscale level developed scaling concepts and understanding of typical behaviour. The continuum level was heavily parameterized from experimental data to make process simulations work. However, the emerging theme here is that computer simulation is beginning to break down many of these division.

The quest for a virtual material requires modelling of polymer behaviour, and the identification of the relevant degrees of freedom at each length and timescale.[6] More specifically we need to understand and represent adequately 'structuring' on all scales. Examples of such structuring elements which are eventually linked to properties are: Electronic structure and torsion potentials, monomer stiffness and persistence length, monomer types and miscibility, crystallinity, liquid crystallinity and orientational texture, reactions, flow fields, pressure, and spatial constraints.

While early models have advanced our understanding of the polymer structuring behaviour, they have sometimes also shaped it to such an extent that it is hard to let go of those models when they are found to be inadequate. Polymer crystallization is an example where the advance in our understanding has been hampered by nucleation models which have after all been an adaptation of theories for simple liquids which were already simplifications even there. Alternatives, such as those proposed by Sadler and Gilmer,[7] recently also Doye and Frenkel,[8] and even more radical approaches[9] are only slowly gaining acceptance. Fortunately the field has been opened up by experimental methods accessing short time and length scales.[10] The question boils down to the way in which statistical mechanics and kinetics can be applied to polymers. Quantities in question such as an interface are often not easily definable as the molecules are larger than the interfacial width, or the definitions (such as a sharp interface for nucleation and a diffuse interface for spinodal decomposition) are inadequate.

3 Molecular Modelling (Electronic and Atomistic Levels)

Molecular modelling has seen huge successes since it's inception about half a century ago,[11] and this trend is set to continue. In fact one could argue that the good times have only started in polymer molecular modelling. The reason for this is that three key elements of successful polymer modelling are being resolved: (i) accurate force fields (ii) the ability to generate realistic conformations and ensembles also for longer chains, and (iii) the ability to deal with different timescales and sample rare events. As for the force fields, faster computers and better algorithms mean that in recent years *ab initio* quantum mechanical methods have been used to calculate precise potentials, which can be accessed either directly or indirectly *via* force fields. As a result, quantitative thermodynamics and dynamics calculations have become a standard for a wide range of systems.[12]

As for the second point, the problem is related to the decrease of entropy with chain length: the chance of placing a chain into an existing sea of others decreases strongly with chain length. Clever Monte Carlo methods have solved this problem for linear polyethylene chains, and are being generalized and extended to other systems.[13] One example is the recoil-growth configurational bias MC.[14] It uses a biased chain growth mechanism to avoid excluded volume overlaps (the effect of the bias is later removed *via* the acceptance criteria), as well as some sort of look-ahead feature which explores the space in the vicinity of the growing chain. Similar to CBMC it can be used to calculate phase equilibria and chemical potentials in polymer systems. A very effective method for equilibrating conformations of polydisperse melts is the so-called end-bridging algorithm.[15] It uses drastic chain reconnections but enforces that a certain molecular weight distribution is kept. Remarkably, its efficiency increases with chain length, and can therefore be used to equilibrate very long chains. A third approach is to use the 'generic' efficiency of lattice models to generate equilibrated lattice chains with the correct end-to-end distance and density, and then reverse map these mesoscale chains to the atomistic level.[16] Further details about the lattice models are described in the mesoscale section.

The classic example for the third point above is the diffusion of gases and solvents in polymers. Whereas the case of small molecule diffusion (up to methane) in rubbery polymers has been addressed pretty exhaustively by straight Molecular Dynamics methods, diffusion rates of larger molecules and in glassy polymers remain outside the reach of such methods. The key to solving these issues lies broadly in the exploration of complex energy landscapes, which depend nonlinearly on penetrant and polymer interactions. However, efficient methods of sampling the transition paths have been devised,[17] and promise to take molecular modelling a significant step further.

While research into further improvements in molecular modelling methods is still very much alive, the field has at the same time reached a mature state. Expert commercial software packages have turned molecular modelling into a readily accessible tool with obvious benefits as well as pitfalls. The analogy with sophisticated instruments such as electron microscopes may be drawn. While it is obvious that one would not want to rebuild a microscope for every research project, one still needs to understand the way it works, and be able to change enough of the settings in order to do useful research. At the same time, a maturing technology is not the same as an old hat. A wealth of phenomena can be explored by research, which focuses on the application of the method rather than writing computer code. One examples of such an approach used MSI's Cerius2 software to study the 'dual-mode' sorption behaviour of small molecules in polymers.[18] It has been known experimentally for some time that the solubility below the glass transition temperature can be described by a superposition of a Langmuir adsorption and a simple linear dependence on pressure according to Henry's law. However, it had never been possible actually to confirm the proposed adsorption in free volume cages directly. A detailed molecular simulation of penetrant molecule behaviour in polymer above and below T_g, however, could indeed show clearly that the small molecule adsorbs to the internal hole surface only below T_g.

Despite these successes, molecular modelling will never be able to address phenomena, which involve length scales of hundreds of nanometers and more, such as the phase separation of block-copolymers, or very long timescales, such as mechanical properties related to sub T_g relaxations. Therefore more and more effort is devoted to the development of mesoscale models linked in a well defined way to the atomic level.

4 Mesoscale Modelling

An area which has emerged particularly strongly over the last few years, and is expected to remain growing for some time is that of 'mesoscale modelling', encompassing anything between the atomic level and the continuum descriptions. The behaviour at the mesoscale is determined in a stochastic way by some underlying faster motions. The timescale of the fast motions is so much smaller that all the associated degrees of freedom can relax in any given observation interval on the mesocale. The classic example of mesocale behaviour is the Brownian particle. While the motion as observed under a microscope is described as a random walk, it is actually the result of atomic motions kicking the particle at a rate of about 10^{21} times per second.

While this area of coarse grained polymer chain models has been studied for many years as outlined in the introduction, the field has attracted a lot of attention in recent years. A number of factors have contributed to this strong growth, including: (i) better integration of theory and modelling by means of advances in numerical algorithms, (ii) realisation that actual materials and processes can be simulated (*i.e.* the front factors in the coarse grained models can be accessed rather than just the exponents), (iii) significant advances incomputer speed and memory, (iv) experimental methods allowing the relevant time and length scales to be investigated (synchrotron, probe microscopy *etc.*), and (v) a strong industrial need for better microstructural control of materials for 'tailored properties'.

An example of is the mesoscale model known as MesoDyn,[19] a dynamic density functional method for the phase separation of block-copolymers. It basically brings together the Gaussian chain model, classical density functional theory, mean field theory, advanced numerical integration methods, and parallel computing techniques to represent a realistic dynamics of diffusive phase separation. While most parts were well established on their own, putting everything together created one of the most realistic models to date of complex self-assembling systems. It lacks however a proper description of hydrodynamics. This is the advantage of the dissipative particle dynamics (DPD)[20] method. Originally designed to simulate fluid droplets containing many atoms, the method has been extended to polymers, which are represented as strings of droplets. The particles undergo dissipative, momentum conserving collisions with a soft repulsive interaction. This means that there is no hard core radius, *i.e.* particles can actually go through each other. While this seems natural in the case of fluid drops, it presents a problem for simulating entangled polymers. Nevertheless, the method has been very successful in dealing with mesoscale phenomena in complex fluids containing polymers due to the large time steps allowed by the soft interaction, and the built-in hydrodynamics.

Another example is that of lattice chain models. Simple square lattice models were established by Flory as a vehicle for calculating configurational entropies *etc.*, and used later in the simulations of the qualitative behaviour, *e.g.* of block copolymer phase separation.[21] More sophisticated models such as the bond fluctuation model,[22] and the face centred cubic lattice chain model[23] have been developed in recent years. They retain the detail required to yield quantitative predictions of observable properties at the mesoscale, such as chain interdiffusion and surface roughness. In these models moves are kept local, and acceptance of attempted moves is decided by a Metropolis Monte Carlo scheme. Although equilibration is not as efficient as with non-local moves, and sampling is not as efficient as in biased schemes, the advantage is that the polymer dynamics, *e.g.* the crossover from Rouse to reptation dynamics, is reproduced correctly. Hence the processes such as the welding of polymers can be studied in detail.

5 Microscale Models

Strong microstructural features are found in many materials, and polymers are no exception. Their description often requires a level in between the mesoscale and the continuum. The length and timescale, *e.g.* of spherulitic crystallization, and liquid crystal polymer texture variations are (still) beyond the reach even of mesoscale models. The way forward here has been to abandon explicit chain models, but retain the important spatiotemporal variations and interactions by means of the implementing constitutive equations in finite difference, finite element or fluid dynamics schemes. Thereby structures develop and evolve in the simulation, providing detailed insight into the polymer behaviour. There are many examples of phenomena emerging from this approach. The implementation of the Johnson-Segalman model showed the formation of shear banding in polymer fluids.[24] The pom-pom model of branched polymers has led to quantitatively correct descriptions of branched polymer rheology.[25] Director models of liquid crystalline polymers have shown the strong interaction between disclinations and shear fields which can lead to disclination multiplication, and the log-rolling behaviour observed in many experiments.[4]

The emerging theme in this field is to establish an ever-stronger link with the molecular scale on the one hand and with the full scale processing simulations on the other. Further improvements in the underlying equations will be made based on mesoscale and molecular scale models, and the simulations coupled with molecular simulations in regions of singular behaviour (*e.g.* disclination cores). As for processing simulation, the development of Lagrangian flow solvers allows the spatiotemporal interactions to be dealt with correctly and efficiently (see also the chapter by McLeish). This should result in future in much more accurate processing descriptions. However, structuring due to, *e.g.*, crystallization will remain a difficult issue for some time to come.

6 Hierarchical and Multiscale Modelling

These ‘buzz words’ include anything that connects one scale to the next, starting from a straight calculation of a constitutive equation parameter from molecular modelling to some more sophisticated schemes of mapping and reverse mapping of scales. The basic idea and attraction is depicted in Figure 2:

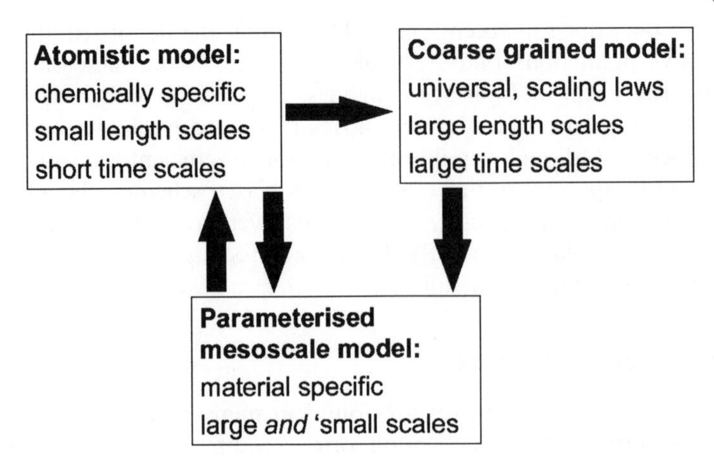

Figure 2 *From universal behaviour/scaling laws to specific materials modelling through the scales*

one would like to deal with the range of scales yet still retain the information about the chemically specific system.

In a way one could regard molecular modelling as hierarchical since it uses force fields which may well have been parameterized from *ab initio* calculations. The question then is what the equivalent of the force field is at the next level up; *i.e.* can the connection be defined in a rigorous way? There is a formal strategy in principle. According to the so-called projection operator formalism interactions at the coarse grained level are derived by integrating the Boltzmann factor of the detailed model over all variables to be eliminated. If timescales between detailed and coarse-grained variables do not overlap the effect of the elimination can be described *via* simple friction factors. However, in practice the approach is usually more or less empirical and different procedures have been developed for different models.

The approach used for the bond fluctuation model is the following: Take an atomistic chain model and determine the distribution function of end-to-end distance for small segments of, *e.g.*, two to five monomers. Also compute the distribution function of angles between two of the segments. Adjust the bond length and bond angle potentials in the lattice model to match the distributions obtained from molecular simulations. This defines the effective bond length of the coarse grained model and its potentials.

In the fcc lattice chain model[23] the approach has been slightly different, since it has a fixed bond length. First one determines the most likely end-to-end distance of segments of two to five monomers. Then for a bulk system, one can calculate the occupancy required on the lattice to match the experimental density, and choose the monomer-to-bead ratio, which gives occupancy of 70–90% in order to represent bulk dynamic behaviour. Finally a bond angle potential is adjusted to achieve the correct mean squared end-to-end distance. The time step can then be calibrated against diffusion or viscosity data, or from Molecular Dynamics simulations.[16]

In the MesoDyn model one uses the following rationale for the mapping of atoms to beads. MesoDyn is based on a Gaussian chain model, and hence the structure of the individual chains is encompassed in the single chain structure factor. For Gaussian chain block copolymers, the structure factor shows a single peak, whose position depends on bead size and chain architecture. This can hence be varied to fit the experimental or molecular simulation data, thereby fixing the atom-to-bead ratio.[26]

The question about these empirical schemes is whether they can be generalized or automated in any way. At the moment it seems that for each specific material, the process needs to done from scratch, involving a lot of human interaction and judgement. Also with the diversity of models and empirical schemes, parameterization is strongly model dependent – the bead size and interaction parameters in MesoDyn, the bond-fluctuation or the fcc lattice chain model of the same system will all be different. The situation hence seems confusing. Any emerging efforts aimed at some more coherence in this area would be beneficial.

7 Knowledge-based Modelling: QSPR/QSAR Methods and Neural Networks.

The area of knowledge-based modelling includes a range of tools, which have in common that they produce some form of a link between a physical model of a (polymer) material on the one hand and a property on the other. In contrast to physical modelling, however, this link is a purely mathematical mapping from a set of inputs to an output quantity. The input is a set of so-called descriptors of the material, which can be any quantity that can be calculated directly from a given molecular structure. Typical descriptors include thermodynamic properties (such as system energies), proportions of ingredients, or functions representing molecular shape (molecular shape area, number of rotable bonds, *etc.*). The output can be any property of the material, *e.g.* its modulus. Commonly these methods are known as QSAR (quantitative structure activity relations) or QSPR (quantitative structure property relations). The former is due to the widespread use of these methods in correlating molecular structures with chemical activity. However, the methods have been found to apply equally well to many other materials including polymers.[27]

Several factors contribute to a successful application of the method. The first is that good training data sets are required to determine the correlation. The second is that relevant descriptors must be found. In the polymer case, group contribution and group interaction models have been very successful.[28] The third is that the mapping needs to be adequate, *i.e.* neither overfitting the data nor losing any of the detail required. There are many different approaches here including simple regressions, multivariate analysis, genetic function approximations and neural networks.[29] The first three approaches yield mathematical formulae linking inputs and outputs. In a neural network the mapping is also largely of pre-set mathematical form but depends on a large number of parameters, the network parameters. In simple neural networks the mathematical form transforming input to output is basically a linear combination of simple functions such as hyperbolic tangents (the parameters of which then are the network parameters). Recently Bayesian neural networks have been used, *e.g.* the Gaussian Process Model,[30] which naturally balance the competing constraints of trying to explain experimental data whilst keeping model complexity to a minimum. In any case, (i) considerable datasets are required to arrive at reliable predictions, and (ii) the predictive power outside the range of the datsets which were used to derive the correlators or train the neural net is small. Another drawback is that one is unable to play 'what if' scenarios that go beyond the limits of the underlying database, or benefit from better insight into the relevant processes. While the advent of molecular modelling promised to change all that in the eighties it became clear not very long after, that the promise could not be kept. Computational times turned out to be prohibitive, and results only qualitative, except if parameters from extensive databases were put in along the way. One could argue that one might as well have stuck to knowledge-based modelling.

The strength of knowledge-based approaches has been demonstrated in many applications of industrial relevance, which have proven to be too complex or too costly to handle by physical modelling. While physical modelling is becoming more powerful, it remains the case that the predictive power and speed of QSAR can in many cases still beat the physical model. This advantage may well remain in future due mainly to two factors. First, faster, automated experimentation techniques (known as 'high throughput experimentation'[31]) deliver more data and hence larger training sets. Second, better descriptors can be obtained from advances in physical modelling and computer speed. For example, the free volume distribution of a polymer can be determined from large amorphous structure models and correlated as one of a set of descriptors with permeation measurements. Likewise the descriptors of the mesoscale phase separated structure of block copolymers can be mapped to mechanical properties.

8 Future Developments

The above sketch of past and present, and the discussion points raised have already pointed towards what the future might hold. Below are some further points I believe to be important.

8.1 Atomistic Modelling

Atomistic scale modelling is going to mature further with a wider range of tools and engines as well as ever improving force fields. We will be able to handle long chain polymers, and specific interactions between all components of interest in a relatively routine way. This will more and more allow us to make quantitative predictions, derive parameters for the mesoscale, and use the techniques as mature instruments for researching truly complex situations with confidence. However, a range of important question will remain tough nuts to crack for a while, especially where they involve many length and time-scales, such as in the case of the glass transition, and the modelling of polymer glasses, which cannot simply be referred to the mesoscale.

8.2 Mesoscale Modelling

Mesoscale modelling will continue to develop rapidly into a standard technique alongside quantum mechanical and atomistic modelling. As the last years have seen a tremendous development in modelling activity, an ever-increasing number of models and model variants has emerged. More comparisons and benchmarking will be needed. It is likely that a range of mesoscale simulation methods and engines such as MesoDyn and DPD will remain in use side by side, with our understanding in which is the 'best tool for the job' improving.

There is a huge potential in exploiting the combination of mesoscale modelling and nanoscale technologies, as the complete nanoscale processes can be simulated directly at the same length scale. Nanolithography for example has reached a stage at which the relevant etching and dissolution processes can be simulated directly by mesoscale models. The same holds for block-copolymers of course. This should lead to many applications from information storage, molecular sieves to foods, and biochemical engineering.

8.3 Hierarchical Modelling

Coarse graining and general scaling ideas have been extremely successful in polymers, helping to understand a wide range of their behaviours, and still helping to understand new phenomena. However, this has been done at the expense of being 'limited' to qualitative predictions. What is required in future are parameterized quantitative models for chemically specific systems. This can be achieved through careful hierarchical modelling. The presently used empirical parameterizations which work for a very limited range of systems need to be more firmly based, and gradually replaced by systematic parameterization routes integrating the faster/shorter degrees of freedom. The result should be a better leverage and control of structuring elements for a wide range of materials. The ability to tailor properties by using just a limited set of monomers and polymers will also improve as a consequence.

An essential part of a successful multiscale modelling strategy will be to use several scales in one simulation, where the decisions about the required level of modelling are taken by the computer and not by human intervention. This will require considerable computer science input. Despite this increasing complexity 'appropriate modelling' will remain important, *i.e.* distilling the essentials of behaviour at each length scale.

8.4 Processing and Properties

Despite the fact that many of the atomistic and mesoscale advances can help with the design and understanding of materials and processes, they generally still fall short of being able to predict key properties resulting from a whole process. It is sobering to realise that it still presents a challenge to predict the mechanical properties of an injection molded plastic bucket from the knowledge of chemical formula and the processing parameters alone. Even fundamental aspects involved in that process such as polymer crystallization in flows remain poorly understood, since they involve a complex range of structures and dynamics. There is huge potential in improving our understanding of such complex structuring during processing in order to control the microstructure and hence properties. However, more than qualitative work seems still some way off. Also the general question remains whether modelling will ever really be able to deal with the huge range of time and length scales involved.

8.5 Interdisciplinarity

Polymer modelling demands and enables interdisciplinarity. As mentioned above, the early split into atomistic biological polymer modelling, and coarse-grained physical modelling has already been largely overcome. In the last few years the fields of polymers and colloids have started to converge and are often addressed by terms such as soft matter, or: polymeric and soft materials. A recent model by Murat and Kremer[32] actually treats whole polymer chains as soft ellipsoidal particles, one could say soft colloids. Also, methods such as Lattice-Boltzmann and DPD are sufficiently flexible to allow a wide range of objects, from molecules *via* polymers and liquid crystal director fields to colloidal particles to be represented. Applications of mesoscale dynamics methods will include chemical engineering issues such as reaction blending as well as the simulation of cell morphology and functionality.[33]

8.6 The Way We Understand and Teach Polymers

The wider use of modelling and its presentation *via* the availability and power of computer graphics is going to change the way we understand polymers and their behaviour, and by implication the way in which we do science and develop materials. Simulations will be used more widely in education to illustrate polymer behaviour on all length scales. Successful future applications of polymer materials depend not only on the latest scientific advances being discussed by a few dozens of people world-wide, but on getting the appreciation of a much larger group of scientist, engineers, managers, technicians and teachers. Modelling can greatly help in this respect.

8.7 Physical *vs* Knowledge-based Modelling

Coming back to the issue raised in the introduction: physical modelling to replace experiment and knowledge-based modelling to couple experiment and simulation more closely: the two opposing trends? My conclusion is that they are in fact complementary: improvements in the physical modelling will lead to better models and descriptors for correlation based approaches, while more experimental data correlated to a range of models should lead to improved simulations. Also both the knowledge-based approach and the direct simulation approach have and will remain to have particular applications. The real advantage could lie in a combination of both approaches, *e.g.* in the use of neural network modelling as an additional engine to help with the complex scale of parameterization necessary in hierarchical modelling.

Whatever the outcome will be between these emerging trends, it is a fact that modelling is making an ever stronger contribution to our understanding of the complex behaviour of polymers, and is changing the way in which new polymer materials are developed.

References

1 P.J. Flory, 'Principles of Polymer Chemistry', Cornell University Press, 1953.
2 W. Kuhn, *Kolloid Z.*, 1936, **76**, 258; 1939, **87**, 3.
3 E.A. Colbourn, 'Computer Simulation of Polymers', Longman Scientific, Harlow, 1994.
4 G. Goldbeck-Wood and A.H. Windle, *Rheol. Acta*, 1999, **38**, 548.
5 M. Doi and S.F. Edwards, 'The Theory of Polymer Dynamics', Clarendon Press, Oxford, 1988.
6 A. Uhlherr A and D.N. Theodorou, *Current Opinion in Solid State & Mater. Sci.*, 1998, **3**, 544.
7 D.M. Sadler and G.H. Gilmer, *Polymer*, 1984, **25**, 1446.
8 J.P.K. Doye, *Polymer*, 2000, 41, 8857.
9 B. Heck, T. Hugel, M. Iijima and G. Strobl, *Polymer*, 2000, 41, 8839.
10 P.D. Olmsted, W.C.K. Poon, T.C.B. McLeish, N.J. Terrill, A.J. Ryan, *Phys. Rev. Lett.*, 1998, **81**, 373.
11 D. Frenkel and B. Smit, 'Understanding Molecular Simulation', Academic Press, San Diego, 1996.
12 H. Sun and D. Rigby, *Spectrochemica Acta*, 1997, **A53**, 1301.
13 D.N. Theodorou, 'Hierarchical Modelling of Polymers', SIMU Newsletter, 2000, **1**, 19.
14 A. Consta, N.B. Wilding, D. Frenkel and Z. Alexandrowiz, *J. Chem. Phys.*, 1999, **110**, 3220.
15 P.V.K. Pant and D.N. Theodorou, *Macromolecules*, 1999, **32**, 5072.
16 K.R. Haire, T.J. Carver and A.H. Windle, *Comp. Theor. Polym. Sci.*, 2001, **11**, 17.
17 M.L. Greenfield and D.N. Theodorou, *Macromolecules*, 1998, **31**, 7068.
18 P.M. Hadgett, Ph.D. Thesis, University of Cambridge, 2000.
19 N.M. Maurits, A.V. Zvelindovsky and J.G.E.M. Fraaije, *J. Chem. Phys.*, 1998, **109,** 11032.
20 R.D. Groot and P.B. Warren, *J. Chem. Phys.*, 1997, **107**, 4423.
21 S. Kumar, 'Computer Simulation of Polymers', Longman Scientific, Harlow, 1994, p228.
22 K. Binder and W. Paul, *J. Polym. Sci. B: Polym. Phys.*, 1997, **35**, 1.
23 T.J. Carver and A.H. Windle, *Comp. Theor. Polym. Sci.*, 2001, in press.
24 P. Espanol, X.F. Yuan and R.C. Ball, *J. Non-Newt. Fluid Mech.*, 1996, **65**, 93.
25 N.J. Inkson, T.C.B. McLeish, O.G. Harlen and D.J. Groves, *J. Rheology*, 1999, **43**, 873.
26 B.A.C. van Vlimmeren, N.M. Maurits, A.V. Zvelindovsky and J.G.E.M. Fraaije, *Macromolecules*, 1999, 32, 646.

27 J. Bicerano, 'Predictions of the Properties of Polymers from their Structures', Marcel Dekker Inc, New York, 1993.
28 D. Porter, 'Group Interaction Models', Marcel Dekker, New York, 1995.
29 H.K.D.H. Bhadeisha, *ISIJ International*, 1999, **39**, 966.
30 C.K. Williams, *Neural Computation*, 1998, **10**, 1230.
31 J.M. Newsam, S.M. Levine, D. King-Smith, D. Demuth and W. Strehlau, *Abs. Papers Am. Chem. Soc.*, 2000, **219**, 13-MTLS, Part 2.
32 M. Murat and K. Kremer, *J. Chem. Phys.*, 1998, **108**, 4340.
33 J.G.E.M. Fraaije, *SIMU Newsletter*, 2000, **1**, 13.

19

Present and Future Developments in SR Technology for Polymer Science

Greg P. Diakun and Nick J. Terrill

CLRC DARESBURY LABORATORY, DARESBURY, WARRINGTON, CHESHIRE WA4 4AD, UK

1 Introduction

The length scales associated with polymer structures are ideal for study employing X-ray scattering techniques. This can be subdivided into two regions, wide angle X-ray scattering (WAXS) which probes atomic distances (1–20 Å) revealing the local conformation of the monomer units in the semi-crystalline state, and small angle X-ray scattering (SAXS) which probes length scales (20–1000 Å) detailing the long-range order present in many polymers. In SAXS, X-rays are sensitive to regions with different electron densities and the technique provides information on crystallite size, unit cell size, thickness of the interface between microphases, or between the crystalline and amorphous regions. Thus crystallisation and phase separation processes are the most commonly studied areas using this technique. The high intensity of X-rays produced from synchrotron radiation sources enables dynamic X-ray scattering experiments to be performed on materials contained within sample environments that mimic processing conditions and mechanical function. The aim of this paper is:

- To briefly describe the facilities available at the SRS Daresbury Laboratory for the study of polymer materials and to summarise the work carried out on the micro focus X-ray beamlines at the ESRF where high intensity X-rays are produced in spot sizes of 10 μm^2 or smaller.
- To describe new stations that will be available over the next two years on the SRS.
- To provide information on the replacement synchrotron, DIAMOND, and types of insertion devices that may be used for SAXS/WAXS experiments on the new machine and to provide some suggestions for possible beamlines on such a facility.

- To describe the types of detector available now and future developments.

2 Current Facilities at Daresbury Laboratory for the Study of Polymers

2.1. X-ray Scattering Facilities.

The beamlines used for X-ray scattering measurements have several principal components – a monochromator, a mirror, various slits and the X-ray camera. As X-rays cannot be focused in a manner comparative to light, beamlines use instead Bragg reflection from crystals, and mirrors set at grazing angles of incidence. The radius of curvature of the mirror and/or crystal determines the point of focus, these elements being controlled by precision motors with the optical elements isolated from external sources of vibration. The crystal monochromator is used to select the X-ray wavelength required by Bragg diffraction, from the broadband of synchrotron radiation and may also focus the beam.

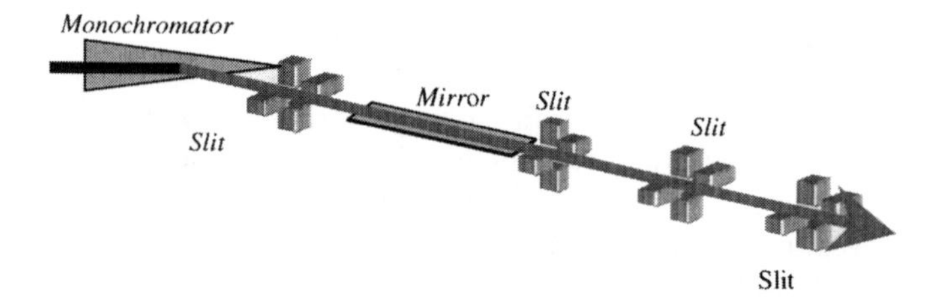

Figure 1 *Generic schematic of the NCD beamlines 2.1, 8.2 and 16.1*

The SRS has three small angle scattering stations, which covers the majority of requirements of the polymer communities. All are based on a similar optical geometry (Figure 1) as outlined below and summarised in Table 1. The three stations operate at a fixed wavelength, with the first optical element being the monochromator. This is used to collect a large amount of white radiation horizontally and monochromate and focus. The choice of monochromator is Ge (111) for its broad rocking curve resulting in increased flux output. The Ge (111) planes are asymmetrically cut (approx. 10.5 °) reducing the angular spread of the X-rays, so that the output beam cross-section is reduced by *ca.* 10 times, *i.e.* the width of the reflected beam is compressed relative to the incident beam. This can be calculated from the following formula (1)

$$A = \frac{\sin(\theta - \alpha)}{\sin(\theta + \alpha)} \quad (1)$$

where θ is the Bragg angle for a specific wavelength and α is the asymmetric cut of the crystal. The monochromator is then bent to provide horizontal focussing. This is followed by a quartz mirror, which provides focussing in the vertical plane. A series of grouped slits are positioned after the mirror to eliminate parasitic scatter, as this influences the resolution range of the station at a particular camera length.

The three fixed wavelength stations described above include variable camera lengths, which allow access to a wide range of d spacings, (see Table 1).

The high X-ray flux and collimation of these beamlines allows time-resolved experiments down to the 10's of microseconds time frames to be performed. They also enable experimenters to utilise new sample environments and use simultaneously, several experimental techniques on one sample. An excellent overview of recent developments in this field is given in Bras and Ryan.[1] Experience has demonstrated that by combining experimental techniques across a wide range of spatial and chemical probes an understanding of complex processes can be obtained. Below are examples of the techniques in use at the SRS.

2.2 SR FTIR Microscopy

The use of FTIR is commonplace in most research laboratories throughout the world for the investigation of molecular bonding and reaction kinetics. An advantage of synchrotron IR is the highly collimated and polarised nature of the beam. This affords the opportunity to examine variation in molecular orientation in very small samples. Using this technique it was possible to measure the angular dependence of the absorption of the polarised beam across the spherulite providing information on orientation and possibly even the location of the crystalline domains. It was also possible to follow the thermal crystallisation of PEO by monitoring the IR spectrum over a narrow range. These absorption spectra were recorded, from a 15 μm spot and include the CH_2 vibrations around 1460 cm^{-1} and around 1350 cm^{-1}. Marked band sharpening is observed during this process and band splitting is also seen during the later stages, see Figure 2(b). This is a rapidly developing technique where reaction kinetics and morphology meet.[2]

2.3 Stress, Strain and Structure

Tensile testing is a fundamental method for looking at polymers to provide information on their mechanical properties. Their behaviour will be influenced by formation of defects in the bulk, as well as conformational changes to the molecular chains. SAXS/WAXS is an excellent tool to correlate these properties to the amount of strain. The results shown in Figure 2(c) demonstrate combined tensile measurements with morphological studies to investigate the properties of polymers under strain. The Minimat Materials Tester provides the experimenter with a method of investigating polymers and matching their mechanical properties to their morphology. These types of measurement can provide critical information for the design of materials for new applications.[3]

Table 1 *Characteristics of stations used for SAXS/WAXS on the SRS*

Station	*Source*	*Wavelength*	*Spot Size* $(H \times V)$ *mm*^{2}	*d spacing Å*	*Principle Operating Mode*
2.1	Bending Magnet	Fixed 1.54 Å	2.5 × 0.5	10–2000	SAXS 1D and 2D
8.2	Bending Magnet	Fixed 1.54 Å	3 × 0.3	6–800 1–6	SAXS/WAXS 1D
16.1	Wiggler	Fixed 1.41 Å	3 × 0.8	10–1000 1–5	RAPID SAXS 2D and SAXS/WAXS 2D
6.2*	Multipole Wiggler	Variable 0.7–2.5 Å	*Ca.* 2.5 × 0.6	Dependent on wavelength	RAPID 1D SAXS/WAXS
14.2**	Multipole Wiggler	Fixed 1.2Å and 1.5 Å	0.4 × 0.3	~1.5–500	30cm MAR Image Plate /ADSC Quantum 4R

* Station operational in 2001.
** Station operational late 2000.

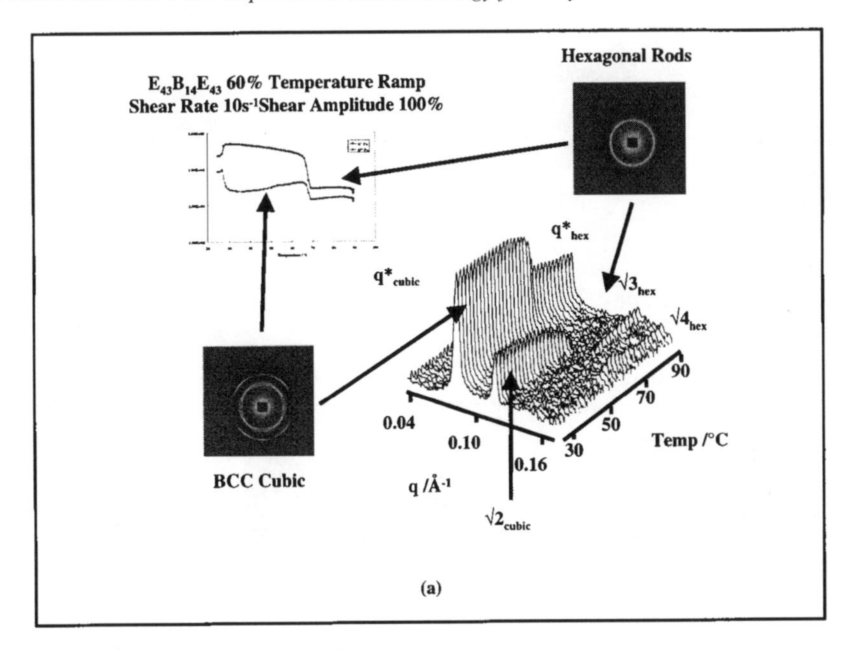

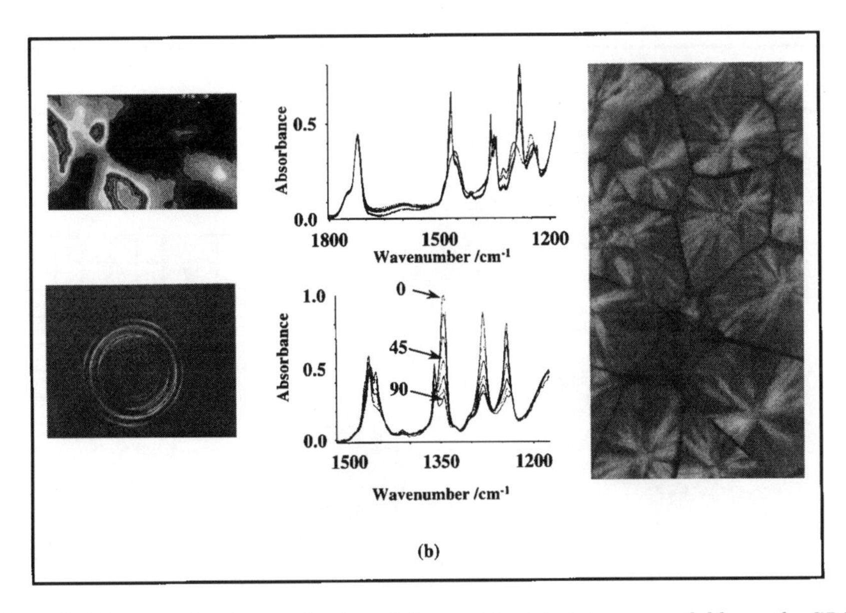

Figure 2 *Diagram showing a selection of the combined techniques available on the SRS. (a) Rheological studies using an in situ parallel plate rheometer on 16.1/2.1.*[7] *(b) Synchrotron FTIR using station 13.1a to investigate poly(oxyethylene) spherulites.*[2] *(c) Tensile testing of polyethylene using 2D SAXS/WAXS on 16.1.*[3] *(d) Reflectivity carried out on 16.2 on thin films of block copolymers comprising poly(oxyethylene)-poly(oxybutylene).*[4] *(e) Simultaneous SAXS/WAXS and Raman using a holoprobe device on 8.2.*[5] *(f) Extrusion studies on polypropylene carried out on 16.1*[6]

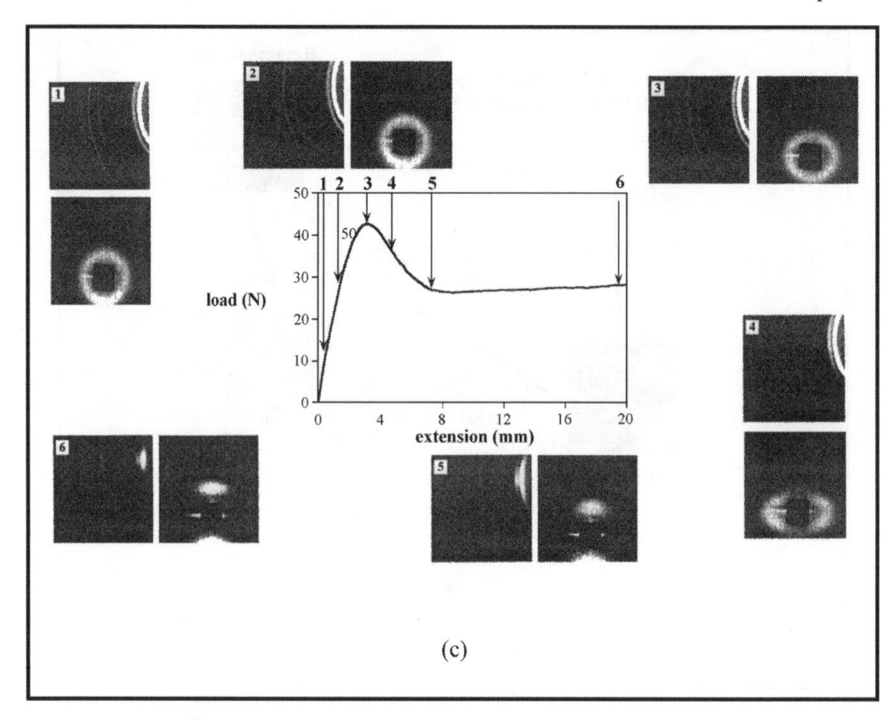
1
2
3
4
5
6
load (N)
extension (mm)

(c)

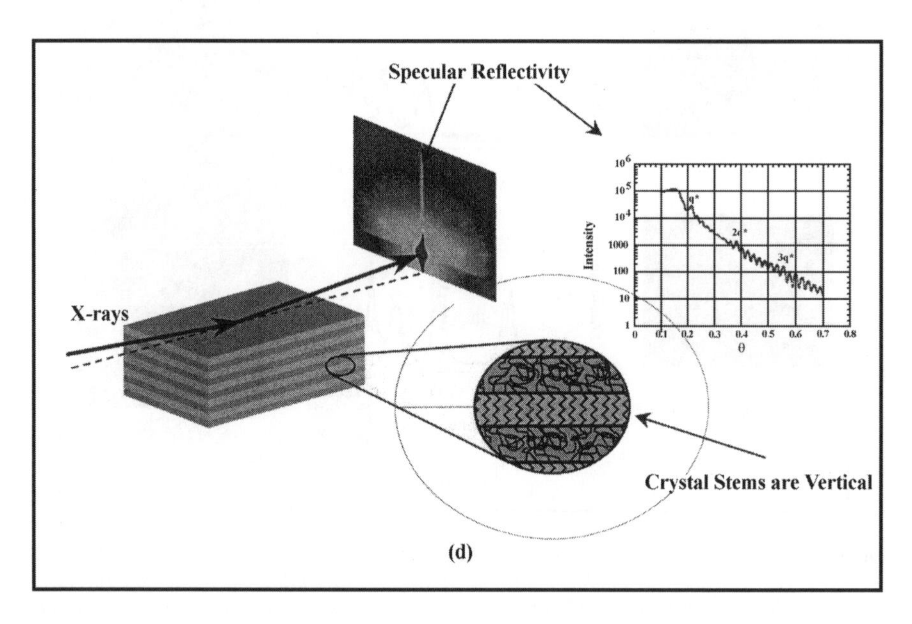
Specular Reflectivity
X-rays
Intensity
θ
q*
2q*
3q*
Crystal Stems are Vertical

(d)

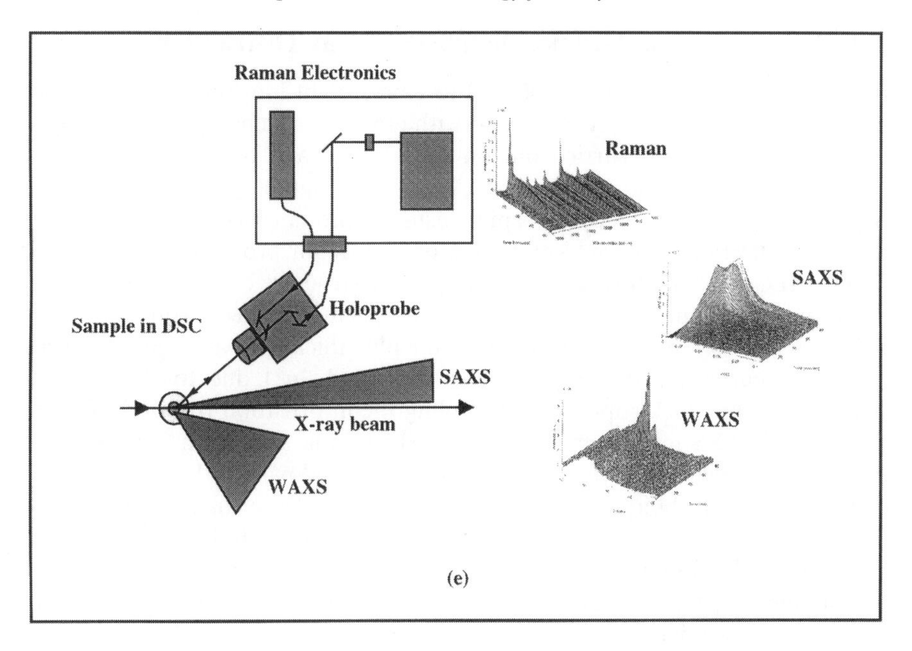

Raman Electronics
Raman
SAXS
Holoprobe
Sample in DSC
SAXS
X-ray beam
WAXS
WAXS

(e)

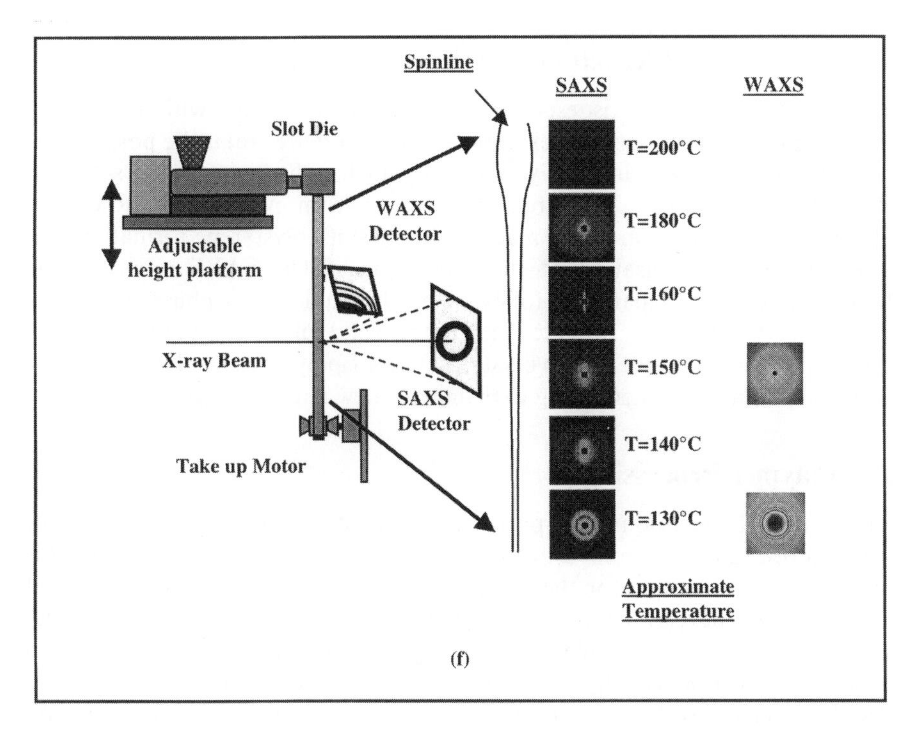

Spinline
SAXS
WAXS
Slot Die
T=200°C
WAXS
Detector
T=180°C
Adjustable
height platform
T=160°C
X-ray Beam
T=150°C
SAXS
Detector
T=140°C
Take up Motor
T=130°C
Approximate
Temperature

(f)

2.4 Reflectivity and Grazing Incidence X-ray Diffraction

A technique, which is being exploited to gain insight into the morphology of polymer films, is reflectivity combined with grazing incidence X-ray diffraction (GIXD). This work is carried out on station 16.2 and operates routinely at wavelengths of 1.3 and 0.5 Å. The example in Figure 2(d) is of a triblock copolymer EBE, produced by spin casting onto a silicon substrate. The specular reflection provides knowledge on the depth profile of the material. The GIXD gives us information relating to the structure of the polymer block at or near the surface of the film. The spacing between fringes on the reflectivity curve corresponds to the total film thickness, and superimposed upon this curve are Bragg peaks (q^*, $2q^*$, and $3q^*$) due to the lamellar structure of the block copolymer. Coupling the information from GIXD with that from WAXS measurements performed on bulk samples it is possible to establish the orientation of the rods in the structured part of the polymer film. This combined technique is excellent for probing the surface layers of thin films, common throughout the packaging and coatings industries. The next stage in the development of this technique will be to temperature profile the morphological properties of these and other polymer films to gain insights into their thermal behaviour.[4]

2.5 Structure and Reaction Kinetics

SAXS/WAXS/RAMAN is especially useful when dealing with chemically induced phase transitions. The example shown in Figure 2(e) is the polymerisation of solvent styrene into polystyrene in which polyethylene is in solution. Polyethylene is soluble in styrene but insoluble in polystyrene. RAMAN allows the determination of the reaction kinetics of polystyrene formation and monitors the crystallisation of the polyethylene. The SAXS monitors the liquid–liquid phase separation followed by the liquid–solid phase transition, whilst the WAXS also observes the liquid–solid phase by monitoring the appearance of peaks due to the crystallisation of polyethylene. These are very valuable parameters when trying to define any new manufacturing process.[5]

2.6 Polymer Processing

The study of extrusion is a prime example of utilising the high flux of synchrotron radiation to investigate real industrial problems. Melt spinning is an important technique for the production of polymer fibres and films. Real time structure development can be followed by using both small and wide-angle scattering. The example shown in Figure 2(f) is an investigation into the crystallisation process. The results indicate density fluctuations occur in the SAXS pattern prior to crystallisation in the WAXS pattern. These results are causing groups to rethink nucleation theory. Better understanding of the origins of the morphology formed will lead to better, higher performance products.[6]

2.7 Rheology and Physical Properties

The microstructural response of polymers to applied shear deformation is of considerable interest commercially, as these materials have numerous industrial applications. For example, it is important to understand the response of the polymer systems during the production process. For this reason a wide variety of shearing devices have been developed with combined SAXS capability to study *in situ* shear induced phase behaviour of polymers. The example shown in Figure 2(a) is of a triblock copolymer E43B14E43 (E = ethylene oxide, B = butylene oxide). The ethylene oxide is crystallisable whilst the butylene oxide always remains amorphous. By carrying out simultaneous SAXS and rheology measurements it is possible to pick out the phase transition from BCC to hexagonal rods on raising the temperature at a constant shear rate and amplitude. Knowing where the phase transitions are in these systems is crucial to the manufacturing industry where failure to understand these parameters can be very costly.[7]

2.8 Microfocus SAXS

Micro-focus SAXS/WAXS is an exciting new technique only made possible by the advent of third generation sources. On these facilities, utilising an undulator beamline, high intensity focused X-ray beams of 1 micron or less are achievable. One of the drawbacks of going to such small beam dimensions is the loss in resolution in the small angle, *e.g.* for a 2 μm beam resolution is *ca.* 150 Å, for a larger beam size of 10 μm a resolution of 670 Å can be achieved.[8]

The use of such small beams allows X-ray scattering patterns on single polymeric fibres with diameters down to a few microns to be collected in a few seconds. Thus scanning experiments across a fibre become feasible. It also allows study of individual spherulites or deformations near the edge of crack tips. Below are examples of work carried out on ID13 at the ESRF.

Polymer materials that are used to contain food are required to be resistant to shock and gas pressure. The physical properties of the final product depend largely on the details of the manufacturing process. By probing the variation of molecular structure in the container wall, C. Martin and co-workers (Keele, UK) have been able to provide a connection between the local order of the polymers and the manufacturing process.[9] (Figure 3(a).)

High performance polymers are not a human invention and nature is still doing better than technology in many cases.

Dragline spider silk has aroused considerable interest due to its excellent mechanical properties, for example stability, elasticity and low weight. A. Bram and co-workers (ESRF) have succeeded in recording X-ray diffraction patterns from a single spider dragline of less than 10 μm diameter[10] (Figure 3(b)). These results allow the elastic properties of the fibres to be linked to the molecular architecture of the polymer chains.

Figure 3(c) shows the SAXS pattern of the centre of a PTMS (poly[tetra-

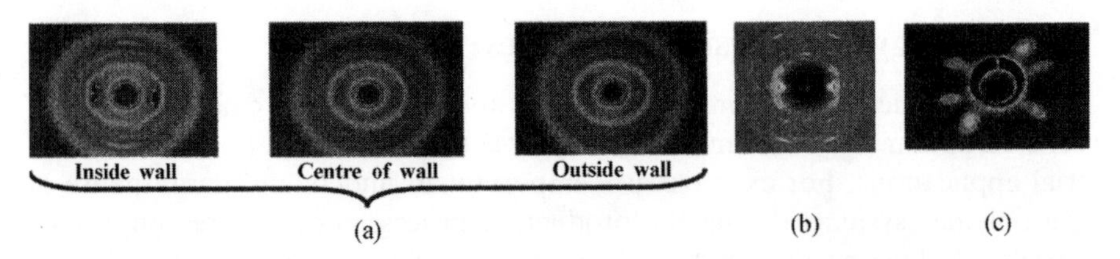

Figure 3 *Examples of Microfocus experiments carried out on ID13 at the ESRF. (a) Patterns from the wall of a PET soft drinks bottle.*[9] *(b) Single Spider Silk fibre.*[10] *(c) Centre of a PTMS spherulite*[11]

methyl-p-silphenylene siloxane]) spherulite recorded with a 2 μm beam.[11] The six peaks indicate the presence of three oriented fibrils in the probe volume.

The experimental study of micron size domains in minute samples opens the way for the detailed analysis of material properties in terms of theoretical models accounting for the molecular architecture of matter.

Time resolved experiments are still very difficult because the flux is relatively low compared with conventional SAXS instruments. In addition, keeping the X-rays illuminated at the same spot on the sample while it is perturbed is very difficult.

2.9 X-ray Photon Correlation Spectroscopy

X-ray Photon Correlation Spectroscopy (XPCS) is another area that is benefiting from the construction of third-generation synchrotron radiation sources. This technique investigates the dynamics of condensed matter at the molecular level.[12–14] This developing procedure draws on the knowledge gained from dynamic light scattering and applies it to the X-ray region. Due to their shorter wavelengths X-rays produce far fewer multiple scattering events and this helps to alleviate many of the problems encountered with other better established varieties, *e.g.* light and neutrons. This has the added benefit of being able to access larger areas of phase space than the other techniques. Thus XPCS has the potential to have a major contribution to our understanding of the motions within many polymeric systems. Further advances in accelerator technology will open up new opportunities in this area, as fourth generation sources will provide even higher brilliance and coherence of X-rays.

3 Future Development

3.1 SRS – Station 6.2

Station 6.2 brings together three techniques optimised for materials processing. These are 1D SAXS/WAXS, X-ray diffraction and XAS (X-ray absorption spectroscopy). This will be a tuneable facility and equipped with state of the art detectors constructed along similar lines to the RAPID 2D detector for station 16.1. This will enable count rates of *ca.* 10 MHz for both 1D small and wide angle experiments exceeding the existing capabilities of the detectors on station 8.2 by *ca.* 40 and 500 respectively. This has huge advantages as in many experiments it is necessary to attenuate the X-ray beam to avoid overloading the detectors. In addition, because of the variable wavelength it will also be possible to carry out anomalous scattering experiments, something which is new to the repertoire of facilities offered to the polymer community. The beamlines optics comprises a vertically focussing collimating mirror followed by a double crystal sagittal bent monochromator and then a plane focusing mirror. It is designed for very rapid, tuneable, combined SAXS/WAXS experiments. Experiments envisaged on this station include:

- Polymer synthesis to enable the structural kinetics of polymers to be studied in real time.
- Polymer processing to permit study of the effects of shear and extension flow upon the crystallisation kinetics of polymers. The variable wavelength is vital because the longer wavelengths enhance scattering from thin films (50–100 μm) or single fibres, whereas shorter wavelengths will improve penetration of *in situ* processing cells.
- Time resolved studies of ionomers and polyelectrolytes, which find applications in areas such as lightweight batteries and engineering materials. Ionic aggregation has been investigated by the use of anomalous SAXS but as yet no time-resolved studies have been conducted. As station 6.2 will be tuneable it should be possible to study the processes *in situ* by a combination of ASAXS/AWAXS to probe the spatial organisation of the ionic groups within the polymer and the local structure of the ions.
- Industrial crystallisation processes by SAXS/WAXS to probe crystal size and shape (SAXS) and polymorphism (WAXS).
- High pressure studies using the short wavelengths available on the station to penetrate the sample cells. Many soft materials undergo high pressure deformation during processing and the consequences of this are poorly understood.

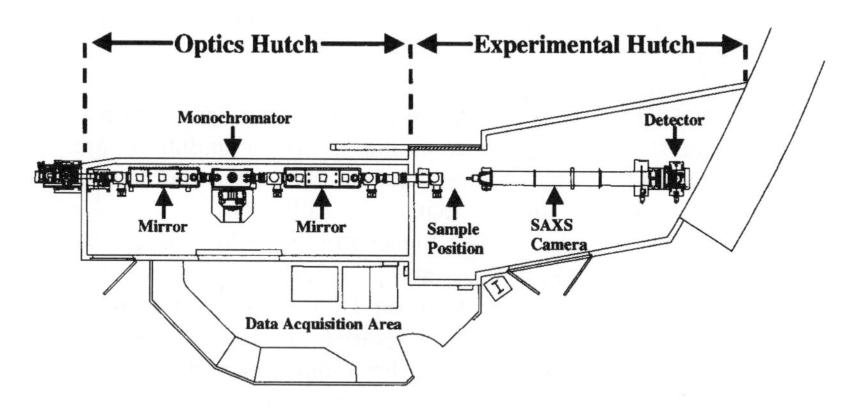

Figure 4 *Schematic Layout of Station MPW 6.2*

3.2 SRS – 14.1 Fibre Diffraction (High Resolution)

Station 14.1 is located towards the edge of the wiggler 14 fan therefore it will receive less high-energy radiation. Nevertheless it should provide over 10 times the flux of the protein crystallography station 7.2 which is used at present for fibre diffraction studies. The station will operate with a Si (111) monochromator at 1.2 Å and 1.5 Å and initially will use a 30 cm MAR Image Plate. This will be replaced with an ADSC Quantum 4R detector when available. The latter is a 2×2 CCD array with an active area of 180 mm^2 and a readout time of 3 seconds. The spot size at the sample will be no larger than 0.3 (V) $\times$ 0.4 (H) mm^2 and the flux is expected to be *ca.* 5×10^{12} photons s^{-1}. A fibre diffraction camera, enabling wide-angle scattering studies of biological macromolecules and non-biological polymers is under development. As an interim measure, the Keele camera from station 7.2 will be fitted and used for fibre studies.

3.3 DIAMOND

It has been identified that there is a need to replace the current high intensity X-ray Synchrotron Radiation Source (SRS) at Daresbury early in the next decade with a new facility (DIAMOND). This new third generation synchrotron facility will be up to 10,000 times brighter than its predecessor and is necessary to help maintain the UK's position in many fields from materials research to the life sciences. It will be commissioned in about five years' time and the project to design and build is currently estimated to cost around £175m. The project is being jointly funded as part of a partnership between the Wellcome Trust, UK and French Governments. The specifications of the new machine have yet to be finalised with key parameters such as energy, number of cells and emittance still to be decided. The final machine energy is likely to be between 3.0 GeV and 3.5 GeV. This paper therefore provides ideas for possible beamlines on DIAMOND for polymer research. However, whatever is finally built will be dependent upon user demand and input. The purpose here is to provide the polymer community with ideas and to encourage feedback concerning anticipated requirements.

3.3.1 Basic Beamline Options. There are primarily two basic insertion device options relevant to the polymer community, (a) an undulator and (b) a multipole wiggler. These offer different characteristics; the undulator provides a narrow central cone of radiation that is generally less than 1mrad in width whereas the multipole wiggler offers higher flux but with a larger beam size and divergence. The community has the opportunity to influence the options of insertion devices it wishes to utilise on DIAMOND but it must have a scientific direction in place to justify its request. It must be remembered that at this point in time no decision has been made regarding allocation of beamlines to either areas of science or technique. It may well be the case that a beamline will be constructed for multiple uses, *e.g.* Small and Wide Angle

Scattering, Powder Diffraction and EXAFS. Thus it is very important that the direction the science will take in the next 10 years is considered now. Now would be an excellent time to discuss the communities' requirements with the scientific staff at Daresbury Laboratory. The purpose of the options presented here is to provide a flavour of what may be possible for polymer science on DIAMOND. It is not designed to direct your ideas in any way.

Working with the two basic insertion devices described earlier the following virility curves give a sense of the performance of two generic stations that it is felt might be applicable to the polymer communities needs. There is a perception that undulators provide fixed wavelength radiation. It is now possible with modern technology to tune undulators to the desired energy, and this can be achieved in real-time (seconds) whilst the storage ring is operational. The examples shown in Figure 5 are tuning curves for an in-vacuum undulator and an out of vacuum undulator. The curves show the odd order harmonics for each insertion device starting with the third harmonic on the left. The tunings curve depicts the available energy by varying the gap of the undulator.

An International Working Group at Daresbury Laboratory concluded that in-vacuum undulators have now reached a level of maturity such that they can be seriously considered from the outset in new project proposals. Magnetic, vacuum, RF shielding, alignment and magnetic measurement all seem sufficiently well resolved to allow use of higher harmonics up to the 11th given sufficient time for shimming. If there is a real requirement for high flux experiments then recourse to Multipole Wigglers is still an option but does not give the same degree of flexibility as an undulator, *e.g.* microfocus experiments.

3.3.2 Combined Technique Ideas. What follows are ideas for beamlines that in-house staff have generated. They may not be relevant or appropriate to future needs of the polymer community but are included to stimulate thought and discussion.

3.3.2.1 Combined SAXS/FTIR beamline. FTIR and X-ray diffraction are common techniques used to evaluate the chemical and structural characteristics of modern materials. Simultaneously measurements have been performed on SRS station 8.2 using a commercial FTIR instrument coupled to a SAXS/WAXS beamline.[15] However, for experiments where spatial resolution is required, FTIR microscopes utilising the IR radiation from a synchrotron can produce 10 μm spot sizes with a *ca.* 30-fold increase in brightness over conventional instruments. Thus the complementary X-ray and IR microscopy beam sizes, offered by third-generation sources, presents a unique opportunity to design a combined instrument which will exploit these properties. It is envisaged that such a beamline would have wide ranging applications in many areas such as biology, chemistry, materials and environmental science as well as polymer science. Experiments possible on this station might include:

- Combined structural and chemical evaluation of many block-copolymer systems during processing – to examine which mechanism dominates the process.

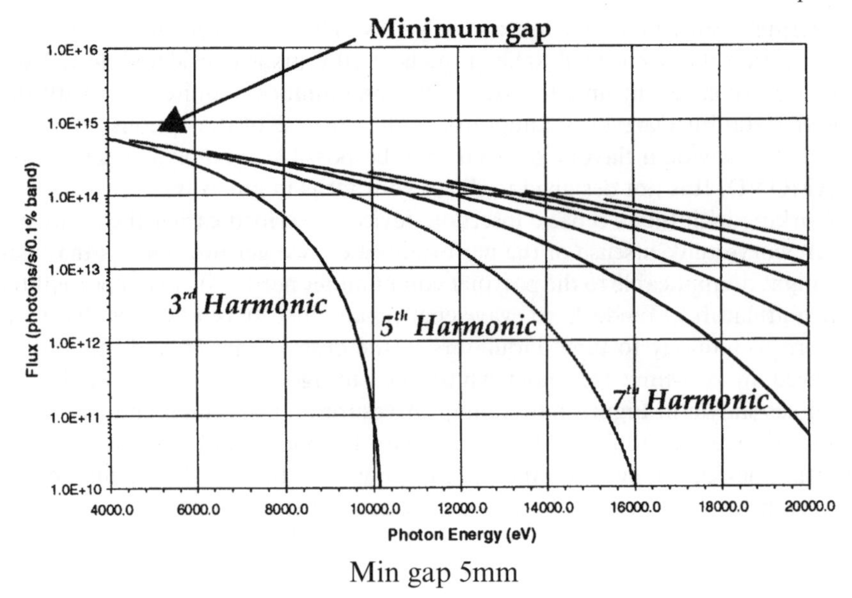

Min gap 5mm

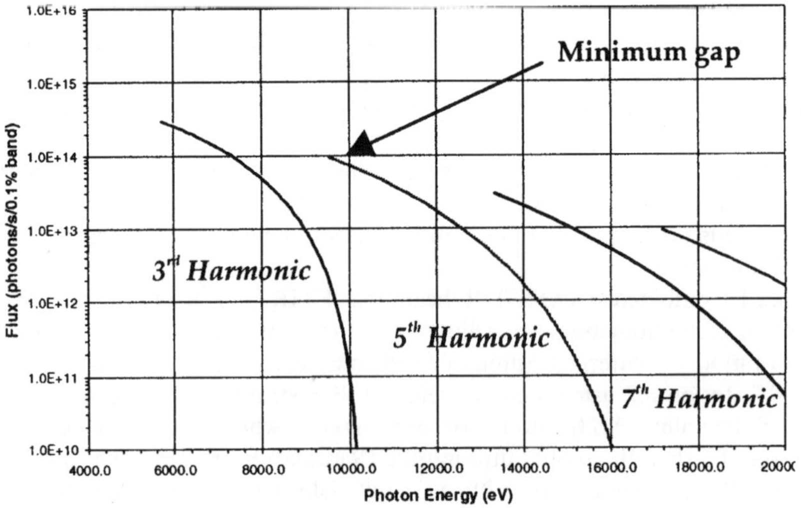

Min Gap 10mm

Figure 5 *Tuning curves for selected Undulator Options*

- Probing spherulite formation both chemically and morphologically – to understand the correlated information.
- Examination of extruded and other processed materials with compatible beam sizes in the same area simultaneously – microprobing both techniques across domains as it would be valuable to assign order both chemically and structurally.

This unique combination of techniques would make it feasible to follow a process dynamically and obtain information not only on the backbone framework of a system but also to gain insights into the orientation of chemically and structurally important moieties.

3.3.2.2 Combined SAXS/NMR beamline. This idea couples one of the most utilised analytical techniques, NMR, to SAXS for a more detailed investigation of dynamic macromolecular structures. NMR can be used to examine biopolymers and lipids to extract information concerning their molecular arrangements and chemical environments. Coupled directly with a structural technique, such as SAXS, a very powerful analytical tool can be created for the examination of dynamics in large-scale structures. The NMR machine that is envisaged to be used initially would be of the old design of split magnets, not the more modern superconducting variety, which would lead to some very challenging technical problems. As the old design is to be used the information obtained will only be qualitative; data collection times will not be fast by Synchrotron standards but it will allow a general overview of experiments to be observed.

3.3.2.3 Combined SAXS/XANES beamline. In a typical SAXS experiment, using a triangular Ge (111) monochromator to select wavelength and focus the beam, the overall bandpass from the bent monochromator is of the order of 1–2% at 10 keV. This results in an energy spread of *ca.* 100–200 eV, more than sufficient for a X-ray Absorption Near Edge Structure (XANES) experiment. XANES can provide useful information on the Oxidation State of the element of interest and geometry of the surrounding atoms. This would allow the experimenter to follow not only the phase changes in the polymer system but also to investigate changes to the metal atom participating in the reaction mechanism. One could envisage a station where the SAXS data was collected simultaneously with a XANES pattern. To perform such an experiment the beam would have to be focused horizontally at the sample, to provide sufficient resolution for the XANES detector, which would reside in the SAXS beamstop. The station would require a compromise in set-up such that the beam is defocused slightly for the XANES experiment to give an improved resolution at the SAXS detector. This arrangement already exists on our SAXS/WAXS beamlines where focusing is usually part way down the camera to afford the best compromise between SAXS and WAXS resolution. It would also be possible, by varying the angle of the beam off the monochromator, to access different wavelengths and also different energy spreads.

3.3.2.4 Combined SAXS/Circular dichroism beamline. Biological macromolecules, such as proteins, carbohydrates and nucleic acids, are composed of many optically active or chiral units that exhibit large Circular Dichroism (CD) signals. CD spectroscopy has therefore been used extensively in the study of proteins, where asymmetric carbon atoms in their amino acid 'backbone' give rise to a CD spectrum. The shape of the spectrum depends on the protein's secondary structure content and allows the proportions of helix, beta structure, turns and random to be determined.

CD is an excellent method for determining the secondary structure content of proteins in their native state, but it is limited by the fact that much of the information is located at wavelengths (below 200 nm) where the light output from conventional Xenon lamps diminishes markedly. In contrast, the flux obtained from synchrotron light remains high at these wavelengths. Also, the inherent polarisation of synchrotron radiation makes it the ideal light source for CD experiments.

Time-resolved CD is a good method for monitoring the development of secondary structure as proteins fold. Proteins are first unfolded by adding denaturants like urea or guanidinium hydrochloride. Folding is then initiated by rapidly diluting the denaturant using a stopped-flow apparatus. Unfortunately, the limited wavelength range offered by conventional CD instruments means that the information obtained is limited, and the development of individual secondary structure types cannot be monitored.

Daresbury scientists have used station 13.1b to collect CD data on the folding of proteins over the wavelength range from 190–240 nm.

Coupling this with SAXS measurements in the Guinier region gives a measure of the change in size of the system as the protein collapses to fold.

A cross beam experiment would be possible with the sample held in a cell with appropriate windows for both SAXS and CD and also different path lengths for their respective incident beams.

Using such a station configuration it might be feasible to follow the formation of micelles and any subsequent phase transitions whilst at the same time following the secondary conformational organisation of the individual polymer chains.

3.4 Detectors

Detectors used for time resolved non-crystalline diffraction experiments on a synchrotron X-ray source should have most if not all of the following characteristics:

- Two dimensions.
- Large size ($>10^4$ mm^2).
- Reasonable spatial resolution ($> 500 \times 500$ pixels with a clean Point Spread Function).
- High dynamic range ($>10^4$).
- Low noise level (<few photons mm^{-2}).
- Rapid read out (smallest possible inter-frame dead time).

A final requirement arises if the stroboscopic technique is to be used, where the experiment is repeated many times and the resulting data summed to provide enough photon statistics. In this case the detector must be capable of being synchronised to the experiment.

Finally, despite significant limitations and a somewhat unfashionable image, the combination of excellent noise performance, high efficiency, reasonable spatial resolution, large active area and rapid frame rate, means that photon counting gas detectors will remain the workhorse for time resolved non-crystalline diffraction for some time to come.

For a more detailed description of the detectors employed at Daresbury Laboratory see Lewis.[16]

3.4.1 Future Developments in Detector Technology. The challenge of new developments in detector technology is to meet the joint needs of the new science that can be envisaged and the increased flux levels produced by third generation sources. The real test will be the evolution of devices with very low noise performance at rapid frame rates, which can detect weak features that need to be measured in 'cutting edge' experiments. The only route to low noise, at present, is photon counting and it is developments in this area that are likely to create the most possibilities for time resolved non-crystalline diffraction.

There are several types of photon counting detector front-ends that can now handle the desired count rates, ranging from the new types of gas detectors (Microstrip gas chambers and MicroGaps) to silicon arrays. Crucial to the success of these detectors is the design and fabrication of a readout and memory system which can handle both the count rate and framing rate of the next generation of time resolved diffraction experiments.

The most immediate solution (2 years away) is the GFS upgrade to the RAPID readout system.[17] The idea behind this system is that by segmenting the active area of the detector it is able to perform many more operations significantly faster. The predicted performance is a count rate in excess of 10^8 s^{-1}, which is certainly sufficient for many experiments. In the longer term truly pixelated detectors, based on readout electronics on a silicon chip, will be necessary to cope with the increased X-ray fluxes.

The massively parallel nature of such a system means that the detector can have an enormous overall count rate capability whilst that of individual pixels can be quite modest. The challenge lies in creating a system with either sufficient read speed or onboard memory capacity to permit rapid time resolved work.

4 Conclusion

The opportunity that DIAMOND affords for new science is only bounded by ones imagination and all we have tried to do in this paper is to provoke thought in this key area. The polymer community should also be considering what experiments they can perform that take place in the microsecond regime to take advantage of the flux levels that will become available on DIAMOND and other third generation sources in the future. Paramount to realising the full potential of the new source will be detector development. At present many Synchrotrons around the world are let down in this area as current detector technology cannot cope with the flux they deliver.

Table 2 *Detectors either available on the SRS at Daresbury Laboratory or under development*

Detector System	*Data Type*	*SAXS/WAXS*	*Dynamic Studies*	*Count Rates*	*Local Count Rate Limits*
Quadrant	1-D Data	SAXS	Yes	250–400 kHz*	3 kHz mm^{-2}
INEL	1-D Data	WAXS	Yes	20 kHz*	5 kHz mm^{-2}
Area	2-D Data	SAXS/WAXS	Yes	250–800 kHz*	3 kHz mm^{-2}
Image Plate	2-D Data	SAXS/WAXS	No	100,000 photons†	
MAR 165 CCD	2-D Data	WAXS	No	67,000 photons†	
Photonic Science XIDIS CCD	2-D Data	SAXS/WAXS	Yes	9,000 photons$^{\partial}$	
RAPID	2-D Data	SAXS	Yes!	40 MHz	1 MHz mm^{-2}
		FUTURE DETECTORS			
RAPID II	1-D Data	WAXS	Yes	10 MHz	1 MHz mm^{-2}
RAPID SAXS	1-D Data	SAXS	Yes	10 MHz	1 MHz mm^{-2}
Gas Microstrip	1-D Data	WAXS	Yes	>10 MHz	> 1 MHz mm^{-2}
POLO	2-D Data	SAXS/WAXS	Yes	–	

* Depending on Local Count rates. Limited by the speed of the readout electronics.
† Saturation point of the readout instrumentation.
$^{\partial}$ Saturation point of the CCD.

References

1 W. Bras, A.J. Ryan, *Adv. Colloid Interface Sci.*, 1998, **75**, 1.
2 N.J. Terrill, M. Tobin, G.P. Diakun, E. Towns-Andrews, in preparation.
3 M.F. Butler, A.M. Donald, W. Bras, G.R. Mant, G.E. Derbyshire, A.J. Ryan, *Macromolecules*, 1995, **28**, 6383.
4 'Daresbury Annual Report' 1997–1998, 40–41.
5 G.K. Bryant, H.F. Gleeson, A.J. Ryan, J.P.A. Fairclough, D. Bogg, J.G.P Goossens, W. Bras, *Rev. Sci. Instrum.*, 1998, **69**, 2114.
6 N.J. Terrill, J.P.A. Fairclough, E. Towns-Andrews, B.U. Komanschek, R.J. Young, A.J. Ryan, *Polymer*, 1998, **39**, 2381.
7 J.P.A. Fairclough, S. Turner, S.M. Mai, C. Booth, A.J. Ryan, I.W. Hamley, S.M. King, A.J. Gleeson, N.J. Terrill, J. Pople, in preparation.
8 M. Muller, C. Czihak, M. Burghammer, C. Riekel, *Proc. SAS 99*, XIth Int. Conference on Small Angle Scattering, 1.
9 C. Martin, A. Mahendrasingam, W. Fuller, J.L. Harvie, D.J. Blundell, J. Whitehead, R.J. Oldman, C. Riekel, P. Engstrom, *J. Synchrotron. Radiation*, 1997, **4**, 223.
10 A. Bram, C.I. Branden, C. Craig, I. Snigireva, C. Riekel, *J. Appl. Cryst.*, 1997, **30**, 390.
11 J. Magill, Univ. of Pittsburgh in collaboration with C. Riekel.
12 L.B. Lurio, D. Lumma, M.A. Borthwick, P. Falus, S.G.J. Mochrie, J.F. Palletier, M. Sutton, *Synch. Rad. News*, 2000, **13**, 28.
13 S.G.J. Mochrie, A.M. Mayes, A.R. Sandy, M. Sutton, S. Brauer, G.B. Stephenson, D.L. Abernathy, G. Grübel, *Phys. Rev. Lett.*, 1997, **78**, 1275.
14 A.C. Price, L.B. Sorensen, S.D. Kevan, J. Toner, A. Poniewierski, R. Holyst, *Phys. Rev. Lett.*, 1999, **82**, 755.
15 W. Bras, G. Derbyshire, D. Bogg, J. Cooke, M.J. Elwell, B.U. Komanschek, S. Naylor, A.J. Ryan, *Science*, 1995, **267**, 996.
16 R. Lewis, 'Time Resolved Diffraction', Clareddon Press, Oxford ,1997, Chapter 10, p. 229.
17 R.A. Lewis, C.J. Hall, W. Helsby, A. Jones, B. Parker, J. Sheldon, *Society of Photo Optical and Instrumentation Engineers Proc. SPIE*, 1995, **2521**, 290–300.

20

The Impact of Neutron Scattering Techniques on Polymer Science

Jeffrey Penfold

ISIS FACILITY, CCLRC,
RUTHERFORD APPLETON LABORATORY,
CHILTON, DIDCOT, OXON. UK

1 Introduction

The use of neutron scattering techniques, both small angle neutron scattering (SANS) and neutron reflectometry, have been central to the development of our understanding of polymers at a microscopic level in bulk (solutions and melts), and at interfaces and in thin films. The power of these techniques and their extensive application stems primarily from the vastly different scattering powers of hydrogen and deuterium. Through this difference, H/D isotopic substitution provides a selectivity and sensitivity at an atomic scale resolution, which is more difficult to obtain with other techniques.

Using this approach SANS has been used to measure the dimension of the Gaussian coil structure of a single chain in melts, solution and blends, provided an affirmation of the screened excluded volume model, and a verification of scaling laws in polymer solutions, determined the structure of diblock copolymer aggregates, and established the relationship between the micro and macroscopic deformation in rubber elasticity.

Neutron reflectivity has established the structure of polymer layers at interfaces (air–solution and solution–solid), determined the surface ordering in block copolymer films, and revealed the nature of inter-diffusion in mixed polymer films.

SANS has historically been the domain of cold neutrons on reactor based sources, and impressive instrumentation, such as notably D22 at the Institute Laue Langevin,[1] is currently available. The more recent emergence of neutron reflectivity as a probe of surface structure[2] has been very much linked to the development of pulsed neutron sources; and much of the initial impact of this technique has arisen from pulsed source instrumentation, such as the CRISP and SURF reflectometers at ISIS.[3,4]

In this paper we will concentrate on the diffraction techniques (SANS and reflectometry), and hence static measurements. However, it should be pointed out that through inelastic scattering, aspects of polymer dynamics are accessible. In particular, it has been possible to access single chain dynamics in bulk systems, deformation and relaxation of polymer melts under shear, shed new light on viscoelasticity in polymer melts, and obtain direct information on polymer reputation and particle fluctuations.

In the Current State of the Art we will review some of the recent SANS and reflectivity data from ISIS, which also serve to point to future directions and opportunities. Recent reflectivity measurements, on the adsorption of polymers and polymer/surfactant mixtures at interfaces, surface ordering in block copolymer systems, time dependent inter-diffusion at polymer–polymer interfaces, and the contribution of capillary waves to interfacial widths, will be described. The use of SANS to investigate the dynamic of trans-esterification of polyester blends, the deformation of copolymers with novel morphologies, and the use of diffraction techniques to determine the structure of polymeric electrolytes, will be presented.

For the future there is a clear trend towards the study of complex multi-component systems, the nature of complex interfaces (such as the liquid–liquid interface), the use of complex environments (shear flow, pressure, confinement, *in situ* electrochemistry), and the increasing realisation that the timescales of some interesting time-dependent phenomena are already accessible. These trends are all closely linked to improving the understanding and ultimate control of ever more realistic but complex materials; in sensors, adhesion, surface engineering, advanced composite materials, intelligent packaging, lubrication, paint, inks, coatings, detergents, foam and gel stabilisation, and fuel additives.

Potential developments in instrumentation and the neutron source at ISIS (Second Target Station optimised for cold neutrons) will be discussed, and the longer term potential of the European Spallation Source, ESS, presented.

2 Current State of the Art

2.1 Neutron Reflectometry

Polymer–polymer interfaces are an important area of study since the interfacial behaviour is fundamental to the bulk properties of the system. This is particularly true when two or more polymers are mixed to form a blend, but the interface also plays a dominant role in areas such as adhesion, welding, surface wetting and mechanical strength. To understand fully polymer behaviour in such applications, the interface must be characterised at a microscopic level. Through deuterium labelling the interface between otherwise indistinguishable polymers can be studied, and neutron reflectivity provides unprecedented detail on interfacial width and shape.[5] In addition to the inherent interdiffusion between polymers at a polymer–polymer interface, the interface is further broadened by thermally driven capillary waves. Capillary waves produce a measured interfacial width which varies logarithmically with film thickness. Xiao *et al.*[6] measured the interfacial width of the polymer for polystyrene (PS) and poly (methyl methacrylate). Using deuterated polystyrene (d-PS) and polystyrene film thickness in the range 20 to 5000 Å they confirmed the logarithmical dependence on film thickness (see Figure 1), to provide a reliable way to extract the intrinsic interfacial width from such measurements.

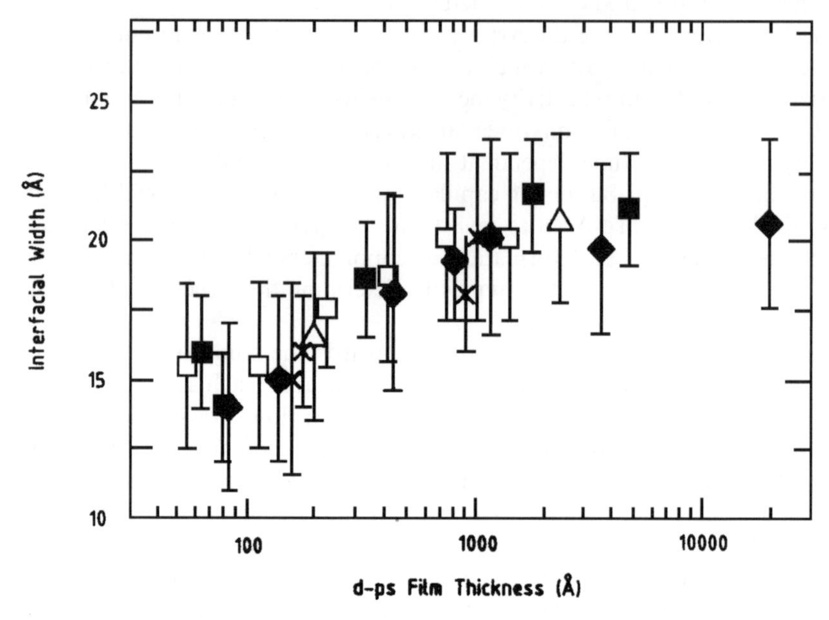

Figure 1 *Interfacial width parameter, Δ, for an interface between d-PS and PMMA, as a function of PS film thickness, the solid line is a fit to the theoretical prediction*

The time dependence of the early stages of polymer interdiffusion at interfaces is indicative of the diffusion process. The normal approach to study such interdiffusion by neutron reflectivity is to use an "anneal/quench" cycle; where the sample is heated for a given time above the glass transition temperature (T_g) of the polymer, then quenched rapidly to room temperature, after which the reflectivity profile is measured. This has proved to be highly effective for a number of systems, but is difficult to apply when T_g is ~ room temperature, or for small molecule ingress into a higher molecular weight polymer layer.

The ability to measure simultaneously over a wide Q range at fixed geometry (angle of incidence) using the white beam time of flight method on a pulsed source, and the fluxes that are now available on the SURF reflectometer at ISIS, mean that *in situ* real time reflectivity measurements are now possible. Bucknall *et al.*[7] have pioneered this approach for the measurement of polymer-polymer interdiffusion with a series of ground-breaking experiments with measurement times down to $\leq$ one minute for carefully optimised sample geometries. The reflectivity profile for such polymer bilayers is characterised

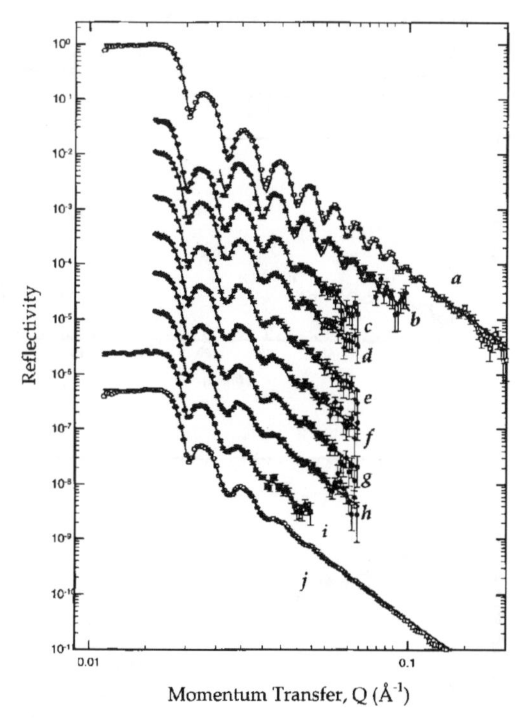

Figure 2 *Reflectivity profiles for h-PS/d-PS bilayer. Measurements made in real time at 115c (points), solid lines are fits to the data (as described in the text). Mean time for profiles plotted are (b) 5 mins, (c) 32, (d) 63, (e) 93, (f) 124, (g) 155, (h) 186, (j) 212, measured for 6 min count times, (a) is before annealing*

by a series of interference fringes who period is inversely related to the deuterated polymer film thickness which comprises half of the bilayer (see Figure 2). These fringes damp with increasing Q, and the rate of damping with Q is directly related to the polymer–polymer interfacial width. As the interdiffusion process proceeds the angle of incidence is altered to capture the region of reflectivity most sensitive to the interfacial width. For the h-PS/d-PS bilayer Bucknall *et al.* were able to show that this approach was equivalent to the "quench/anneal" cycling and measured reflectivity profiles at five minute intervals. The data was consistent with a symmetrical error function interface characterising the polymer interdiffusion. Below the reptation time, τ_R, the interfacial width followed a $t^{1/4}$ behaviour and above τ_R as $t^{1/2}$ (classical Fickian diffusion), consistent with theoretical predictions.

Using a specially designed cell, based in the type of cell used for studying the liquid–solid interface, they have demonstrated that the same approach can be used to study small molecule ingress. A similar series of reflectivity profiles for the diffusion of oligomeric-styrene (Ost) into high molecular weight deuterated polystyrene (d-PS) were obtained. Measurements were again made at five minute intervals, and during the course of the experiment the grazing angle of incidence was decreased from 0.8 to 0.5° to move the window of measurements

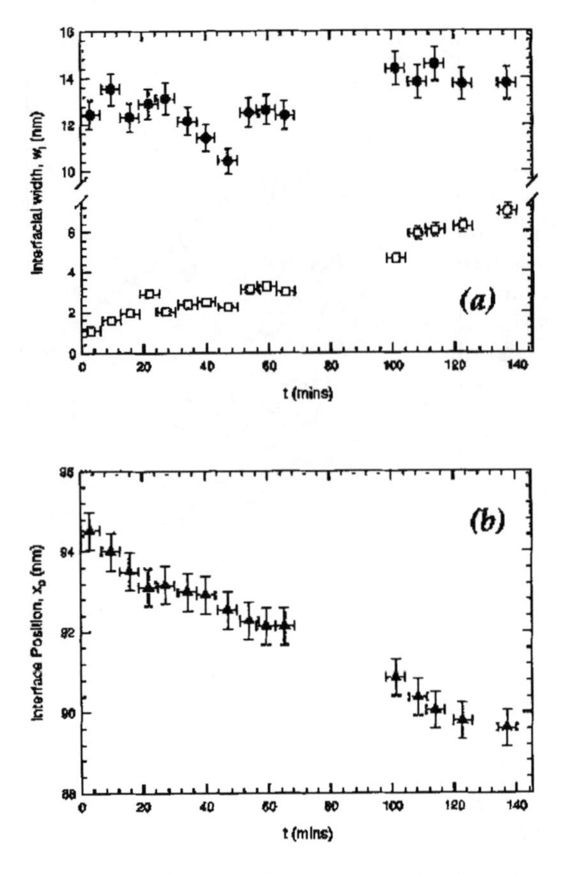

Figure 3 *Variation of (a) interfacial width (w_1, w_2) and (b) position of the interface (x_0) as a function of annealing time for OSt/d-PS*

to lower Q as the interface progressively broadened. In this case the data was consistent with a highly asymmetrical interfacial profile, and could be described by two error functions of width w_1, w_2 which were discontinuous about the interface centre x_o. The time dependence of these parameters are shown in Figure 3.

The width of the d-PS side of the interface is larger than that on the OSt side, confirming that the oligomer diffusion is much faster than the polymer diffusion. Both widths increase with time, and as expected the interface moves further into the d-PS with increasing time. More recent measurements suggest that time slices as short as ~ 10 seconds are now accessible.

Water soluble polymers are frequently incorporated in aqueous surfactant solutions in many domestic and technological applications (in formulations such as shower gels and hair shampoos), as viscosity modifiers, stabilisers and deposition aids. Water soluble polymers often interact strongly with surfactants in aqueous solution, giving rise to a rich pattern of behaviour in properties such as surface tension. The bulk properties of a variety of polymer/

surfactant mixtures have been extensively studied. In contrast, there is no sound theoretical basis for understanding surface tension behaviour in such mixtures. Neutron reflectivity measurements, where surfactant, polymer and water can all be separately deuterium labelled, provided an opportunity to determine adsorbed amounts and obtain information about the surface structure in such complex mixtures.[8,9] Recent measurements[10] on a mixture of dodecyltrimethyl ammonium bromide, $C_{12}TAB$, and sodium polystyrene sulfonate, NaPSS, have revealed an unexpected rich pattern of behaviour with surfactant concentration, which is not evident from the surface tension data. The surface tension data for this mixture shows typical polymer/surfactant behaviour, with two break points at two different surfactant concentrations. These are usually attributed to two bulk phase changes, the first corresponding to the formation of micelles on the polymer (cac), and the second to the formation of free micelles (cmc). The unusual feature in this system is the the cac and cmc are separated by ~3–4 decades in concentration. The neutron reflectivity measurements show that at low surfactant concentrations (~cac) a simple monolayer ~20 Å thick is formed at the air–water interface; comprising of surfactant and polymer. The polymer volume fraction in the layer was estimated to be ~25%. At higher surfactant concentrations the reflectivity profiles change dramatically, corresponding to a more complex structure being formed at the interface (see Figure 4). The transition from a simple single monolayer to a more complex structure occurs over a narrow range of concentration, and in a region where the surface tension data shows no changes.

Measurements with different 'contrasts', d-$C_{12}TAB$ / NaPSS / cma, d-$C_{12}TAB$ / NaPSS / D_2O and h-$C_{12}TAB$ / NaPSS / D_2O, enable the detailed

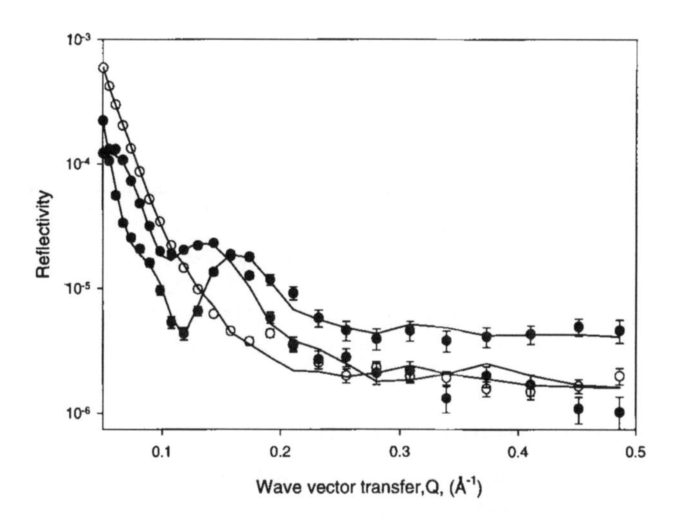

Figure 4 *Reflectivity profiles for 2.3×10^{-2} M $C_{12}TAB$/140 ppm NaPSS (•) d-$C_{12}TAB$/NaPSS/cma, (o) d-$C_{12}TAB$/NaPSS/D_2O, (Δ) h-$C_{12}TAB$/NaPSS/D_2O: solid lines are fits to data using the layered structure described in the text*

structure of the surface layer to be determined. Preliminary analysis shows that at the highest surfactant concentration measured (2.3 x 10^{-2}M) the structure corresponds to layers of polymer with surfactant attached (either in monomeric or micellar form), ordered parallel to the interface.

Extensive neutron reflectivity studies on surfactant adsorption at the air–water interface[11] show that a surfactant monolayer is formed at the interface. Even for concentration >> cmc, where complex sub-surface ordering of micelles may exist,[12] the interfacial layer remains a monolayer. This is in marked contrast to the situation for amphiphilic block copolymers, where recent measurements by Richards *et al.*[12] on polystyrene polyethylene oxide block copolymers (PS-b-PEO) and by Thomas *et al.*[13] on poly(2-(dimethylamino)ethylmethacrylamide-b-methyl methacrylate) (DMAEMA-b-MMA) show the formation of surface micelles at a concentration <bulk cmc. This is shown in Figure 5 for the DMAEMA–MMA block copolymer, where an abrupt change in thickness is observed at a finite concentration, and signals the onset of surface micellisation.

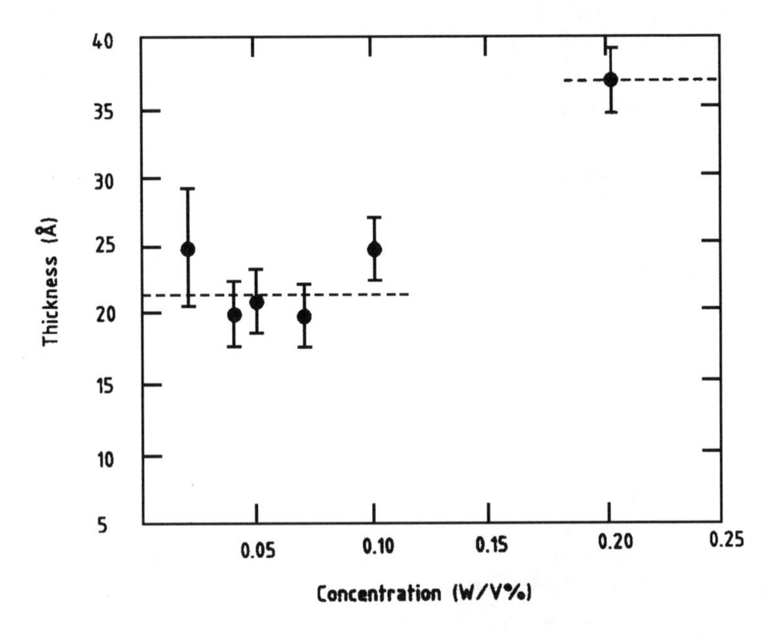

Figure 5 *Variation of thickness of DMAEMA b MMA copolymer layer at the air-water interface with concentration; lines are a guide to the eye only*

2.2 Small Angle Neutron Scattering

SANS has played a major role in physical polymer science since its possibilities were first demonstrated some years ago. The initial interest was in the determination of the size of the polymer molecule, as a means of exploring theories of molecular configuration in the solid state and in solution, and required data at high and low values of Q. Results from recent experiments on

LOQ at ISIS emphasise the importance of the intermediate Q range for investigating polymer–polymer thermodynamics, molecular configuration and inter-molecular interactions. Polyesters are generic polymers which are widely used in a variety of applications, such as fibres, packaging, video and computer tapes. They are prone to a random scission and recombination reactions, known as trans-esterification, which takes place in the melt. Trans-esterification does not reduce the average molecular weight of the polymer, but it does alter the molecular weight distribution. Such changes in molecular weight distribution can influence melt viscosity and crystallisation behaviour in undersirable ways. Furthermore, and importantly in this case, where two different polyesters are melt mixed, trans-esterification leads to co-polymer formation. Although information on compositional changes can be obtained using other techniques such as NMR, kinetic data can only be obtained by SANS. Such data is obtained by following the changes in SANS from a deuterated polyester mixed with a hydrogenous one. The data can be interpreted straightforwardly in terms of an activation energy, the measurements were made on PET/PBT mixtures.[15] The value determined for the activation energy show that in this case the combination trans-esterification is a random process, and is not driven by the polymer end groups.

A recent SANS study on LOQ by McLeish *et al.*[16] has demonstrated the contribution that SANS can make to the understanding of flow properties in novel polymer. Although producing highly monodisperse polymers will improve the strength of the final product, it often gives rise to poor processing, as their flow properties are prone to instabilities in the extremes of shear and extensional flow. It has long been known that random branching of polymers improves such processing. This provides a route to the sensitive control of flow properties, but requires an understanding of the molecular physics of entangled branched polymer. McLeish *et al.* have made a series of rheological and SANS measurements on some monodisperse H-shaped poly-isoprenes, in order to advance that understanding. The rheological measurements provided indirect confirmation of their theoretical predictions that under strong flow the molecules disentangle from their awkward branches rapidly, by pulling the chain ends into the centre tube previously occupied by sections of the cross-bar. The SANS measurements (see Figure 6), with deuterium labels at the extreme ends of the arms, provided a direct confirmation of this mechanism. Furthermore the data showed that a small amount of the arm material was being drawn into the central tube at lower levels of strain than expected from the initial theoretical predictions, providing an impetus for a refinement to the theory.

2.3 Crystalline Diffraction

The ability of neutron diffraction to probe the location of light atoms in the presence of heavier ones is particularly important in the study of lithium batteries, where the synthesis of new cathode materials in an important aspect of the development of high power density, lightweight cells for laptop

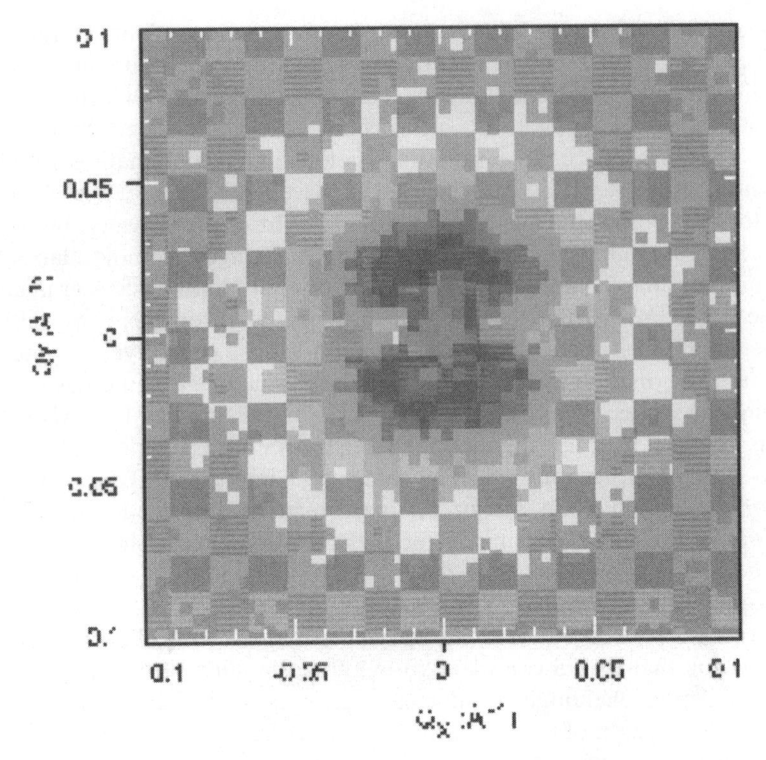

Figure 6 *SANS data for H-shaped polyisoprene, for a stretch of 2.3 times original length*

computers, mobile phones, and a variety of other applications. A detailed knowledge of the crystal structure is required to understand the dynamics of the Li^+ intercalation mechanism including the effects of repeated charging/discharging to determine those factors which limit the ultimate lifetime of the battery.

The recent example of the *ab initio* structure determination of the polymer electrolyte Poly (ethylene oxide)$_6$:$LiAsF_6$ by Bruce *et al.*[17] is a notable example of the complex structures that can be determined from powder diffraction on a pulsed neutron source. Polymer electrolytes consist of salts dissolved in solid high molecular weight polymers, and represent a unique class of solid coordination compounds. Their importance lies in their potential in the development of truly all-solid-state rechargeable batteries. The structure of the 6:1 complex is particularly important, as it is a region where the conductivity increases markedly. The structure of the complex is distinct from all known crystal structures of PEO:salt complexes (see Figure 7). The Li+ cations are arranged in rows, with each row located inside a cylindrical surface formed by two PEO chains, with the PEO chains adopting a previously unobserved conformation. Furthermore the anions are located outside the PEO cylinders and are not coordinated with the cations.

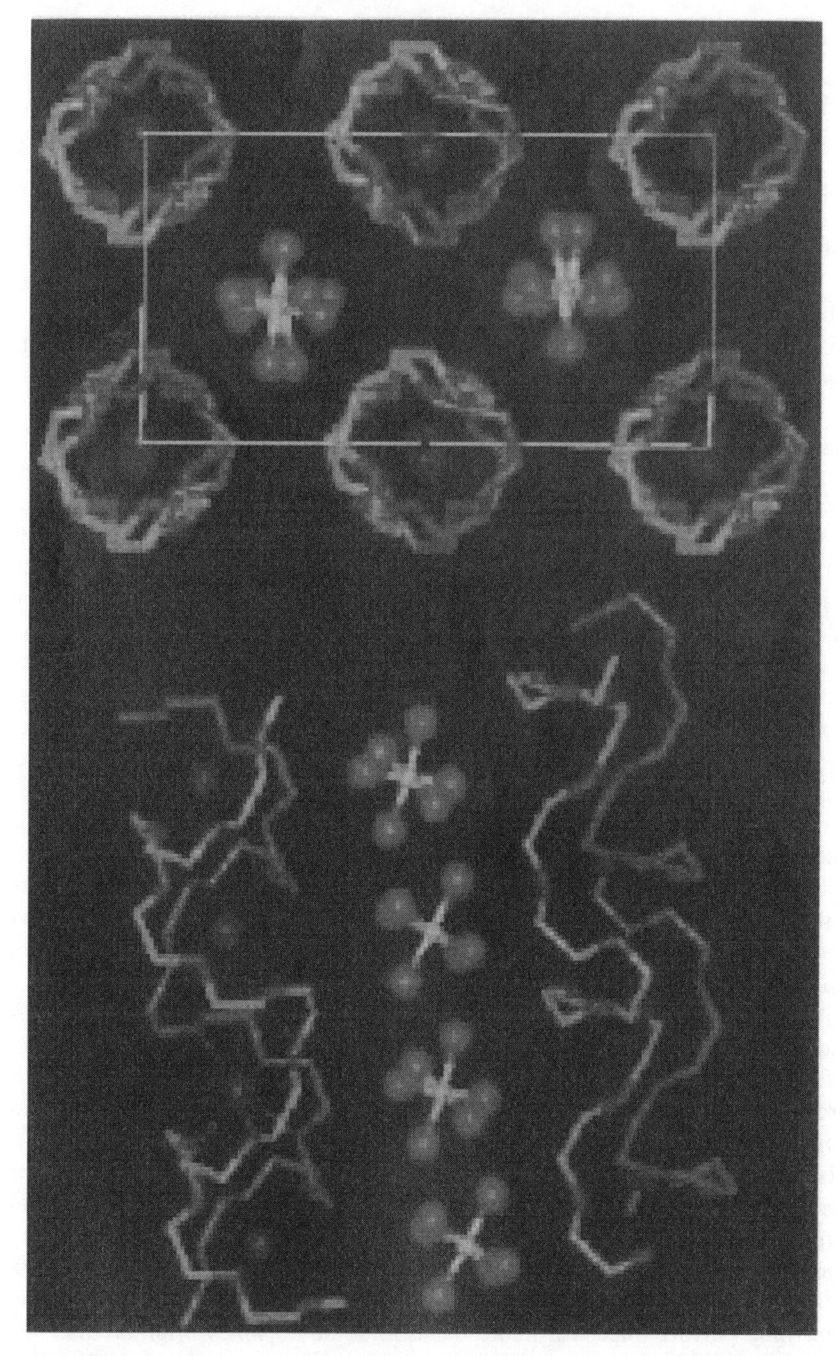

Figure 7 *Structure of poly (ethylene oxide)$_6$: $LiAsF_6$, thin lines show coordination around Li^+ cation, for clarity hydrogen ions are not shown*

3 Future Potential

3.1 Development of Pulsed Neutron Sources

Although much of the pioneering work and reputation of neutron scattering has been based on reactor sources, in recent years through primarily the contribution from ISIS, the potential of accelerator based pulsed neutron sources is being increasingly recognised.

The advantages of neutrons in scattering studies, selectivity through isotopic labelling, the magnetic interaction, the ability to probe atomic motion, and the penetrability of the probe, are well established and documented. The effective and growing exploitation of ISIS over a broad range of condensed matter research has now powerfully demonstrated the specific benefits of the time-of-flight technique in neutron scattering on a pulsed source. Such benefits include improved instrumental resolution, simultaneous wide coverage of both momentum and energy transfer, intrinsically low backgrounds, the ease of use of fixed geometries, the high fluxes of epithermal neutrons, and the broad spectral range available. Developments at ISIS have had a particular impact on high resolution powder diffraction studies, magnetic inelastic scattering, and the use of reflectometry to study surfaces and interfaces.

ISIS is particularly optimised for thermal/epithermal neutron production. Recent developments have however demonstrated that substantial gains in performance over the existing target station at ISIS can be achieved specifically for cold neutrons.[18] By optimising both target and moderator configurations on a second target station, at a low source repetition rate (10hz) (see Figure 8), for cold neutrons and high resolution instrumentation, performance enhancement in the range 10–20 times can be achieved.

This will have a particular impact on instrumentation such as SANS, reflectometry, neutron spin echo, high resolution diffraction and microvolt resolution spectroscopy, and provide exciting opportunities in the technologically significant areas of soft matter, polymers, biological sciences, and advanced materials. The detailed scientific case for the Second Target Station at ISIS is described in detail elsewhere.[19]

With the longer term in mind, a number of European countries have produced a specification and scientific case for a European Spallation Source,[20] and design studies on key components and concepts are in progress. The source specification is based on a proton accelerator producing an average proton beam power of 5 MW at a repetition rate of 50 Hz with a proton pulse length of 1 μs or less. This would feed two target stations, a 50 Hz target station able to accept up to 5 MW of beam power, and a 10 hz target station able to accept 1 MW and optimised for cold neutrons. The anticipated neutronic performance will make the ESS some 30 times brighter than ISIS, and coupled with anticipated developments in instrumentation will enhance data collection rates by up to three orders of magnitude in some fields. An initial reference suite of 44 instruments has been specified.

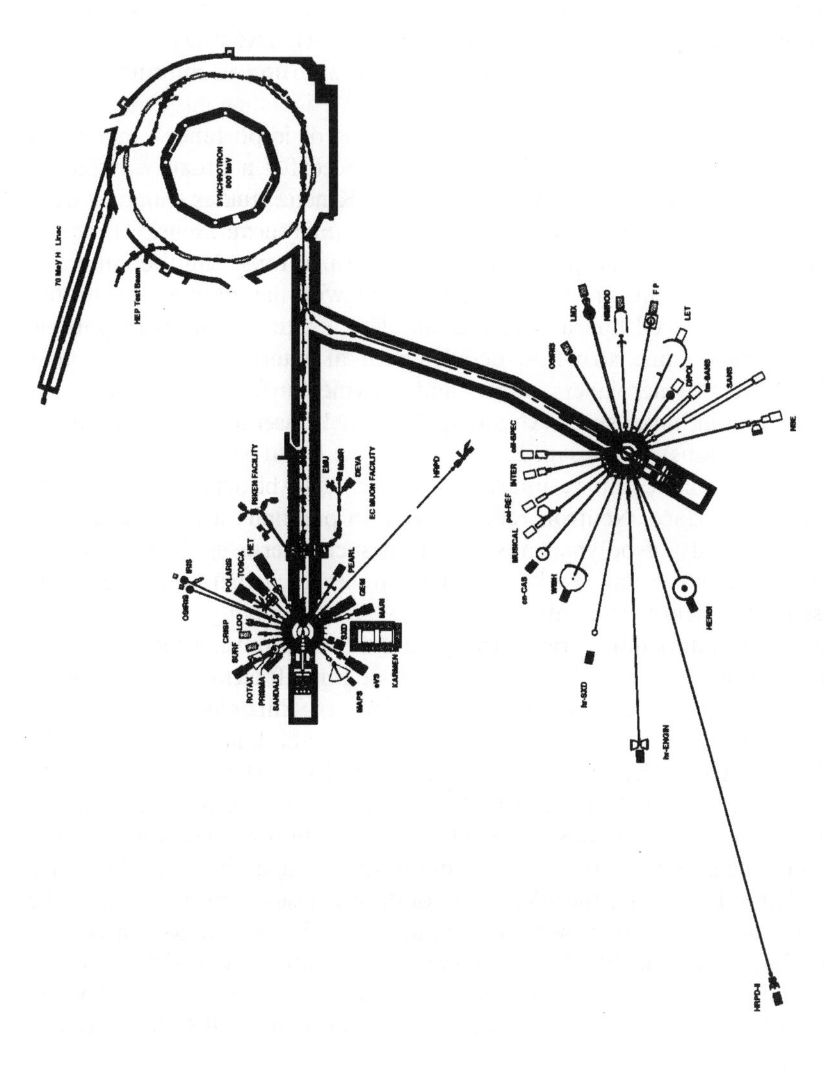

Figure 8 *Schematic representation of incorporation of low frequency Second Target Station at ISIS*

3.2 New Scientific Opportunities

The anticipated developments in neutron sources and instrumentation will provide exciting new opportunities for the study of polymers, soft matter, and complex fluids. Rather than try to cite specific detailed examples, broad areas of potential exploitation and current trends will be highlighted. The examples described in the earlier section of this paper indicate a distinct trend towards the study of complex multi-component or multi-phase systems, the use of complex environments (flow, pressure, confinement), the study of complex interfaces (for example, liquid–liquid interface), and non-equilibrium *in situ* studies.

In many of these systems the dimension scales of importance range from molecular to meso-scale, and this dictates the need for a broad wavelength band and a greater emphasis on cold neutrons. Kinetic studies (ranging from the kinetics of chemical reactions, to probing dynamic surface tension) require a broad spectral range and a higher flux of cold neutrons. Multi-component and multi-phase systems are often only tractable with the 'parametric' studies possible with enhanced flux and resolution. The greater use of H/D isotopic labelling to probe fine structure details, both at interfaces using specular reflectivity (in polymer layers, films, and polymer/surfactant complexes at interfaces), and in bulk using SANS (copolymer self assembly, polymer melts, blends and solutions), is envisaged.

One of the more complex interfaces so far relatively unexplored is the liquid–liquid interface. Neutron reflectivity is, in principle, a promising technique for the study of polymer adsorption at such interfaces. Until recently, apart from the pioneering work of Richardson *et al.*,[21] there has been little progress in that field with neutron reflectivity. This is mainly associated with the experimental difficulties arising from the need to establish a thin layer (~ a few microns) of either the oil or aqueous phase for sufficient transmission of the neutron beam at grazing incidence. However, Zarbkhsh *et al.*[22] have recently made impressive progress in developing a methodology for studing the liquid-liquid interface by specular neutron reflectivity. A thin wetting film of a suitable alkane on a hydrophobically treated silicon surface is frozen when it has thinned to a few microns. Contact is then made with the aqueous sub-phase, and the system is brought to equilibrium. Using this approach they have started to investigate the adsorption of different block-co-polymers at the oil–water interface.[23] Some initial results in Figure 8 shows a comparison of the distribution of the block co-polymer polybutadiene-polyethylene oxide (PB-PEO) at the hexadecane–water and air–water interfaces. It shows that in this case the oil–water interface is more diffuse, and that the PB block extends into the oil phase.

In the context of complex interfaces, *in situ* measurements, such as *in situ* electrochemistry, are now becoming established, and a recent example of this development is the work of Hillman *et al.*[24] on the structure of electro-active polymer bilayers. Higher fluxes, and the ability to use smaller sample areas will greatly facilitate the further development of this area of application.

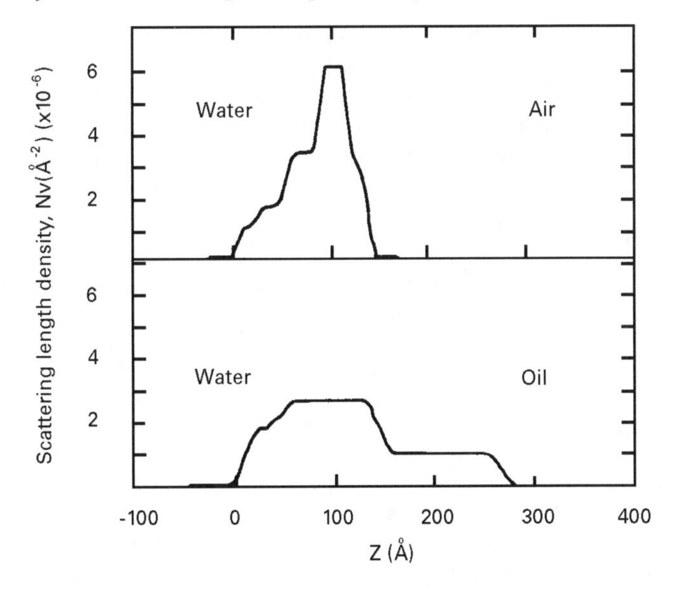

Figure 9 *Distribution of the block co-polymer PB-PEO at the (a) air–water and (b) hexadecane–water interfaces, derived from specular neutron reflection measurements*

Neutron specular reflectivity is essentially a 'depth profiling' technique, however, 'off-specular' scattering at grazing incidence provides information about lateral structure and correlations on a length scale comparable to light scattering. This has been exploited recently by Jones *et al.*[25] to investigate dewetting in polymer films. This is an area that will be used increasingly in combination with more direct imaging methods, such as atomic force microscopy, AFM, Brewster angle microscopy, BAM, and Scanning near-field optical microscopy, SNOM; but will nevertheless remain essential for 'buried' interfaces. Indeed, it is anticipated that increasingly specular neutron reflection will be used in conjunction with other techniques, which include X-ray scattering, ellipsometry and Raman spectroscopy.

The nature of polymer chains in confined geometries, thin films, and structured surfaces for example, is of much current interest. Grazing incidence SANS (using the evanescent wave to control depth sensitivity) and transmission SANS on free standing thin films[26] in combination with H/D labelling, are likely to make a vital contribution to our understanding of conformation in confined geometries.

In the application of SANS to bulk polymeric systems, the anticipated developments in instrumentation and flux will provide the opportunity to extend the work of McLeish *et al.*,[16] and probe polymer conformation under extrusion and in flow. Measurements as a function of position will enable complex spatial distributions of velocity and stress within an extruder to be mapped. Measurements with partial deuterium labelling will allow further development of the structure/rheology relationship in polymer processing.

The greater fluxes available, and current developments in 'white beam' polarisers will stimulate the routine incorporation of polarisation analysis, for the separation of coherent and incoherent scattering in hydrogenous material.

In conclusion, the potential developments in pulsed neutron sources and instrumentation provide exciting opportunities for the future study of polymers.

References

1 Neutron Beam Facilities at the high flux reactor, available to users, Institute Laue Langevin, Grenoble, France, (1994).
2 J. Penfold, R.K. Thomas, *J. Phys. Condens. Matter*, 1990, **2**, 1369.
3 J. Penfold, R.C. Ward, W.G. Williams, *J. Phys. E*, 1987, **20**, 1411.
4 J. Penfold, R.M. Richardson, A. Zarbakhsh, J.R.P. Webster, D.G. Bucknall, A.R. Rennie, R.A.L. Jones, T. Cosgrove, R.K. Thomas, J.S. Higgins, P.D.I. Fletcher, E. Dickinson, S.J. Roser, I.A. McLure, A.R. Hillman, R.W. Richards, E.J. Staples, A.N. Burgess, E.A. Simister, J.W. White, *J. Chem. Soc., Faraday Trans.*, 1997, **93**, 3899.
5 T.P. Russel, *Mat. Sci. Rep.*, 1990, **5**, 1.
6 C. Xiao, M. Sferazza, R.A.L. Jones, D.G. Bucknall, J.R.P. Webster, J. Penfold, *Phys Rev Lett.*, 1997, **78**, 3693.
7 D.G. Bucknall, S.A. Butler, J.S. Higgins, *Macromolecules*, 1999, **32**, 5453.
8 A. Creeth, P.G. Cummins, E.J. Staples, L.J. Thompson, I. Tucker, J. Penfold, R.K. Thomas, N. Warren, *Faraday Disc*, 1997, **104**, 245.
9 D.J. Cooke, C.C. Dong, J.R. Lu, R.K. Thomas, E.A. Simister, J. Penfold, *J. Phys. Chem. B*, 1998, **102**, 4912.
10 J. Penfold, R.K. Thomas, D. Taylor, E.J. Staples, I. Tucker, unpublished results
11 J.R. Lu, R.K. Thomas, J. Penfold, *Adv. in Coll. Int. Sci.*, 2000, **84**, 143.
12 J.R. Lu, E.A. Simister, R.K. Thomas, J. Penfold, *J. Phys. Chem.*, 1993, **97**, 13907.
13 S.W. An, T.J. Su, R.K. Thomas, F.L. Baines, N.C. Billingham, S.P. Armes, J. Penfold, *J. Phys. Chem. B*, 1998, **102**, 387.
14 P.F. Dewhurst, M.R. Lovell, J.L. Jones, R.W. Richards, J.R.P. Webster, *Macromolecules*, 1998, **31**, 7851.
15 S.C.E. Backsona, R.W. Richards, S.M. King, *Polymer*, 1999, **40**, 4205.
16 T.C.B. McLeish, J. Allgaier, D.K. Bick, G. Bishko, P. Biswas, R. Blackwell, B. Blottiere, N. Clarje, B. Gibbs, D.J. Groves, A. Hakiki, R.K. Heenan, J.M. Johnson, R. Kant, D.J. Read, R.N. Young, *Macromolecules*, 1999, **32**, 6734.
17 G.S. MasGlashan, Y.G. Andreev, P.G. Bruce, *Nature*, 1999, **398**, 792.
18 *ISIS Facility Annual Report 1997-98*, RAL-TR-1998-050, 1999.
19 *A Second Target Station at ISIS, RAL-TR-2000-032*, 2000.
20 *ESS, A next generation source for Europe*, Vol 1–**3, 1997**.
21 T. Cosgrove, J.S. Phipps, R.M. Richardson, *Colloids Surf.*, 1992, 62, 199.
22 A. Zarbakhsh, J. Bowers, J.R.P. Webster, *Meas. Sci. Technol.*, 1999, 10, 738.
23 A. Zarbakhsh, J. Bowers, J.R.P. Webster, *Experimental report RB10222 in reference 18*.
24 A.R. Hillman, A. Glidle, R.M. Richardson, S.J. Roser, P.M. Saville, M.J. Swann, J.R.P. Webster, *J. Am. Chem. Soc.*, 1998, 120, 12882.
25 M. Sferrazza, M. Heppenstall-Butler, R. Cubitt, D.G. Bucknall, J.R.P. Webster, R.A.L. Jones, *Phys. Rev. Lett.*, 1998, 81, 5173.
26 T.P. Russell, P. Lambooy, J.G. Barker, P. Gallagher, S.K. Satija, G.J. Kellogg, A.M. Mayes, *Macromolecules*, 1995, 28, 782.

21

Scanning Probe Microscopy on Polymers: Recent Developments and Perspectives

Georg Krausch

LEHRSTUHL FÜR PHYSIKALISCHE CHEMIE II AND BAYREUTHER ZENTRUM FÜR KOLLOIDE UND GRENZFLÄCHEN (BZKG), UNIVERSITÄT BAYREUTH, 95440 BAYREUTH, GERMANY

1 Introduction

Since its invention in the early 1980's, scanning tunnelling microscopy (STM)[1] and a still growing family of related scanning probe microscopies (SPM) have become routine tools for surface analysis in different areas of science and technology.[2] In all of these techniques a nanoscopic probe is scanned in close proximity to the sample surface under study and some physical property is recorded as a function of the lateral position of the probe. The resulting images can have ultra-high spatial resolution. In the following, we shall summarise some recent developments regarding the application of SPM to problems in polymer science and sketch aspects of ongoing and future work in the field. We shall neither try to be comprehensive, nor present an unbiased view of the field, the latter being due to the author's own interests and experience.

Among the various techniques, we shall only discuss STM, scanning force microscopy (SFM or AFM),[3] and scanning near-field optical microscopy (SNOM or NSOM),[4] as those can well be considered the most prominent members of the SPM family. Among the three, we shall further focus on SFM, as most of the work done on polymers has indeed made use of this method. However, a brief note on STM and SNOM seems appropriate.

2 Scanning Tunnelling Microscopy (STM)

Scanning tunnelling microscopy (STM) requires conducting samples, which establishes certain limitations for its use on polymeric samples. Impressive work has been done by Rabe and others on the molecular ordering of rather short chains adsorbed from solution on to conductive substrates.[5–7] As soon as (non-conductive) chain molecules exceed a certain length and form coil-like structures, STM becomes increasingly difficult to use. However, in the area of conducting or semiconducting polymers STM can indeed be helpful to study electrical or electro-optical properties with high spatial resolution. Experiments along these lines are still in their beginnings. The IBM-group in Rueschlikon has started investigations on polymeric materials used for light emitting devices.[8] They were able not only to measure the I/U-characteristics on a local scale but also to use the STM tip as a nanoscopic electrode to locally excite electro-luminescence. So far the results indicate significant lateral differences in the electrical and electro-optical properties of the samples under study, clearly justifying the necessity of local experiments. However, since the current is used to control the tip/sample distance, local variation in the conductivity of the specimens result in different indentation depths and thereby different areas of contact. This drawback can be circumvented, if alternative tip/sample distance control mechanisms are used. First experiments using shear-force tip/sample distance control[9,10] for metallic tips are presently being carried out in our laboratory.

3 Scanning Near-field Optical Microscopy (SNOM)

The idea of using nanoscopic light sources (or detectors) in the optical near field to overcome the Abbe-diffraction limit in optical microscopy dates back to the beginning of the twentieth century, when Synge and Einstein discussed the potential of holes in badly silvered mirrors as nanoscopic optical apertures.[11,12] However, near-field microscopy in the visible became possible only when piezo-electric actuators were introduced in the mid-eighties.[4] Meanwhile, sub-wavelength spatial resolution is indeed realised in various, commercially available scanning near-field optical microscopes (SNOM). Yet, the technology of SNOM-tips is far from being satisfactory in terms of reproducibility and stability.[13] Aside from these inherent problems, confocal laser scanning microscopy (LSM) has become a mature and powerful competitor to SNOM. LSM offers a somewhat poorer spatial resolution (150–250 nm), which however is realised reproducibly on an everyday basis. In terms of SNOM, the last decade therefore was characterised by high hopes and various enthusiastic attempts. It is probably fair to say, however, that as yet no major breakthrough in the field of polymers has been achieved using this technique. In my opinion, the greatest potential for SNOM experiments can be seen in the investigation of isolated biomolecules adsorbed on suitable surfaces. Fluorescence labelling is well established in biology and complex (far-field) optical techniques (such as, *e.g.* non-radiative energy transfer) are well established. Here, single molecule spatial resolution together with the simultaneous detection of topography and fluorescence intensity goes significantly beyond the potential of other optical techniques.

Regarding synthetic polymers, attempts have been made to use SNOM both for local modification ('lithography')[14,15] and characterisation of (electro-)

optical properties in semiconducting materials.[16] In the latter case, the nanoscopic SNOM probe can also be used as a sub-wavelength sized detector for electro-luminescence in polymeric LED's. The central goal of these experiments is to study the mechanisms of breakdown of the devices at as early a time as possible. Therefore, high spatial resolution is needed. One has to realise, though, that the SNOM probe has to be brought into the optical near field of the light emitting region, which requires special LED schemes, using, *e.g.*, semitransparent metal electrodes. First experiments along these lines have recently been performed in our laboratory.

4 Scanning Force Microscopy

Among the scanning probe techniques applied to polymer physics problems, SFM is certainly the most prominent one. The technique has matured to an extent that today most major polymer laboratories have access to a commercial SFM instrument. It is therefore fair to state that SFM has become a standard analytical technique for the investigation of polymer surfaces.[17] This development is closely related to some important technological improvements. The implementation of reliable and stable non-contact or intermittent-contact imaging conditions (the most prominent being the *TappingMode™* by Digital Instruments, Santa Barbara) have significantly reduced sample damage during imaging even on very soft samples. (A quite impressive application of this approach was the recent demonstration by Herminghaus and co-workers, who showed that droplets of simple liquids could be imaged in *TappingMode™*.[18]) It further became possible to establish material contrast along with the topography image by various 'nano-mechanical' imaging schemes.[19] Since the SFM cantilever and the polymeric sample establish a complex mechanically coupled system, some information on the mechanical properties of the polymeric material can be inferred from the cantilever response. On this theme and again using *TappingMode™*, one can detect the phase difference between excitation of the cantilever and its actual movement. Other techniques systematically vary the force between tip and sample and thereby get access to elastic properties of the polymer and as well as to adhesion forces between polymer and tip[20] (*e.g.* the *Pulsed Force Mode* by Witec, Ulm). Presently, commercial SFM's are being equipped with heating stages enabling SFM imaging at well-controlled elevated temperatures up to some 150 to 200 °C. Thereby it is possible to study kinetic processes and phase transitions at a given spot on the sample in real time. Imaging in liquids and various gas and vapour atmospheres has recently become state of the art as well. As an example, imaging in water or in aqueous solutions not only considerably reduces the (capillary) forces between tip and sample, it also is of utmost importance for the imaging of biopolymers under *in vivo* conditions.

Among the various applications of SFM in polymer science, the imaging of heterogeneous synthetic polymeric materials (blends, copolymers, hybrid materials, *etc.*) is probably most widespread and can be considered the easiest task, in particular when the materials are (at least partly) glassy at the imaging temperature. It is in this area of polymer science that SFM can already be considered a standard analytical technique. The ubiquity of SFM's and the fact that even the non-expert is able to produce 'nice' SFM images within reasonably short times bring the danger of a loss in scientific quality. Indeed, there appears to be an information gap between the 'SFM-community' on the one hand and SFM users within the polymer community on the other hand, regarding possible artefacts and well-known problems related to the technique. A sound understanding of the rather complex dependence of, *e.g.*, the imaging conditions on the resulting SFM images in *TappingMode™* and related techniques is indispensable for reproducible and reliable scientific work. As a simple example we quote the (well-known and well-understood) fact that a *TappingMode™* topography image of a heterogeneous polymer sample *does not* necessarily reflect the 'true' surface topography, in particular when the material exhibits a significant phase contrast between different constituents.[21] Here, it will be important in the near future to demonstrate some of the most important aspects on polymeric samples and make the results visible to the respective community.

There are other, more recent, areas of polymer science where the use of SFM is not as established but where important insights are to be expected in the near future. As an example, Magonov and co-workers demonstrated that the combination of high resolution imaging and temperature control enables the study of crystallisation processes on local scales and in real time. The potential to establish different melting temperatures for different types of aggregates adds an important contribution to scattering experiments in an established area of polymer science, which only recently has attracted renewed attention. Along the same lines one could apply SFM to explore the microscopic processes related to the effect of nucleating agents used, *e.g.* for polyolefin crystallisation.

Growing interest in SFM imaging can also be seen in biology and biophysics, where single biomolecules such as DNA or particular proteins can be imaged on a molecular level.[22] As for synthetic polymers, the development of non-destructive imaging modes together with the possibility to image in water or suitable buffer solutions proved to be essential for successful SFM experiments. Meanwhile a more important issue is the immobilisation of the biomolecules on suitable substrates. A compromise has to be found between stable imaging conditions (which require strong fixation) and a minimum influence on the biological activity (requiring loose binding to the substrate). One of the major problems one faces in biological experiments of this type is the purity of the liquids in use. In principle it is possible to study interactions between different biomolecules in real time, *e.g.* the binding of a protein to specific sites of an immobilised DNA molecule. Presently, the limit in spatial resolution is given both by the radius of curvature of the tip and the stability of the molecules on the substrate surface. In view of real-time biological experiments, a new generation of scanning force microscopes is presently being developed, which are expected to enable faster image collection. SFM imaging with an image collection rate of one image per second seems feasible in near future.

Another, quite impressive biological application of SFM is the possibility to measure forces on single molecules. Based on suitable chemical modification of the SFM tip, Gaub and others have been able to chemically bind a molecule between the tip and a substrate.[23] If the tip is withdrawn from the surface, one can measure the forces involved in stretching chains or unfolding proteins. Obviously this so-called 'force-spectroscopy' is not yet a standard procedure and many possible artefacts (*e.g.* multiple binding to the tip) have to be sorted out. However, the experiments performed so far have demonstrated the general feasibility of the approach and an increasing number of scientists have started work along these lines.

Finally, we shall briefly discuss a very recent development extending SFM beyond its standard use as a surface imaging technique. Magerle has recently shown that an iteration of SFM imaging and step-wise erosion of the sample can be used to establish real-space images of bulk morphologies with a few nm spatial resolution along all three dimensions.[24] If together with the surface topography some material contrast is imaged by SFM, this experiment reveals a stack of curved surfaces, each containing additional information on the material distribution on this surface. If the same area is imaged after each erosion step, one can reconstruct the three-dimensional material distribution by suitable image processing software. This approach has been demonstrated on a poly(styrene-*block*-butadiene-*block*-styrene) (SBS) triblock copolymer exhibiting a cylindrical micro domain morphology. *TappingMode™* phase contrast has been used to determine the styrene/butadiene ratio and plasma etching was used to erode the specimen. Given that no other three-dimensional real space imaging technique with comparable spatial resolution is available, this so-called 'nanotomography' has the potential to become a routine analytical technique for the study of heterogeneous polymeric materials. It should be noted that the approach is neither limited to *TappingMode™* SFM as the imaging technique, nor to plasma etching as the means to erode the sample. Any SPM method combined with some suitable erosion technique should in principle enable three-dimensional imaging.

5 Conclusion

SPM techniques have found their way into polymer science and without doubt, SFM has done so to the largest extent. The polymer community has welcomed the novel techniques with great enthusiasm, which is probably due to the fact that we all like to see colourful images of our samples. In some cases, the enthusiasm has led to some questionable over-interpretation of SFM results, which deserves a word of warning. As an example, it is somewhat questionable when conclusions on the bulk morphology of, say, a block copolymer sample are simply drawn from SFM images of its surface structure. Here, the limits of a surface sensitive technique are certainly reached. A recent application of the (SFM-based) nanotomography approach described above has been used to demonstrate that surface morphology and near-surface morphology do not necessarily relate in a straightforward manner.[25] It is therefore fair to take an optimistic view of the future of SPM techniques in polymer science: When used with the care and understanding that every complex experimental technique deserves, SPM techniques will certainly play an important role as a complementary analytical technique for polymers in the future. This holds both for synthetic polymers and for biopolymers.

Acknowledgements

The author is grateful for financial support through the Deutsche Forschungsgemeinschaft (SFB 481) and the Bayreuther Zentrum für Kolloide und Grenzflächen (BZKG).

References

1 G. Binnig and H. Rohrer, *Helv. Phys. Acta,* 1982, **55,** 726.
2 R. Wiesendanger, *J. Vac. Sci. Tech. B,* 1994, **12,** 515.
3 G. Binnig, C. F. Quate, and C. Gerber, *Phys. Rev. Lett.,* 1986, **56,** 930.
4 D. W. Pohl, W. Denk, and M. Lanz, *Appl. Phys. Lett.,* 1984, **4,** 651.
5 A. Stabel, R. Heinz, J. P. Rabe, G. Wegner, F. C. D. Schryver, D. Corens, W. Dehaen, and C. Süling, *J. Phys. Chem.,* 1995, **99,** 8690.
6 R. Heinz and J. P. Rabe, *Langmuir,* 1995, **11,** 506.
7 R. Heinz and J. P. Rabe, *Langmuir,*1995, **11,** 2857.
8 D. G. Lidzey, S. F. Alvarado, P. F. Seidler, A. Bleyer, and D. D. C. Bradley, *Appl. Phys. Lett.,* 1997, **71,** 2008.
9 E. Betzig, E. L. Finn, and J. S. Weiner, *Appl. Phys. Lett.,* 1992, **60,** 2484.
10 R. Toledo-Crown, P. C. Yang, Y. Chen, and M. Vaez-Iravani, *Appl. Phys. Lett.,* 1992, **60,** 2957.
11 E. H. Synge, *Phil. Mag.,* 1928, **6,** 356.
12 D. McMullan, *Proc. RMS.,* 1990, **25/2,** 127.
13 H. Heinzelmann and D. W. Pohl, *Appl. Phys. A,* 1994, **59,** 89.
14 G. Krausch and J. Mlynek, *Microelectronic Engineering,* 1996, **32,** 219-228.
15 R. M. Nyffenegger and R. M. Penner, *Chem. Rev.,* 1997, **97,** 1195.
16 M. H. P. Moers, H. E. Gaub, and N. F. V. Hulst, *Langmuir,* 1994, **10,** 2774.
17 S. S. Sheiko, *Adv. Poly. Sci.,* 2000, **151,** 61.
18 T. Pompe, A. Fery, and S. Herminghaus, *Langmuir,* 1998, **14,** 2585.
19 G. Krausch, M. Hipp, M. Böltau, J. Mlynek, and O. Marti, *Macromolecules,* 1995, **28,** 260-263.
20 H. U. Krotil, T. Stifter, and O. Marti, *Rev. Sci. Inst.,* 2000, **71,** 2765.
21 A. Knoll, R. Magerle, and G. Krausch, submitted for publication, 2000.
22 H. G. Hansma and L. Pietrasanta, *Curr. Opin. Chem. Biol.,* 1998, **2,** 579.
23 M. Rief, F. Oesterhelt, B. Heymann, and H. E. Gaub, *Science,* 1997, **275,** 1295.
24 R. Magerle, *Phys. Rev. Lett.,* 2000, in press.
25 M. Konrad, A. Knoll, G. Krausch, and R. Magerle, *Macromolecules,* 2000, **33,** 5518.

Molecular Machines

22

Mechanically-Linked Macromolecules

Guy J. Clarkson and Dave A. Leigh*†

CENTRE FOR SUPRAMOLECULAR AND
MACROMOLECULAR CHEMISTRY,
DEPARTMENT OF CHEMISTRY, UNIVERSITY OF WARWICK,
COVENTRY CV4 7AL, UK

1 Introduction

Given the high cost of introducing new chemical building blocks, it seems likely that the polymers of the new century will be derived primarily from the ever-decreasing pool of existing cheap monomers. Future demand for new materials with enhanced properties will therefore have to be largely satisfied by assembling these building blocks in new ways and using novel types of polymer architecture to provide improved characteristics and give 'added value'. Such thinking has fuelled much activity in the dendrimer and hyperbranched polymer fields in recent years[1] and has also led to the investigation of 'supramolecular polymers',[2] macromolecules where the monomer units are held together by noncovalent interactions such as hydrogen bonds (Figure 1). The use of weak, reversible interactions between the components allows them to assemble with a high degree of structural integrity – 'mistakes' in the architectures are automatically corrected – but it also makes them fragile, and disrupting the equilibrium conditions under which they were formed (by changing the solvent or temperature, for example) often results in their destruction. Other types of novel molecular-level architectures which are increasingly attracting interest include those based upon mechanically interlocked subunits, catenanes (interlocked molecular rings) and rotaxanes (consisting of a thread circumscribed by one or more macrocycles),[3,4] the idea being that incorporating flexible or co-conformationally switchable linkages into polymer backbones could generate a range of highly unusual and potentially useful properties at the macroscopic level including improved flexibility, solubility, processibility, compatibility, novel rheological properties,

† Present address: Department of Chemistry, University of Edinburgh, UK.

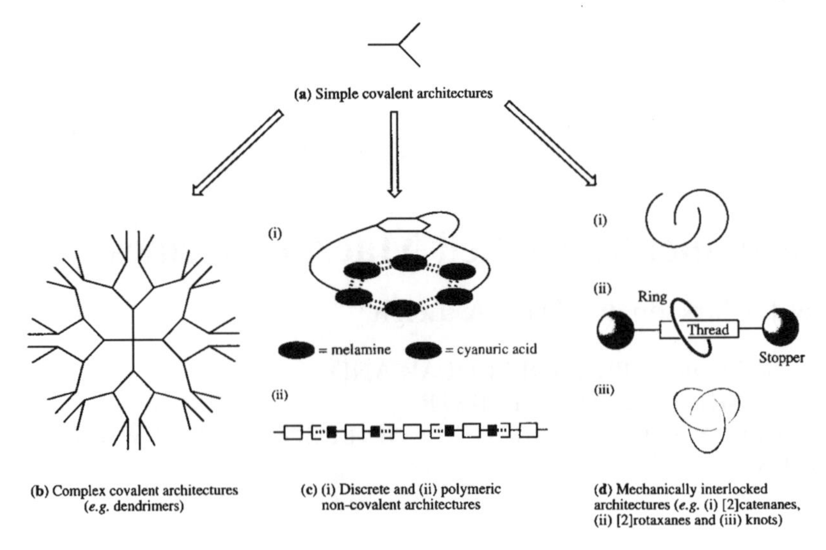

Figure 1 *The evolution of synthetic molecular level architectures*

variable surface properties, low temperature stability, shock absorbance and elasticity *etc.*

Whilst polyrotaxanes (where the thread is part of a polymer chain) have been synthetically accessible for several years and their structure–property relationships are beginning to be understood,[5] the same cannot be said for macromolecules derived from catenanes (Figure 2). In general, catenane syntheses have thus far been too low yielding or labour-intensive to provide the quantities of materials necessary to make them realistic candidates for exploitation in macromolecules. However, recent advances in synthetic routes to catenanes – including their synthesis in one step from cheap, commercially available reagents and their assembly under thermodynamic control – together with a greatly improved understanding of their dynamic properties, look set to change this situation.

2 Current State of the Art

The most impressive catenane synthesis so far is unquestionably that of the longest linear catenane, the [5]catenane 'Olympiadane' (Figure 3).[6,7] Olympiadane was synthesised using a [3]catenane as the template on which to clip, in a stepwise manner, smaller tetracationic cyclophanes. Using ultrahigh pressure conditions the yield is 30% (also formed are branched [5]-, [6]-, and [7]catenanes in a combined yield of >54% yield).[7] Whilst impressive given the complexity of the molecular level architecture, these interlocking yields are still much too low to envisage preparing a wholly mechanically interlocked linear polymer of type **G** in this way. Therefore effort is currently being directed at preparing more modest, but more practically significant, mechanically-linked architectures such as **A**.

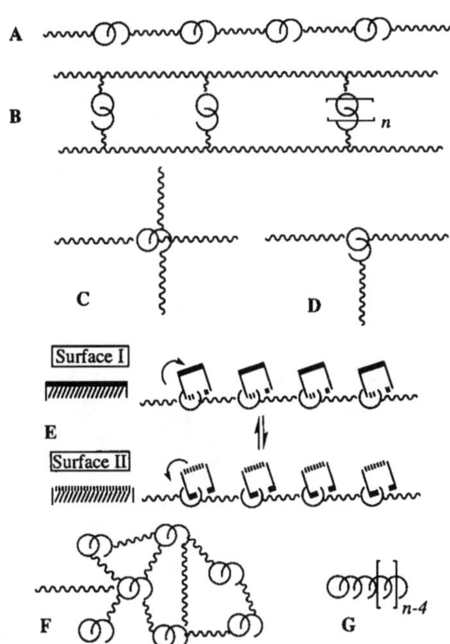

Figure 2 *Possible types of polymers based upon catenane architecures*: **A** *Linear mechanically-linked polymers;* **B** *mechanically cross-linked polymers;* **C** *stars with a flexible, potentially switchable, catenane core;* **D** *combs with catenane linkages;* **E** *main chain catenanes with switchable surface properties;* **F** *highly mechanically-crosslinked networks;* **G** *poly- or [n]catenanes. Only examples of types* **A** *and* **F** *have thus far been prepared*

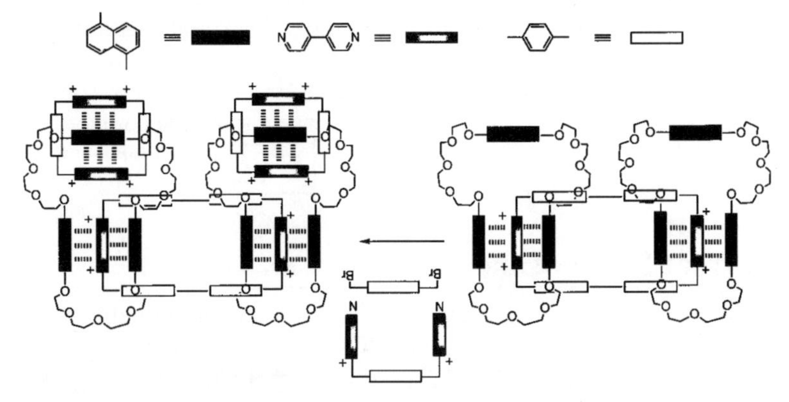

Figure 3 *Synthesis of 'Olympiadane'' – a [5]catenane, from a [3]catenane. The bracketed prefix indicates the number of interlocked components*

The first linear polymers containing [2]catenanes in their backbones (mechanically linked main chain polymers, architecture **A**) were based on the Sauvage transition metal-coordinated catenanes.[8] Polycondensation reactions were carried out between a diacid and a catenane diol to give a poly[2]catenane with Mn (600 000) and Mw (4 200 000) based on polystyrene standards (Figure 4).

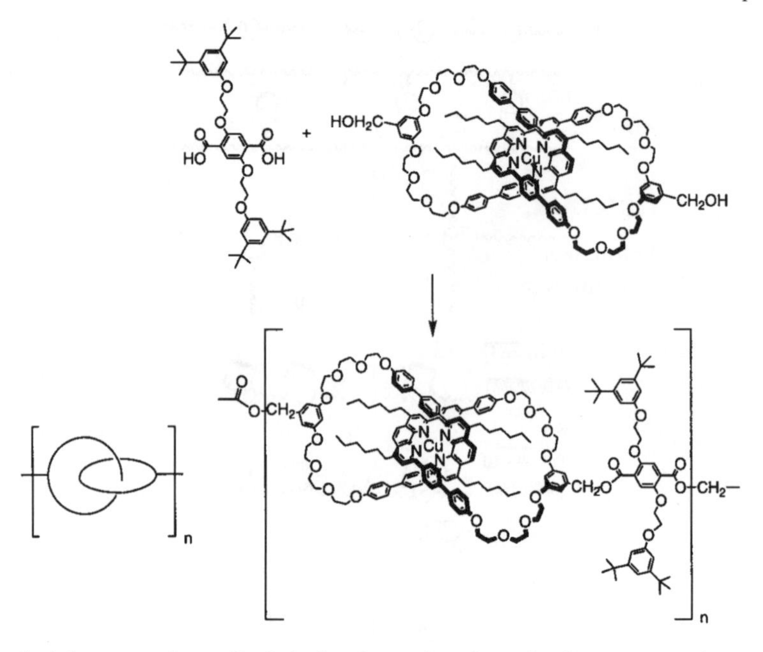

Figure 4 *A linear mechanically linked polymer based on the Sauvage metal-containing catenanes*

Similarly, the Stoddart catenanes based on π-electron rich-π-electron poor aromatic systems have been incorporated into polymer backbones (*e.g.* Figure 5).[9] These systems give a degree of polymerisation of ~17.

Unfortunately, neither the metal- or paraquat-based systems have yet been

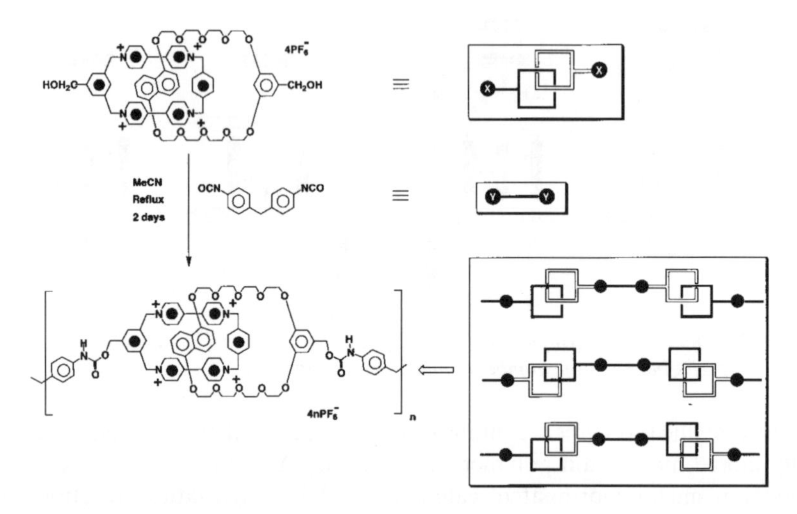

Figure 5 *A linear mechanically linked polymer based on the Stoddart paraquat-based catenanes*

incorporated into 'conventional' polymer systems, nor are they amenable to comparison with analogous non-mechanically interlocked polymer systems so the true value or property effects of introducing such linkages have not yet even begun to be investigated.

3 Future Potential

In order to exploit mechanically interlocked units in macromolecular structures, several challenges must first be overcome:

(i) Routes to catenane and rotaxane architectures must be developed that are (a) *cheap* and (b) *easily accessible.*
(ii) The property effects of incorporating mechanical linkages into *conventional* polymers must be established by comparing the physical, rheological, structural and mechanical properties of *mechanically* and *non-mechanically* interlocked polymers with each other. This, again, requires the availability of catenane and rotaxane building blocks on a routinely large scale.

Fortunately, recent (and no-doubt future) advances in synthesis mean there is real promise for the bulk production of such molecules. Hydrogen bond-directed synthesis allows the preparation of both catenanes and rotaxanes in a single chromatography-free step from very cheap, commercially available materials (Figure 6).[10]

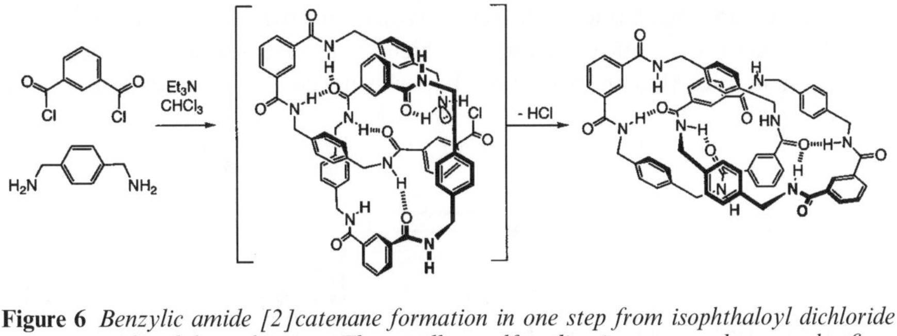

Figure 6 *Benzylic amide [2]catenane formation in one step from isophthaloyl dichloride and xylylene diamine. The smallest self-replicating system known; the first-formed macrocycle templates the formation of an identical second one through it by hydrogen bonding*

Alternatively, the use of reversible, thermodynamically-controlled synthetic methods – similar to those used to make the 'supramolecular polymers' discussed in Figure 1 – can be used to assemble mechanically interlocked molecules in near quantitative yields (Figure 7).[11] However, unlike the supramolecular systems, once formed (after removal of the catalyst) the mechanically interlocked structures are *robust* and cannot be destroyed without breaking a (strong) covalent bond!

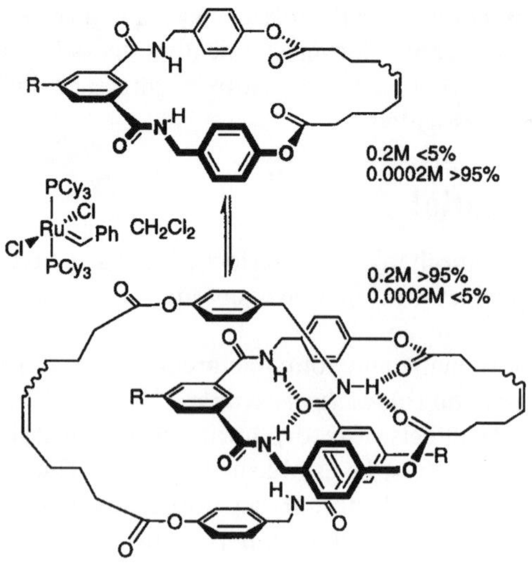

Figure 7 *Organic 'Magic rings'. Interconversion between macrocycle and catenane under thermodynamic control*

By incorporating such motifs into 'handcuff'-shaped monomers, it should be possible to combine some of the characteristics of supramolecular polymers – self-correcting, thermodynamically-controlled assembly – with the hallmark properties and stability of standard polymers with covalent backbones (Figure 8). Such monomers would, in the presence of a catalyst, spontaneously assemble to form mechanically linked polymers of precise length and architecture defined by the concentration at which they were prepared. Indeed, the various structures (linear, branched, hyperbranched *etc.*) would be intercon-

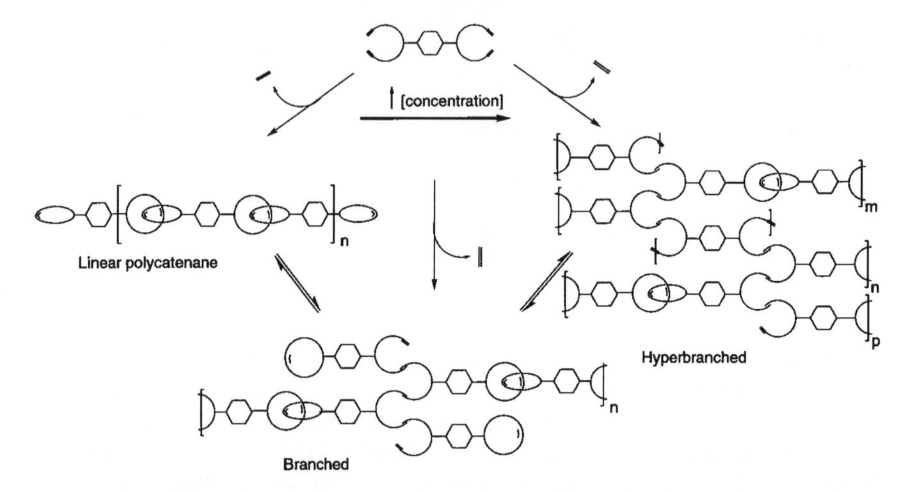

Figure 8 *'Magic handcuffs'. Interconversion between mechanically-linked polymer architectures under thermodynamic control*

vertable by just changing concentration *but only in the presence of the catalyst*! In it's absence, the polymers would be permanent and stable.

These types of high yielding, ready accessible synthetic methods should give access to all the different types of macromolecular architectures shown in Figure 1 (and many others). A systematic study will then identify the possible benefits that can be achieved by using the special properties available only through mechanical bonding at the molecular level.

4 Conclusions

The recent improvements in methods for [2]catenane synthesis mean that interlocked structures are now becoming readily available as building blocks and cores for novel macromolecular materials. Many new architectures and polymer systems will undoubtedly be developed over the next few years and the effects of mechanical, rather than covalent, linkages on bulk and surface properties will begin to be evaluated.

Acknowledgements

This work was supported by the TMR initiative of the European Union, the EPSRC and the ICI Strategic Research fund. DAL is an EPSRC Advanced Research Fellow (AF/982324).

References

1 (a) G.R. Newkome, C.N. Moorefield and F. Vögtle. '*Dendritic Molecules: Concepts, Syntheses, Perspectives*', VCH, Weinheim, 1996. (b) F. Zeng and S.C. Zimmerman, *Chem. Rev.*, 1997, **97**, 1681.

2 (a) J.P. Mathias, E.E. Simanek, J.A. Zerkowski, C.T. Seto, and G.M. Whitesides, *J. Am. Chem. Soc.*, 1994, **116**, 4316. (b) R.P. Sijbesma, F.H. Beijer, L. Brunsveld, B.J.B. Folmer, J.H.K. Ky Hirschberg, R.F.M. Lange, J.K.L. Lowe and E.W. Meijer, *Science*, 1997, **278**, 1601. (c) F.H. Beijer, H. Kooijman, A.L. Spek, R.P. Sijbesma and E.J. Meijer, *Angew. Chem. Int. Ed. Engl.*, 1998, **37**, 75.

3 For accounts of interlocked molecules see, for example, (a) G. Schill. '*Catenanes, Rotaxanes and Knots*', Academic Press, New York, 1971. (b) C. Dietrich-Buchecker, and J.-P. Sauvage, *Bioorg. Chem. Frontiers,* 1991, **2**, 195. (c) D.B. Amabilino and J.F. Stoddart, *Chem. Rev.*, 1995, **95**, 2725. (d) C.A. Hunter, *J. Am. Chem. Soc.*, 1992, **114**, 5303. (e) F. Vögtle, T. Dünnwald and T. Schmidt, *Acc. Chem. Res.*, 1996, **29**, 451. (f) D.G. Hamilton, J.K.M. Sanders, J.E. Davies, W. Clegg, and S.J. Teat, *Chem. Commun.*, 1997, 897. (g) D.A. Leigh and A. Murphy, *Chem. Ind.*, 1999, 178.

4 For some accounts of the unusual properties of mechanically interlocked architectures see, for example, (a) R.A. Bissell, E. Córdova, A.E. Kaifer and J.F. Stoddart, *Nature*, 1994, **369**, 133. (b) D.A. Leigh, K. Moody, J.P. Smart, K.J. Watson and A.M.Z. Slawin, *Angew. Chem. Int. Ed. Engl.*, 1996, **35**, 306. (c) D.A. Leigh, A. Murphy, J.P. Smart and A.M.Z. Slawin, *Angew. Chem. Int. Ed. Engl.*, 1997, **36**, 728. (d) C. Gong and H.W. Gibson, *Angew. Chem. Int. Ed. Engl.*, 1997, **36**, 2331.

5 The extensive research already undertaken into the synthesis and properties of polyrotaxanes merits discussion in its own right, but is outside the scope of this

manuscript. For a recent review see H.W. Gibson and C. Gong, *Curr. Op. in Solid State and Mat. Sci.*, 1997, **2**, 647.

6 D.B. Amabilino, P.R. Ashton, A.S. Reder, N. Spencer and J.F. Stoddart, *Angew. Chem. Int. Ed. Engl.*, 1994, **33**, 433.

7 D.B. Amabilino, P.R. Ashton, V. Balzani, S.E. Boyd, A. Credi, J.Y. Lee, S. Menzer, J.F. Stoddart, M. Venturi and D.J. Williams, *J. Am. Chem. Soc.*, 1998, **120**, 4295.

8 J.-L. Weidmann, J.-M. Kern, J.-P. Sauvage, Y. Geerts, D. Muscat and K. Müllen, *Chem. Commun.*, 1996, 1243.

9 S. Menzer, A.J.P. White, D.J. Williams, M. Belohradsky, C. Hamers, F.M. Raymo, A.N. Shipway and J.F. Stoddart, *Macromolecules,* 1998, **31**, 295.

10 A.G. Johnston, D.A. Leigh, R.J. Pritchard, and M.D. Deegan, *Angew. Chem. Int. Ed. Engl.*, 1995, **34**, 1209.

11 T.J. Kidd, D.A. Leigh and A.J. Wilson, *J. Am. Chem. Soc.,* 1999, **121**, 1599.

23

Polymers and Soft Nanotechnology

Richard A.L. Jones

DEPT OF PHYSICS AND ASTRONOMY,
UNIVERSITY OF SHEFFIELD, HICKS BUILDING,
HOUNSFIELD ROAD, SHEFFIELD S3 7RH, UK

1 Introduction

The idea of nanotechnology – the manufacture of machines and devices at nanometer length scales – has generated a degree of fervour bordering on the religious, but there is a gulf between what nanotechnology's more extreme evangelists have promised and what has yet been achieved or is even in realistic prospect. Approaches to nanotechnology have used one of two philosophies. In the first, conventional silicon processing technology borrowed from the world of microelectronics allows the fabrication of mechanical components at the micron length scale, which can be directly integrated with electronic logic circuits. This technology – often referred to as micro-electro-mechanical systems (MEMS) – already has widespread applications at market, such as the micro-jet assemblies in ink-jet printers, and the micro-accelerometers used to control automotive airbag systems. Applications of this technology will continue to grow, but it is not clear that it will be possible to scale this approach down from the micron scale to make devices in the nanometer range. Size does matter, and the physics that dominates the operation of devices at the nanometer length scale has qualitative differences from the physics that dominates our macroscopic world. Issues such as friction, dissipation and Brownian motion become limiting at these length scales, particularly in aqueous environments. The other approach, working from the molecular level, is currently further from applications. Here fascinating efforts are being made to make device components, like shuttles and gears using ingenious supramolecular chemistry (see for example ref. 1), but as yet we are a long way from putting these components together to make devices that do something interesting.

On the other hand, perhaps the most compelling argument that sophisticated nano-scale devices are indeed possible is the fact that an intricate and effective nanotechnology does already exist – in biology. Recent advances in molecular biology have moved beyond the stage of characterising the structure of individual macromolecules; now biology is elucidating the way macromolecules are put together to form complex machines, such as ATP-synthase[2] or the nuclear pore complex. In this article I argue that one of the most fruitful approaches to making nano-scale devices is likely to follow from distilling these lessons, and attempting to make devices that use the basic operating principles of cell biology, but that are realised in synthetic components. Just as the mechanisms of cell biology are dominated by macromolecules, so it is polymer science that will provide us with the ingredients for making their synthetic analogues. By making devices that emulate the principles used by cell biology, we may both develop useful technologies and start to understand more deeply the physical principles that underly the operation of cell biology.

These devices will be, like their biological analogues, essentially *soft*. By this I mean that their structures are likely to be held together by weak bonds – hydrogen bonds, hydrophobic/hydrophilic interactions, screened Coulomb interactions and so on. Such devices will thus retain the general features of soft matter – that is to say, they will be *self-assembling* and *responsive to their environment*. Structures built in this way are also intrinsically floppy, rather than stiff, and they operate in an environment which consists of solvent (possibly or preferably water) at constant, relatively high temperature. Being small they will be highly affected by Brownian motion and surface forces, and any motion is in a fluid medium which will be in conditions of extremely low Reynolds number. Thus they will have dynamics characterised by negligible inertia, dominated by viscosity and long range hydrodynamic interactions. Being small devices in an isothermal environment means that one cannot use any kind of heat engine to provide a power source; instead one will have to use molecular motors which can directly convert chemical energy to mechanical energy. The mechanics these soft devices will exploit is thus going to be very different from the mechanics which operate in human scale machinery, so in designing devices we will have to use very different principles to those we use at human sizes. We can't make a nano-machine by simply scaling down a human size machine.

What kinds of devices are we talking about? In many cases, crude devices exemplifying some of these principles already exist. One example is the widespread use of polymers for the encapsulation of various kinds of active ingredient, whether they are drug molecules in pharmaceutical controlled release preparations or flavouring or fragrance agents in consumer products. These are micron or sub-micron scale containers that protect delicate molecules from degradation until they are needed, when they can be released. Currently, although these often make effective products, the techniques for creating the capsules and triggering and controlling the release are still crude. Extensions of ongoing research will extend the range and specificity of the triggering. But perhaps in the future it would be possible to imagine an active vesicle, that could seek out a target like a tumour, move there with its own motive power, and then release its load. What would we need to do before we could achieve such a goal? Other possibilities might include selective valves, which would transport only certain target molecules; if these valves were coupled to an energy source molecules could be transported against a concentration gradient in order to isolate and purify reaction products. Other applications would combine the soft and the hard to yield selective membranes, microfluidic devices and the other components of a 'lab on a chip'.

In this paper, I will first consider some of the general issues that will need to be addressed to achieve such goals. These issues must include how to make the devices in the first place. The advantage of emulating nature in designing a soft nanotechnology is that we can use the natural propensity of soft matter to *self-assemble*, rather than having to build up a nano-device molecule by molecule. It is the *responsiveness* of soft matter that will allow us to design devices that respond to their environment, exploiting phase transitions and conformational changes to provide selective triggers and to do mechanical work. Finally, in order to create truly useful devices, we need to find a way of converting energy from light or chemical energy into useful work – we need to design *molecular motors.* I will then discuss some progress that has been made that illustrates these themes, before speculating on some targets that provide challenging, but possibly achievable goals for future research.

2 Fundamental Issues

A remarkable characteristic shared by biological and non-biological soft matter systems is their propensity to *self-assemble* – that is for many molecules to arrange themselves in ordered structures driven simply by the tendency of systems to lower their overall free energy. In synthetic soft matter one thinks of soap mesophases and the complex structures formed by block copolymers; in biology one has the folding of proteins and the formation of membranes. Attempts to apply the self-assembly paradigm to biology are often criticised for neglecting the fact that biological systems are not at equilibrium. The argument is that, because the lowest free energy state of an organism is not interesting (that is to say not alive!), and that in biological processes there is a continuous input of energy, one cannot think of biological processes in terms of self-assembly. This misses an important point; one can exploit self-assembly in processes other than that of achieving a global energy minimum. In polymer processing one is used to the idea of *locking in* a non-equilibrium structure that is created by self-assembly in different conditions in which it is at equilibrium. One way of doing this is to make a self-assembled structure with a *precursor* material, which is then processed in some way as to lock in the structure. This is a trick that cell biology uses, for example in the assembly of collagen fibres. Likewise, by using encapsulation, one can move towards a local equilibrium in a highly restricted volume; the resulting structures may be stable when they are released from their containers, even if unconstrained they are not at equilibrium. This seems to be how heat shock proteins work as chaperones to promote unimolecular protein folding in circumstances when the lowest free energy state of a finite concentration of proteins would be a biochemically useless polymolecular aggregate. Individual protein molecules are prevented from interacting with other molecules by being confined to a tiny container[3] (itself the product of self-assembly). The action of confining the protein and subsequently releasing it requires an energy input, but the folding of the protein is still an act of self-assembly in an attempt to find a local, highly restricted equilibrium. In summary, self-assembly is not only the only practicable way of creating complex structures in soft matter at the nano-scale; it is also less restrictive than it may seem at first sight. With the help of additional energy inputs, one can use it to make structures that are not at equilibrium.

Another property shared by biological and synthetic soft matter is *responsiveness* to changes in the environment. If the conformation of a macromolecule is at equilibrium with the environment, then when the environment changes the conformation of the macromolecule may change in response. The same applies to macromolecular or supramolecular assemblies. The most familiar example in polymer science is the coil-globule transition undergone by a polymer molecule in response to a change in solvent quality; a polymer in a good solvent has the open conformation of a random walk with excluded volume, but if the temperature changes so that the solvent quality becomes poor then the polymer coil will collapse to a much more compact configuration. The property of responsiveness is most dramatic when the change in environment drives a *conformational transition*; if one has a sharp phase transition driven by an environmental change then near the transition a small change in environment can drive a large response. The responsive quality of soft matter is exemplified at a macroscopic scale by the volume phase transitions undergone by certain polymer gels, but the same principles operate at the molecular level too, for example in the collapse of a grafted polymer brush in response to changing temperature (see Figure 1). The range of environmental changes to which a synthetic hydrophobic polymer will respond is limited to temperature and solvent type.

Polyelectrolytes in aqueous solution offer richer possibilities, including the salt concentration, which controls the range of the screened Coulomb interaction, and the pH, which will control the degree of ionisation of a weak polyelectrolyte. In biological systems, besides these rather generic interactions, there are often conformational changes in response to the binding of very specific ligands. These are crucial to the operation of many biochemical systems; however, they involve no new issues of principle and it should be possible to design analogous, if less finely engineered, interactions in synthetic systems.

The biggest difference between biological systems and current synthetic ones is that biological systems operate with a constant input of energy. This energy arrives in the form of light or chemical energy and must be converted from chemical energy to mechanical energy to power the many cellular and supracellular processes. In human technology, conversion of chemical energy to mechanical energy usually takes place in a heat engine. It is very difficult to make a heat engine at the nanoscale – one simply cannot keep heat confined for long enough. The time taken for an enclosure at a higher temperature than

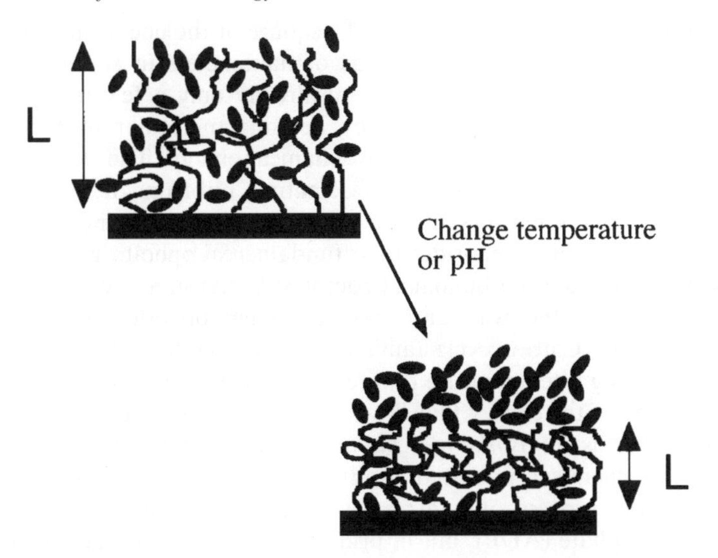

Figure 1 *Responsiveness in a polymer brush. In a good solvent (a) favourable interactions between polymer segments and solvent molecules lead to the chains being stretched – the loss of configurational entropy attributable to the chain stretching is outweighed by the lowering of energy due to the polymer/solvent interaction. If this interaction becomes less favourable (b), due to a change in temperature or pH, for example, the chains reversibly shrink*

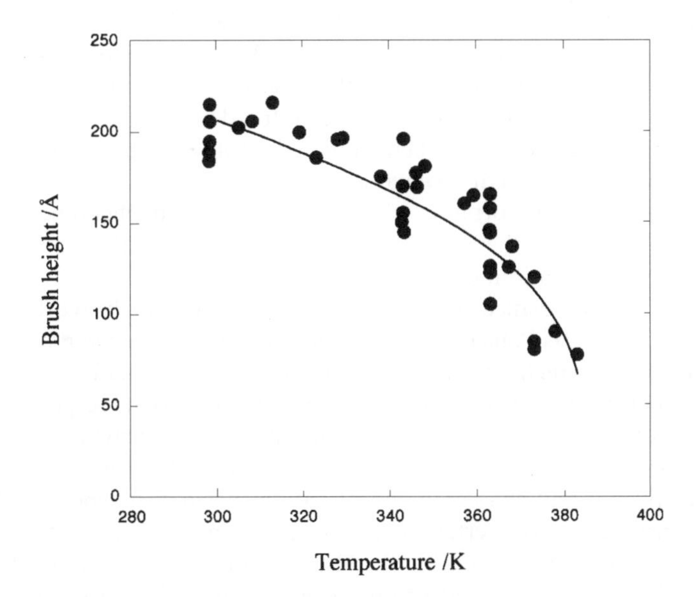

Figure 2 *The height of a polymer brush as a function of temperature, as revealed by neutron reflectometry.*[4] *The brush is composed of end-grafted deuterated polystyrene, and the environment is poly(vinyl methyl ether). As the temperature is increased the interaction between polystyrene and poly(vinyl methyl ether) becomes increasingly unfavourable and the brush height decreases reversibly*

its environment to equilibrate scales as the square of the linear dimension; for enclosures with dimensions of hundreds of nanometers in water or hydrocarbon environments relaxation times are already only a few microseconds. Nature has got round this problem by devising molecular motors, which convert chemical energy directly into mechanical energy. The details of how biological molecular motors work are fascinating, but from the point of view of trying to emulate them in synthetic systems, it is perhaps more fruitful to look beyond the details to consider their fundamental operating principles. In this way one can avoid the biological complexity that needs to be considered when one discusses the way any specific system operates. Heat engines, exemplified by the Carnot cycle familiar from elementary thermodynamics, operate by creating a temperature difference between two reservoirs, and then converting some of the energy that flows from hot to cold into useful work. A molecular motor is an *isothermal motor* driven by differences in *chemical potential.* The chemical potential gradient that powers biological motors derives from the process of hydrolysis of adenosinetriphosphate (ATP) to adenosinediphosphate (ADP), but in principle any chemical reaction that can be kept out of equilibrium could serve as the driving force. Indeed, one does not need a chemical reaction; a simple gradient in composition produces a gradient of chemical potential, and the resulting gradient of osmotic pressure can be used to extract mechanical work. The two-dimensional analogue of such an osmotic motor is a device exploiting the *Marangoni effect.* If a surface has a concentration gradient of an adsorbed species, this leads to a gradient in surface tension; these capillary gradients can cause work to be done. An long-studied example of this phenomenon occurs when crystals of camphor are floated on the surface of water; the surface tension gradients set up by the dissolution of the crystal cause the crystal to 'dance' across the surface. Although the serious study of such motion is well over a century old the subject has recently been rediscovered as a chemical motor.[5] For the presence of this discussion the importance of this long-established phenomenon is its *scalability;* by virtue of the involvement of surfaces it is likely to be more rather than less important in nanoscopic systems.

Moving on to the detailed mechanics of how a motor might work, the key ingredient in many biological motors seems to be the coupling of a cyclic chemical reaction, which continuously interconverts ATP and ADP, with a conformational change in a macromolecule which occurs in response to the binding of ATP. It is the conformational change that provides the *power stroke*; motor proteins such as kinesin are provided with a long lever arm so that the relatively small change in conformation that occurs when the bound ATP is hydrolysed is converted to a substantial movement of the motor. Abstracting away from the specific details, this coupling of a cyclic chemical reaction with a conformational change seems to underly a wide variety of biological molecular motors, and it should be possible to devise a synthetic analogue. An even more generalised description of molecular motors has been provided in terms of Brownian diffusion in the presence of dissipation in a time-varying asymmetric potential,[6] and this again may be the basis for alternative realisations of a synthetic molecular motor. Finally, it is worth commenting that while motor proteins provide the most extensively studied biological motors, it seems unlikely that other mechanisms are not at work in biological systems. Nature probably exploits any mechanism that works, and it would be a surprise not to find that Marangoni or direct osmotic mechanisms are important in some biological systems as well.

3 Synthetic Systems

In the last section, we discussed some of the general principles that are common to all soft matter systems, whether of synthetic or biological origin. Here I will mention some examples of synthetic systems that exploit some of these principles.

Self-assembly as general principle for creating complex morphologies and structures is familiar both for small molecule amphiphiles and for block copolymers. Examples of potentially important functional structures created by self-assembly are *vesicles* and *micelles*, which form convenient containers for localising chemical species and reactions. Vesicles are now much used as vehicles for the delivery of drugs and other active ingredients. Phospholipid vesicles form very simple models for cells and organelles, though it has to be clearly recognised that cell membranes have a very much more complex structure than a simple phospholipid bilayer, In addition to the importance of membrane proteins and pore complexes, the membrane itself is mechanically coupled to layers of macromolecules both inside and outside the cell. It is this coupling that gives the membrane its ability to respond to external stimuli; changes in environment lead to changes in the elastic moduli of the membrane, which in turn can lead to shape transitions of the vesicle. Current synthetic systems are much cruder. They may use phase transitions within the membrane in response to environmental changes, for example to trigger release of the vesicle contents, and they may well incorporate polymers to increase their robustness, but these subtle couplings between the membrane and associated macromolecules have not yet been fully exploited. In addition to producing vesicles from small molecule amphiphiles, it is also possible to exploit self-assembly to create vesicles from block copolymers,[7] which are substantially tougher and less permeable than lipid vesicles.

The volume transition in gels was mentioned in the last section as an example of responsiveness. The volume transition that a gel undergoes when the solvent conditions are changed can be triggered by a variety of stimuli, including temperature, specific ligand concentration or light; these transitions can be exploited in the release of drugs and other active ingredients.[8] Because of the scalability of the volume transition, this phenomenon can be exploited as well in microscopic gel particles as in macroscopic samples of gel. An interesting recent example[9] combines the principle of a pH driven volume transition with the idea of a vesicle as an isolating vessel; a 6.5 micron size particle of a pH responsive hydrogel is encapsulated in a lipid bilayer. Only when the bilayer was broken by the application of an electric field was the gel able to respond to the changed pH of the environment, providing a very rapid, triggerable release of a drug. This structure closely mimics a cell biological structure, the secretory granule.

The collapse of a polymer gel in response to a change in environment can be scaled down to the single chain level. A layer of polymer chains grafted to a surface forms a *polymer brush*, for which the collapse transition can be nicely observed using neutron reflectivity.[10] A pore lined with a responsive polymer brush will form a *selective valve*;[11] for example if the grafted polymer is a weak polybase, in aqueous acidic conditions the brush will be charged and will expand to close the pore, while in basic conditions the brush will be neutral. This principle has been used to create a selective membrane,[12] which shows greatly reduced permeability in acidic conditions.

These simple examples show that the use of self-assembly and responsiveness in biomimetic systems are already well advanced. There has been much less progress towards creating synthetic molecular motors. Both osmotic and Marangoni mechanisms have been used to drive motion at the micron scale or larger. Gradients of concentration have been shown to be highly effective in propelling vesicles *via* a hydrodynamic mechanism,[13] while Marangoni effects driven by temperature can drive fluid flow in microfluidic systems at the micron length scale.[14] For the reasons discussed above, Marangoni effects driven by temperature are unlikely to be effective at nanometer length scales, but surface tension gradients can be induced by other methods, including illumination of photosensitive surfactants[15] and the use of spatially localised electrochemical redox reactions.[16] The principle of coupling a cyclic chemical reaction with a conformational transition has been demonstrated at the macroscopic level;[17] a responsive hydrogel was immersed in a vessel in which a Belousov-Zhabotinsky (B-Z) reaction was being carried out. The chemical oscillation of the B-Z reaction led to a mechanical oscillation of the gel as it successively swelled and deswelled in response to the changing oxidation state of the environment. As was remarked on above, the volume transition of a hydrogel is a phenomenon that can be scaled down to the molecular level. This should offer one route to the direct conversion of chemical energy at the molecular level; we are actively pursuing this possibility at Sheffield.

4 Conclusions

The world of cell biology is full of intricate and ingenious machinery, but the principles it employs are the same that govern the behaviour of synthetic soft matter systems, and these principles can be illuminated by designing synthetic devices that mimic the operation of cellular and sub-cellular structures and devices. I foresee a growing trend in polymer science to seek to use these principles to make increasingly complicated devices at the micron length scale or below. Examples of these devices will include self-motile capsules with increasingly specific triggering mechanisms and membranes and pores that specifically pump chemicals against concentration gradients. High added-value applications will be found in medicine, consumer products, and in many other areas yet to have been imagined.

References

1 C.P. Collier, E.W. Wong, M. Belohradsky, *et al.*, *Science*, 1999, **285**, 391
2 V.K. Rastogi, M.E. Girvin, *Nature*, 1999, **263**
3 K.K. Kim, R. Kim, and S. Kim, *Nature*, 1998, **394** 595
4 M. Sferrazza, R.A.L. Jones, D. Bucknall, *Phys. Rev. E*, 1999, **59** 4434
5 T. Mitsumata, K. Ikeda, J.P. Gong, Y. Osada, *App. Phys, Lett.*, 1998, **73**, 2366
6 J. Prost, J. Chawin, L. Peliti, A. Ajdari, *Phys. Rev. Lett.*, 1994, **72**, 2652
7 D.M. Discher, Y. Won, D.S. Edge, *et al.*, *Science*, 1999, **284**, 1143
8 T. Okano, *Biorelated Polymers and Gels*, Academic Press, San Diego, 1998
9 H. Kiser, G. Wilson, D. Needham, *Nature*, 1998, **394**, 459
10 A. Karim, S. K. Satija, J. F. Douglas, J. F. Ankner, and L. J. Fetters, *Phys. Rev. Lett.*, 1994, **73**, 3407.
11 R. Israels, D. Gersappe, M. Fasolka, V. A. Roberts, and A. C. Balazs, *Macromolecules*, 1994, **27**, 6679.
12 A.M. Mika, R.F. Childs, J.M. Dickson, *J. Memb. Sci.*, 1999, **153**, 45
13 J. Nardi, R. Bruinsma, E. Sackman, *Phys. Rev. Lett.*, 1999, **82**, 5168
14 D.E. Kataoka, S.M. Troian, *Nature*, 1999, **402**, 794
15 J.Y. Shin, N.L. Abbott, *Langmuir*, 1999, **15**, 4404
16 D.E. Bennett, B.S. Gallardo, N.L. Abbott, *J. Am. Chem. Soc.*, 1996, **118**, 6499
17 R. Yoshida, T. Takahashi, T. Yamguchi, H. Ichijo, *J. Am. Chem. Soc.*, 1996, **118**, 5134

Subject Index